Industrial Valves

Industrial Valves

Calculations for Design, Manufacturing, Operation, and Safety Decisions

First Edition

Karan Sotoodeh
University of Stavanger
Oslo, Norway

WILEY

Published by John Wiley & Sons, Inc., Hoboken, New Jersey.
Published simultaneously in Canada.

Library of Congress Cataloging-in-Publication Data
Names: Sotoodeh, Karan, author.
Title: Industrial valves : calculations for design, manufacturing, operation, and safety decisions / Karan Sotoodeh.
Description: First edition. | Hoboken, New Jersey : Wiley, [2023] | Includes bibliographical references and index.
Identifiers: LCCN 2023002088 (print) | LCCN 2023002089 (ebook) | ISBN 9781394185023 (cloth) | ISBN 9781394185030 (adobe pdf) | ISBN 9781394185054 (epub)
Subjects: LCSH: Valves–Design and construction–Mathematics. | Fluid dynamics–Mathematics. | Engineering mathematics–Formulae.
Classification: LCC TS277 .S68 2023 (print) | LCC TS277 (ebook) | DDC 621.8/40151–dc23/eng/20230302
LC record available at https://lccn.loc.gov/2023002088
LC ebook record available at https://lccn.loc.gov/2023002089

Cover Image(s): Rawf8/Adobe Stock Photos; Who is Danny/Adobe Stock Photos
Cover Design: Wiley

Set in 9.5/12.5pt STIXTwoText by Straive, Pondicherry, India

Contents

1

Flow Capacity

1.1 Introduction

Valve flow capacity, also called flow coefficient or capacity index, is defined as the valve's capacity for a liquid or gas to flow through it. The flow coefficient is technically defined as the flow rate of water in US gallons per minute at the temperature of 60 °F with a pressure drop of 1 psi across the valve. Flow coefficient, as shown with parameter C_v, increases by opening the valve to the maximum value when the valve is 100% open. Pressure drop, also called differential pressure, is defined as the difference between the inlet and outlet of the valve; for a valve with a specific size, the greater the differential pressure, the greater the flow rate. These two parameters, flow rate and differential pressure, are brought together by a flow coefficient, which allows the performances of different valves regarding flow rates to be compared. The differential pressure across the valves is calculated by knowing the flow rates, and finally, the flow rates are determined for given differential pressure values. Furthermore, the flow coefficient is an essential parameter for sizing valves. The flow coefficient (C_v) for liquids is calculated from Eq. (1.1).

Flow Coefficient (C_v) Calculation for Liquids

$$C_v = Q\sqrt{\frac{SG}{\Delta P}} \tag{1.1}$$

where:

C_v: Flow coefficient (US gal per minute [gpm]/psi);
Q: Flow rate (US gpm);
SG: Fluid-specific gravity (dimensionless) calculated by Eq. (1.2);
ΔP: Differential pressure across the valve (pound per square inch [psi]).

Note: Water-specific gravity is equal to one.

Industrial Valves: Calculations for Design, Manufacturing, Operation, and Safety Decisions, First Edition. Karan Sotoodeh.

Specific Gravity (SG) Calculation

$$SG = \frac{\rho_{\text{fluid}}}{\rho_{\text{water}}} \tag{1.2}$$

where:

ρ_{fluid}: Density of the fluid passing through the valve;
ρ_{water}: Water density equal to $1\ \text{g/cm}^3$ or $1000\ \text{kg/m}^3$.

To summarize, the flow coefficient calculated by Eq. (1.1) is a parameter obtained by experiments that indicate the flow capacity of incompressible fluid (liquid) during nonchoked and nonflashing flow conditions with a unit of differential pressure across the valve. Choked flow is fluid with a limited mass flow rate due to restrictions like nozzles, orifices, and reducers. Flashing flow refers to the significant evaporation of liquid due to a dramatic pressure drop.

Example 1.1 Calculate the flow coefficient of a control valve that passes 20 US gpm of water with a 4 psi pressure drop in a fully open position.

Answer

$$C_v = Q\sqrt{\frac{SG}{\Delta P}} \rightarrow C_v = 20\sqrt{\frac{1}{4}} = 10\ \text{gpm/psi}$$

Two flow coefficients are commonly used in theory and practice: one American flow coefficient (C_v) and the other metric (K_v) also called the flow factor. The metric version of the flow coefficient or flow factor offers the measurement in terms of cubic meters per hour (m^3/h) of flow in a temperature range of 5–30 °C for a differential pressure of 1 bar. The flow factor (K_v) for liquids is calculated by Eq. (1.3).

Flow Factor (K_v) Calculation for Liquids

$$K_v = Q\sqrt{\frac{SG}{\Delta P}} \tag{1.3}$$

where:

K_v: Flow coefficient ($m^3/h \cdot \text{bar}$);
Q: Flow rate (m^3/h);
SG: Fluid-specific gravity (dimensionless) calculated through the use of Eq. (1.2);
ΔP: Differential pressure across the valve (bar).

The flow coefficient (C_v) for air and gases is calculated based on either Eq. (1.4) or Eq. (1.5), depending on the criticality of the pressure drop. A critical pressure drop indicates an outlet pressure (P_O) that is less than or equal to 50% of the inlet pressure (P_i). A noncritical or subcritical pressure drop is where the outlet pressure is higher than 50% of the inlet pressure.

Flow Coefficient Calculation for Air and Gases (Critical Pressure Drop)

$$C_v = Q \frac{\sqrt{SG\,(T + 460)}}{660P_i} \quad \frac{P_O}{P_i} \leq 0.5 \tag{1.4}$$

where:

C_v: Flow coefficient (Standard cubic feet per minute [SCFM]/psi);
Q: Gas flow rate (Standard cubic feet per minute [SCFM]);
SG: Specific gravity of flowing gas relative to air at 14.7 psi and 60 °F (Note 1);
T: Flowing air or gas temperature (°F); and
P_i: Inlet gas absolute pressure (psia) (Note 2).

Note 1
The specific gravity of some common gases can be found in Table 1.1. SG of air is equal to one.

Table 1.1 Specific gravity of some common gases.

Gas	Specific gravity
Acetylene (ethyne) – C_2H_2	0.90
Air (Note 1)	1
Alcohol vapor	1.601
Ammonia – NH_3	0.59
Argon – Ar	1.38
Benzene – C_6H_6	2.6961
Blast furnace gas	1.02
Butadiene – C_4H_6	1.87

(Continued)

Table 1.1 (Continued)

Gas	Specific gravity
Butane – C_4H_{10}	2.0061
I-Butene (butylene) – C_4H_8	1.94
Isobutene – C_4H_8	1.94
Carbon dioxide – CO_2	1.5189
Carbon monoxide – CO	0.9967
Chlorine – CL_2	2.486
Digestive gas (sewage or biogas)	0.8
Ethane – C_2H_6	1.0378
Ether vapor	2.586
Ethylene (ethene) – C_2H_4	0.9683
Fluorine	1.31
Helium – He	0.138
Heptanes	3.459
Hexane	2.973
Hydrogen	0.0696
Hydrogen chloride – HCL	1.268
Hydrogen sulfide – H_2S	1.1763
Hydrofluoric acid	2.370
Hydrochloric acid	1.261
Isobutane	2.01
Isopentane	2.48
Krypton	2.89
Mercury vapor	6.940
Methane – CH_4	0.565
Natural gas	0.6–0.7
Neon	0.697
Nitric oxide – NO	1.037
Nitrogen – N_2 (pure)	0.9669
Nitrogen – N_2 (atmospheric)	0.9723
Nitrous oxide – N_2O	1.530
Nonane	4.428
Octane	3.944
Oxygen – O_2	1.1044

Table 1.1 (Continued)

Gas	Specific gravity
Ozone	1.660
Pentane	2.487
Propane – C_3H_8	1.5219
Sulfur dioxide – SO_2	2.264
Water gas	0.71
Water vapor	0.6218
Xenon	4.53

Note 2

Absolute pressure measurement considers zero as the reference point, and it is the total pressure of gauge and atmospheric pressure values. Atmospheric pressure represents the pressure within the earth's atmosphere.

Flow Coefficient Calculation for Air and Gases (Noncritical Pressure Drop)

$$C_v = \frac{Q}{1360}\sqrt{\frac{\mathrm{SG}\,(T + 460)}{\Delta P \times P_O}} \quad \frac{P_O}{P_i} > 0.5 \tag{1.5}$$

where:

P_O: Outlet gas absolute pressure (psia);
ΔP: Differential pressure between the inlet and outlet (psia) equal to $P_i - P_O$;
T: Fluid temperature (°F).

Example 1.2 Air with a pressure of 1000 psia is flowing through a globe valve that reduces the pressure to 400 psia. Considering the flow of air through the valve is 400 SCFM, and its temperature is 30 °C, what is the valve's flow coefficient in SCFM/psi?

Answer

Figure 1.1 is a schematic of the valve, inlet, and outlet air including values of pressure, flow rate, and specific gravity.

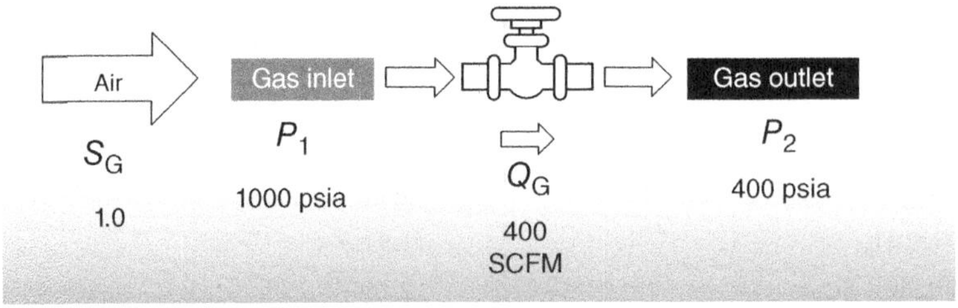

Figure 1.1 A globe valve schematic including inlet and outlet air service, values of pressure, and flow rate.

The service is air, and the outlet pressure is less than half of the inlet pressure indicating a critical pressure drop, so the flow coefficient is calculated using Eq. (1.4). The air temperature is 30 °C, which equals to 86 °F.

$$C_v = Q\frac{\sqrt{SG\,(T+460)}}{660P_i} = 400\frac{\sqrt{1\times(86+460)}}{660\times 1000} = 0.014\,\text{SCFM/psi}$$

Example 1.3 Methane (CH_4) with a flow rate of 100 SCFM and a pressure of 140 psia is passing through a control valve, which reduces the pressure to 110 psia. Assuming the gas temperature is equal to 60 °F, calculate the flow coefficient C_v.

Answer

The outlet pressure is 78.5% of the inlet pressure, so Eq. (1.5) for gases with non-critical pressure drop should be used. The specific gravity of methane is 0.565, per Table 1.1.

$$C_v = \frac{Q}{1360}\sqrt{\frac{SG\,(T+460)}{\Delta P\times P_O}} = \frac{100}{1360}\sqrt{\frac{0.565\,(60+460)}{30\times 110}}$$

$$= 0.0735\times 0.2984 = 0.022\,\text{SCFM/psi}$$

The flow factor (K_v) for air and gases is calculated based on either Eq. (1.6) or Eq. (1.7), depending on the criticality of the flow and pressure drop. A supercritical flow or pressure drop indicates an outlet pressure (P_O) that is less than or equal to 50% of the inlet pressure (P_i). A noncritical pressure drop is where the outlet pressure is higher than 50% of the inlet pressure.

Flow Factor Calculation for Air and Gases (Critical Pressure Drop)

$$K_v = \frac{Q}{257p_i}\sqrt{\rho \times T_i} \quad \frac{P_O}{P_i} \leq 0.5 \tag{1.6}$$

where:

K_v: Flow factor ($m^3/h \cdot bar$);
Q: Gas flow rate (Standard cubic meter per hour [m^3/h]);
P_i: Inlet gas absolute pressure (bara);
ρ: Gas density (kg/m^3);
T_i: Flowing air or gas inlet temperature (°K).

Flow Factor Calculation for Air and Gases (Noncritical Pressure Drop)

$$K_v = \frac{Q}{514}\sqrt{\frac{\rho \times T_i}{\Delta P \times P_O}} \quad \frac{P_O}{P_i} > 0.5 \tag{1.7}$$

where:

P_O: Outlet gas absolute pressure (bara);
ΔP: Pressure loss in valve (bara).

Example 1.4 Isobutane with temperature and pressure values of 20 °C and 10 bar, respectively, passes through a control valve where the pressure is reduced to 2 bar. Calculate the flow factor of the valve, assuming the flow rate of isobutane is 2000 m^3/h.

Answer

$\frac{P_O}{P_i} = \frac{2}{10} = 0.2 < 0.5 \rightarrow$ It is a critical pressure drop, so the flow factor is calculated according to Eq. (1.6). The specific gravity of isobutane is 2.01 (refer to Table 1.1). Considering the air density equal to 1.225 kg/m^3, the isobutane density is calculated as follows:

$$SG_{isobutane} = \frac{\rho_{isobutane}}{\rho_{air}} \rightarrow 2.01 = \frac{\rho_{isobutane}}{1.225} \rightarrow \rho_{isobutane} = 2.46\ kg/m^3$$

The isobutane temperature is converted from °C to °K as follows:

$$T_i(°K) = T_i(°C) + 273.15 = 20 + 273.15 = 293.15\,°K$$

$$K_v = \frac{Q}{257p_i}\sqrt{\rho \times T_i} = \frac{2000}{257 \times 10}\sqrt{2.46 \times 293.15} = 20.90\ (m^3/h \cdot bar)$$

Conversions between the flow coefficient and flow factor are calculated from Eqs. (1.8) and (1.9).

Relationship Between Flow Coefficient and Factor

$$C_v = 1.16 \times K_v \tag{1.8}$$

$$K_v = 0.853 \times C_v \tag{1.9}$$

Example 1.5 Calculate the flow coefficient (C_v) for the control valve in Example 1.4.

Answer

$$C_v = 1.16 \times K_v = 1.16 \times 20.90 = 24.24 \text{ US gpm/psi}$$

1.2 Flow Coefficient Chart and Flow Curve

A flow coefficient chart or flow curve shows the relationship between the opening percentage and the amount or percentage of C_v or flow rate. Figure 1.2 is a flow characteristics chart or curve showing the relationship between the opening ratio of valves and the percentage of maximum flow or flow coefficient. Three flow

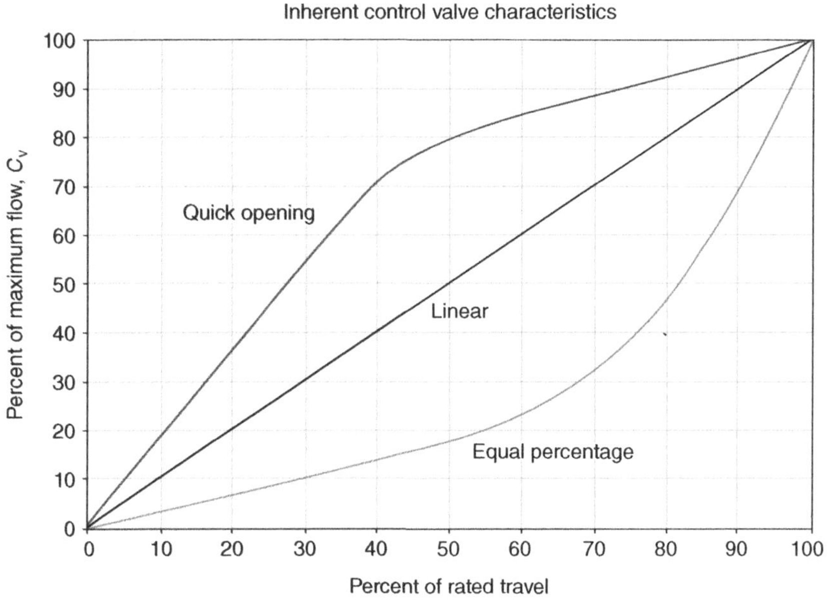

Figure 1.2 Flow characteristics of valves.

curves or characteristics are shown in the figure: quick opening, linear, and equal percentage. Quick opening means that a large amount of flow passes through the valve as it begins to open. A small valve opening, especially in the beginning, can lead to a high amount of flow. Linear flow means that the change in the flow is equal to the opening percentage. For example, 40% of opening a valve passes 40% of the fluid through the valve. An equal percentage means that the valve releases a relatively low amount of fluid at the beginning of the opening stage. For example, a valve that is getting opened by 35% to just start releasing the flow has an equal percentage flow characteristic. Control valves have typically equal percentage and linear flow characteristics, and the quick opening flow characteristic is most commonly used for on/off valves.

Example 1.6 A control valve with a linear flow characteristic has a flow coefficient of 700 at a fully open position. Calculate the C_v value of the valve at 5% and 95% opening percentages.

Answer

Flow rates and C_v values are increased linearly by increasing the valve opening percentage for a linear flow characteristic. Thus, values of flow rate and flow coefficient at any valve opening percentage for a linear flow characteristic are calculated according to Eqs. (1.10) and (1.11), respectively.

Valve Flow Rate Calculation Based on Valve Opening Percentage for a Linear Flow Characteristic

$$q_{vp} = q_{max} \times \text{vp} \tag{1.10}$$

where:

q_{vp}: Valve flow rate linked to the valve opening percentage;
q_{max}: Maximum flow rate passing through the valve at the fully open position;
vp: Valve opening percentage.

Valve Flow Coefficient Calculation Based on Valve Opening Percentage for a Linear Flow Characteristic

$$C_{vvp} = C_{vmax} \times \text{vp} \tag{1.11}$$

where:

C_{vvp}: Valve flow coefficient linked to the valve opening percentage;
C_{vmax}: Maximum flow coefficient of the valve at the fully open position.

$$C_{vvp} = C_{vmax} \times \text{vp} \rightarrow C_{v5\%} = 700 \times 0.05 = 35 \text{ and}$$

$$C_{v95\%} = 700 \times 0.95 = 665$$

Therefore, the flow coefficients of the valve at 5% and 95% opening percentages are 35 and 665, respectively.

Example 1.7 A control valve with an equal percentage flow characteristic has a flow coefficient of 640 at a fully open position. Calculate the C_v value of the valve at 10% and 90% opening percentages, respectively.

Answer

Flow rates and C_v values are not increased linearly by increasing the valve opening percentage for an equal percentage characteristic. Thus, values of flow rate and flow coefficient at any valve opening percentage for an equal percentage flow characteristic are calculated according to Eqs. (1.12) and (1.13), respectively.

Valve Flow Rate Calculation Based on Valve Opening Percentage for an Equal Percentage Flow Characteristic

$$q_{\text{vp}} = q_{\text{max}} \times \alpha^{\text{vp}-1} \tag{1.12}$$

where:

$$\alpha = 50$$

Valve Flow Coefficient Calculation Based on Valve Opening Percentage for an Equal Percentage Flow Characteristic

$$C_{\text{vvp}} = C_{\text{vmax}} \times \alpha^{\text{vp}-1} \tag{1.13}$$

where:

$$C_{\text{vvp}} = C_{\text{vmax}} \times \alpha^{\text{vp}-1} \rightarrow C_{\text{v10\%}} = 640 \times 50^{0.1-1} = 640 \times 0.02957 = 18.93$$

$$C_{\text{vvp}} = C_{\text{vmax}} \times \alpha^{\text{vp}-1} \rightarrow C_{\text{v90\%}} = 640 \times 50^{0.9-1} = 640 \times 0.676 = 432.80$$

Therefore, the flow coefficients of the valve at 10% and 90% opening percentages are 18.93 and 432.80, respectively.

Valves are divided into two major categories based on the closure member motion: linear and quarter-turn. Linear motion valves set off have a closure member that moves up and down with a linear motion to change the flow rate through the valve. Quarter-turn like ball and butterfly valves in which a closure member and the stem rotate 90° between the open and closed positions. The valve flow coefficient chart or flow characteristic for a quarter-turn valve could be provided as a relationship between the flow rate or C_v and the angle of the stem or closure member. It should be noted that the angle of stem and closure member for quarter-turn valves is equal. For example, Figure 1.3 illustrates a couple of flow coefficient charts or curves for a 38″ CL1500 (equivalent to 250 bar nominal pressure) ball valve. The chart here shows the relationship between the angle of stem and flow coefficient, whereas the lower chart illustrates the connection between the opening percentage and the flow coefficient of the valve. Even when the stem angle is 33°, no flow is passing through the valve, meaning that the valve has an equal percentage flow characteristic. In light of the fact that 90° rotation of the stem and closure member corresponds to 100% or full valve opening, using interpolation, a 33° stem angle corresponds to 36.6% opening. The full open C_v of the valve is the one typically provided by valve manufacturers on general arrangement (GA) valve drawings, which is 118,179 gpm in this case. GA valve drawings deliver the dimensions

Tables of estimated C_v values

Flow ratio (%)	Estimated stem angle (°)	Estimated C_v
0	33	3
5	36	666
10	39	1750
15	42	3169
20	44	4797
25	47	6756
30	50	8938
35	53	11,543
40	56	14,454
45	59	17,964
50	62	21,945
55	64	26,836
60	67	32,496
65	70	39,595
70	73	47,962
75	76	58,557
80	79	70,896
85	81	85,623
90	84	1,00,306
95	87	1,12,664
100	90	1,18,179

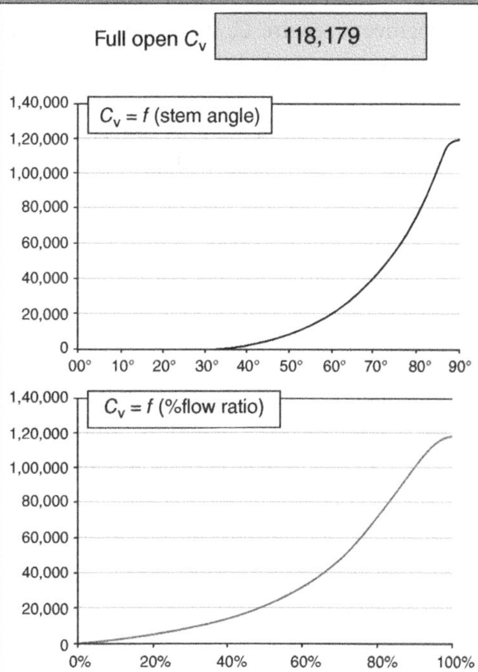

Figure 1.3 Flow curves for a 38″ CL1500 ball valve.

Table 1.2 Flow coefficient (C_V) values in gpm/psi for gate, globe, and check valves in the size range of 1/4″–3″ from one manufacturer.

NPS	DN	Reduced port gate	Full port gate	Globe	Piston check	Ball check	Swing check
1/4″	8	1.7	1.7	1	0.9	0.8	
3/8″	10	4.2	5.7	1.2	1	0.9	
1/2″	15	5.7	8.2	1.5	1.3	1.2	11
3/4″	20	8.2	22	2.4	2.1	1.9	17.6
1″	25	26	34	5.6	5	4.4	31.4
1 1/4″	32	37	60	15			
1 1/2″	40	60	92	21	12.6	10.5	57.6
2″	50	92	200	29	17.4	14.5	80.1
3″	80	200					

of main valve attributions, such as face-to-face or end-to-end, and valve height. Process engineers require C_V values to ensure that valves provide expected flow capacities. It is important to know that the flow coefficient depends on many factors such as the size and type of valves, type of design (e.g. full bore or reduced bore), and valve manufacturer. Table 1.2 provides C_V values in gpm/psi for gate, globe, and check valves in the size range of 1/4″–3″ from one manufacturer. Gate and globe valves are used for stop/start and flow regulation, respectively. Check valves are nonreturn valves that are opened by the fluid flow in one direction and prevent the flow from returning to the upstream side of the valve. NPS and DN in Table 1.2 represent nominal pipe size and dimension or diameter nominal, respectively. NPS and DN are two ways to express the size of the piping and valves. DN is expressed in metric units of a millimeter. For example, DN25 and DN50 imply dimensions/diameters of 25 mm and 50 mm, respectively. DN is the metric equivalent of NPS, and DN is calculated by multiplying the pipe or valve size in inches (NPS) by 25. For example, a 1″ NPS pipe or valve corresponds to DN25.

1.3 Rangeability and Turndown

Rangeability is another flow rate definition for industrial valves used for valve sizing. Rangeability is relevant for the valves used for flow control or throttling, such as control, globe, butterfly, plug, and V-notch ball valves. A valve with higher flow rangeability enables control of the flow over the wider flow range. A control valve is an actuated globe valve used to control the fluid by altering the size of the fluid passage. Direct flow rate control leads to changes in major process quantities such as pressure, temperature, and liquid level. An actuator is a mechanical or electrical device installed on top of industrial valves to operate (open and close) the valves automatically without any need for an operator. Rangeability is defined as the ratio of maximum flow to a minimum controllable flow of a valve installed in a piping system, which is calculated by Eq. (1.14). Thus, if the minimum controllable flow is 10% of the maximum controllable flow, the rangeability of the valve is 100/10 = 10. It is important to note that the minimum controllable flow is not the flow leakage through the valve when it is closed.

Valves are often not required to handle the maximum controllable flow; as an alternative, the maximum operation flow passes through the valve. Thus, the term turndown is equal to or smaller than rangeability and is calculated according to Eq. (1.15). For example, the valve for which the rangeability is calculated may need to have a maximum normal operation flow equal to only 70% of the maximum controllable flow, so the turndown is 70/10 = 7.

Rangeability Calculation

$$\mathrm{R} = \frac{\textit{Maximum controllable flow}}{\textit{Minimum controllable flow}} \tag{1.14}$$

Turndown Calculation

$$\mathrm{T} = \frac{\textit{Maximum normal operating flow}}{\textit{Minimum controllable flow}} \tag{1.15}$$

Example 1.8 The maximum and minimum controllable flows by a control valve are at 95% and 5% of their full opening positions, respectively. Calculate the valve rangeability in two cases of the linear and equal percentage flow characteristics.

Answer

$$R = \frac{\text{Maximum controllable flow}}{\text{Minimum controllable flow}} = \frac{\text{Flow at }95\%\text{valve position}}{\text{Flow at }5\%\text{valve position}}$$

$$R = \text{Linear flow characteristic} = \frac{95\%\ q_{\max}}{5\%\ q_{\max}} = \frac{95}{5} = 19$$

$$R = \text{Equal percentage flow characteristic} = \frac{q_{\max} \times \alpha^{0.95-1}}{q_{\max} \times \alpha^{0.05-1}} = \frac{50^{0.95-1}}{50^{0.05-1}}$$

$$= \frac{0.8223}{0.0243} = 33.84$$

Example 1.9 The flow coefficient of a valve in the fully open position is 640. The pressure drop of the fluid service, which is water, is 1 psi when the valve is fully open. Calculate the valve rangeability, assuming that the valve's maximum and minimum controllable flows are adjusted at 90% and 10% of their full opening positions, respectively, and the valve flow characteristic is linear.

Answer

$$C_v = Q\sqrt{\frac{SG}{\Delta P}} \rightarrow 640 = Q\sqrt{\frac{1}{1}} \rightarrow Q = q_{max} = 640 \text{ gpm}$$

$$10\% \text{ opening position} \rightarrow q_{vp} = q_{max} \times \text{vp} = 640 \times 0.1 = 64 \text{ gpm}$$

$$90\% \text{ opening position} \rightarrow q_{vp} = q_{max} \times \text{vp} = 640 \times 0.9 = 576 \text{ gpm}$$

$$R = \frac{\text{Maximum controllable flow}}{\text{Minimum controllable flow}} = \frac{\text{Flow at } 90\% \text{ valve position}}{\text{Flow at } 10\% \text{ valve position}}$$

$$= \frac{576}{64} = 9$$

1.4 Valve Authority

Valve authority is another identification for selecting and sizing the valves used for flow control, such as control valves. It expresses the ratio between the pressure drop across the valve used for flow control and the total pressure drop across the whole system as per Eq. (1.16).

Valve Authority

$$N = \frac{\Delta P_1}{\Delta P_1 + \Delta P_2} \tag{1.16}$$

where:

N: Valve authority (dimensionless);
ΔP_1: Pressure loss in a valve in a fully open position (bara);
ΔP_2: Pressure drop across the remainder of the system (circuit);
$\Delta P_1 + \Delta P_2$: Pressure drop across the whole system (circuit)

The conventional definition of a control valve authority (N) measures how much the system's pressure drop is related to the control valve. A valve authority value lower than 0.25 indicates that the valve for fluid control has unstable fluid control characteristics, so it is not recommended. Conversely, a valve authority value from 0.5 to 1 provides good to excellent fluid control but with a high pressure drop. So, it results in high energy consumption, which is undesirable. A valve authority between 0.25 and 0.5 provides fair to good control with a reasonable pressure drop. The best engineering practice for control valve selection and sizing is to keep the value of N near 0.5 but not greater than, e.g. 0.4.

Example 1.10 A system has a total pressure drop ($\Delta P_1+\Delta P_2$) of 1.25 bar. If the control valve in the system has a valve authority (N) equal to 0.45, what is the pressure drop across the valve? Assuming that the valve fluid is water with a flow rate of 15 m^3/h, what is the valve flow factor?

Answer

$$N = \frac{\Delta P_1}{\Delta P_1 + \Delta P_2} \rightarrow 0.45 = \frac{\Delta P_1}{1.25}$$
$$\rightarrow \Delta P_1 = 0.5625 \ \text{(Pressure drop across the valve)}$$

$$K_v = Q\sqrt{\frac{SG}{\Delta P}} \rightarrow K_v = 15\sqrt{\frac{1}{0.5625}} = 20 \ \left(m^3/h \cdot bar\right);$$

1.5 Valve Gain

Valve gain is the ratio of flow change to the valve closure member travel (stroke) or valve opening percentage. This parameter is calculated from Eq. (1.17). Gain should never be less than 0.5 to avoid any problem for the valve to control the flow. The gain same as the pressure drop is constant in the linear valve, whereas it increases as the valve opens in equal percentage flow characteristics.

Valve Gain

$$\text{Valve gain} = \frac{\text{Flow rate change}}{\text{Valve opening percentage change}} \tag{1.17}$$

Example 1.11 Three flow rates of 35 gpm, 100 gpm, and 170 gpm correspond to stroke or opening percentages of 42%, 69%, and 90%, respectively. Calculate gain values, and can this valve perform flow control?

Table 1.3 Valve flow rates and corresponding opening percentages.

Flow (gpm)	Stroke (%)	Change in flow (gpm)	Change in stroke (%)
35	42	100 − 35 = 65	69 − 42 = 27
100	69		
170	90	170 − 100 = 70	90 − 69 = 21

Answer

All provided information, including the valve flow rates and corresponding opening percentages, is summarized in Table 1.3:

$$\text{Valve gain \#1} = \frac{\text{Flow rate change \#1}}{\text{Valve opening percentage \#1}} = \frac{65}{27} = 2.41$$

$$\text{Valve gain \#2} = \frac{\text{Flow rate change \#2}}{\text{Valve opening percentage \#2}} = \frac{70}{21} = 3.33$$

The difference between the two gains should be less than 50% of the higher gain value.

$$3.33 - 2.41 = 0.92, \quad 50\% \times 3.33 = 1.65$$

Since 0.92 is less than 1.65, there should be no problem using the valve for flow control. Also, it should be noted that both gain values are larger than 0.5. Thus, the selected valve can perform flow control nicely without any operational problems.

Questions and Answers

1.1 Which answer is correct for a given size of a valve?

A The greater the flow, the less pressure drop
B The greater the pressure drop, the greater the flow
C The lesser the flow, the greater the pressure drop
D The greater the flow, the lesser the pressure drop

Answer

Option B is the correct answer. Increasing the pressure drop or differential pressure across a valve leads to a higher flow rate through the valve.

1.2 What statement is correct about the flow coefficient in the form of a C_V value?

A C_v is a theoretical value that an experiment cannot measure.
B C_v value can only be measured when the valve is fully open.

C C_v value is independent of the type of fluid passing through the valve.
D Metric units are not used for C_v value calculation.

Answer
Option A is incorrect because an experiment can measure C_v. Option B is not correct either since C_v can be measured from fully closed to fully open at any valve position. Option C is wrong because the type of fluid passing through the valve affects the fluid's density, specific gravity, and flow coefficient. Option D is the correct answer.

1.3 Which sentence is correct about valve flow characteristics?
A Opening a valve by 30% leads to passing 30% of the total flow capacity of the valve. This valve has a linear flow characteristic.
B A valve is opened 50%, while the fluid passage from the valve is just 20% of the total flow. This valve has the flow characteristic of quick opening.
C If it is needed to have a small flow passage at the beginning of the opening of the valve, a linear flow characteristic trim is required.
D A valve releases 25% of the total flow capacity at a 27% opening. This valve has a quick opening characteristic.

Answer
Option A is correct, and the valve-opening percentage is equal to the percentage of the flow capacity. So the characteristic is linear. Option B explains an equal percentage flow characteristic since the amount of flow at the half opening is relatively low. Therefore, option B is not correct. Option C is wrong because having a small flow at the beginning of the valve opening is an equal percentage and not linear. Option D is very similar and close to the linear flow characteristic. So it does not address the quick opening characteristic.

1.4 Water with a flow rate of 20 m^3/h is circulated in a piping system, including the valve. Calculate the pressure drop across the valve with a K_v value of 20.
A 0.5 bar
B 1 bar
C 2 bar
D 0.25 bar

Answer
The fluid service is water and liquid. So, Eq. (1.3) is applicable to fluid factor (K_v) calculation as follows:

$$K_v = Q\sqrt{\frac{SG}{\Delta P}} \rightarrow 20 = 20\sqrt{\frac{1}{\Delta P}} \rightarrow \Delta P = 1$$

Thus, option B is the correct answer.

1.5 Find the wrong statement about flow coefficient and flow factor.

A Conversion between the flow coefficient and flow factor gives a larger value to the flow coefficient than a flow factor.

B The value of the flow coefficient on a valve GA drawing is not given for the valve's fully open position.

C The flow factor is the metric version of the flow coefficient.

D A flow coefficient of 2 US gpm/psi is equal to a flow factor of 1.706 m^3/h · bar.

Answer

Except for option B, all options are correct since the flow coefficient values on the GA drawings are provided for the valve's fully open positions.

1.6 Find the correct choice about rangeability and turndown.

A The maximum normal operating flow is 66 gpm, and the minimum controllable flow is 3 gpm. The rangeability is equal to 22 in this case.

B Turndown is always smaller than rangeability.

C If the minimum controllable flow is 2% of the maximum controllable flow, the turndown is 50.

D It is impossible to have a flow rangeability of 1 : 1 for a control valve.

Answer

Option A is incorrect because 22 is turndown and not rangeability. Option B is not correct because turndown is equal to or smaller than rangeability. Option C is wrong because the value of 50 is for rangeability. Option D is the correct answer since globe valves usually provide a flow rate from 5% to 95%, giving a rangeability of 1 : 19.

1.7 Which parameter is related to the valve flow factor?

A C_v

B K_v

C A_v

D SG

Answer

Option B is the correct answer.

1.8 Figure 1.4 illustrates a couple of flow coefficient charts or curves for a 20″ CL1500 (equivalent to 250 bars' nominal pressure) ball valve. Which statement is correct for the valve in this example?

A The valve has a quick opening flow characteristic.

B The maximum flow rate passing through the valve in the fully open condition is 22,970 US gpm.

Tables of estimated C_v values

Flow ratio (%)	Estimated stem angle (°)	Estimated C_v
0	33	1
5	36	166
10	39	430
15	41	775
20	44	1171
25	47	1645
30	50	2192
35	53	2798
40	56	3521
45	59	4328
50	61	5301
55	64	6443
60	67	7740
65	70	9325
70	73	11,126
75	76	13,291
80	79	15,725
85	81	18,186
90	84	20,537
95	87	22,202
100	90	22,970

Full open C_v: 22,970

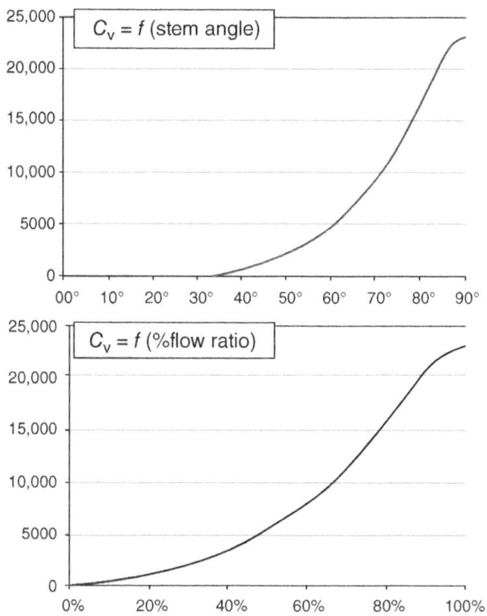

Figure 1.4 Flow curves for a 20″ CL1500 ball valve.

C The flow coefficient value on the valve GA drawing is 22,970 US gpm/psi.

D Half opening position of the valve provides a flow coefficient equal to 2192 US gpm/psi.

Answer

Option A is incorrect because the flow curves indicate an equal percentage flow characteristic. An equal percentage implies that the valve discharges a relatively low amount of fluid at the start of the opening stage. The curve on the top illustrates the relationship between the stem angle and flow coefficient, demonstrating that the flow starts passing through the valve after the stem angle of 33°, equal to 36% valve opening approximately. Option B is not correct either because there is no information available about the type of fluid and its specific gravity as well as the pressure drop across the valve. So, it is not possible to convert the flow coefficient to the flow rate. If the fluid is water and the pressure drop is one psi, option B would be correct. Option C is correct because valve manufacturers provide the value of the flow coefficient in the valve's fully open position on the valve's GA drawings. Option D is not correct because a 50° stem angle corresponds to a 55.55 valve opening percentage, giving a flow coefficient of 2192 US gpm/psi. Thus, option C is the right answer.

1.9 What is the optimum range of valve authority?

A 0.1–0.25
B 0.25–1.0
C 0.5–0.75
D 0.25–0.5

Answer

Option D is the correct answer.

1.10 Calculate the valve authority if the pressure drop across the control valve is 40 kPa and the total pressure drop in the remainder of the system is 60 kPa.

A 0.4
B 0.33
C 1.5
D 0.25

Answer

$$N = \frac{\Delta P_1}{\Delta P_1 + \Delta P_2} = \frac{40}{40 + 60} = 0.4$$

where:
N: Valve authority (dimensionless);
ΔP_1: Pressure loss in a valve in a fully open position (bara);
ΔP_2: Pressure drop across the remainder of the system (circuit);
$\Delta P_1 + \Delta P_2$: Pressure drop across the whole system (circuit).

Thus, option A is the correct answer.

Further Reading

Bahadori, A. (2014). *Natural Gas Processing*, 1e. Austin, TX: Elsevier (Gulf Professional Publishing).

Bolton, W. (2021). *Instrumentation and Control Systems*, 3e. Austin, TX: Elsevier (Gulf Professional Publishing).

Boyes, W. (2010). *Instrumentation Reference Book*, 4e. Oxford: Butterworth-Heinemann.

Nesbitt, B. (2007). *Handbook of Valves and Actuators: Valves Manual International*, 1e. Oxford: Elsevier.

Skousen, P.L. (2011). *Valve Handbook*, 3e. New York: McGraw-Hill.

Smit, P. and Zappe, R.W. (2004). *Valve Selection Handbook*, 5e. New York: Elsevier.

Sotoodeh, K. (2021). *Subsea Valves and Actuators for the Oil and Gas Industry*, 1e. Austin, TX: Elsevier (Gulf Professional Publishing).

2

Valve Sizing

2.1 Introduction

Valve sizing is an essential engineering task that must be done accurately since improper valve sizing is both expensive and inconvenient. If a valve is too small, it cannot pass the required flow and compromises flow assurance and process integrity. On the other hand, an oversized valve increases the cost. Valve sizing is typically conducted based on both theoretical and experimental methods: Valve sizing is split based on the valve's applications, such as isolation valves, nonreturn or check valves, control valves, and safety valves.

2.2 Isolation Valve Sizing

A ball, plug, or gate valve that stops and starts fluids normally has the same size as the pipe it is connected to. A 3″ isolation valve, for example, is typically connected to a 3″ pipe. It is possible for ball valves to have a full or reduced bore, and the bore is defined by the size of the hole inside the ball. Full-bore ball valves in 3-in. sizes have a nominal pipe size (NPS) of 3″. Reduced bore ball valves in 3 in. sizes have a NPS of 3″ × 2″. As an example, a 3″ refers to the size of the valve end connected to the piping, whereas a 2″ refers to the size of the valve bore. Engineers intend to use a reduced bore ball valve as a first choice because of its lower cost unless the reduced bore causes flow assurance problems or undesirably very high-pressure drop. As an example, ball valves on subflare lines which are installed before and after pressure safety valves (PSVs) shall be a full bore to allow a sudden release of fluid from the piping system to the flare line upon overpressurization. Further, some ball valves are subject to the piping injected gadget (PIG) running for cleaning or maintenance. So these valves must have an internal diameter equal to the

Industrial Valves: Calculations for Design, Manufacturing, Operation, and Safety Decisions, First Edition. Karan Sotoodeh.

internal diameter of the connected piping. Using the following example, it is possible to look at how to select a ball valve for stopping/starting the flow in a piping system. This will satisfy the process engineers' need to set the flow rate.

Example 2.1 A ball valve is installed on the water piping that is 4 in. in diameter and 60 °F. According to the process engineer, the valve has a flow rate of 8000 US gallons per minute (gpm). Valve inlet and outlet pressures are 600 psi and 585 psi, respectively. Tables 2.1 and 2.2 provide the flow coefficient values provided by the valve manufacturer for full bore and reduced bore ball valves in size ranges ranging from 1/2″ to 20″. Select the appropriate size for the valve.

Table 2.1 Flow coefficient (C_v) for full-bore ball valves (size ranges: 1/2″–18″).

Valve size									
in.	1/2″	3/4″	1″	1.5″	2″	3″	4″	6″	8″
mm	12	19	25	37.5	50	75	100	150	200
Flow coefficient (C_v)	26	50	94	560	480	1300	2300	5400	10,000
in.	10″	12″	14″	16″	18″				
Mm	250	300	350	400	450				
Flow coefficient (C_v)	16,000	24,000	31,400	43,000	57,000				

Table 2.2 Flow coefficient (C_v) for reduced bore ball valves (size ranges: 3″–20″).

Valve size									
in.	3″	4″	6″	8″	10″	12″	14″	16″	18″
mm	75	100	150	200	250	300	350	400	450
Flow coefficient (C_v)	420	770	1800	2500	4500	8000	12,000	14,000	18,000
in.	20″								
mm	500								
Flow coefficient (C_v)	22,000								

Ball valves in the size range of 2″ and below are not reduced bore typically.

Answer

$$C_V = Q\sqrt{\frac{SG}{\Delta P}} \rightarrow C_V = 8000\sqrt{\frac{1}{15}} = 2066\ \text{gpm/psi}$$

The first ball valve choice is a reduced bore 4″ × 3″ ball valve, which is a cheaper alternative. However, a 4″ reduced bore ball valve has a flow coefficient of 770 gpm/psi as per Table 1.2 which does not meet the required flow coefficient of 2066 gpm/psi. Thus, a reduced bore ball valve is not appropriate in this case. Alternatively, the flow coefficient of a 4″ full-bore ball valve is 2300 gpm/psi, which is compatible with the required flow coefficient. Thus, a 4″ full-bore ball valve is the best choice.

Butterfly valves have gained popularity for flow isolation since they are more compact and lighter than gate and ball valves. A butterfly valve is a reduced bore valve, which is helpful to know. Butterfly valves, which are quarter-turn valves, have advantages over ball valves, gate valves, and plug valves, such as saving weight, space, and costs, as well as the amount of torque required to open and close a butterfly valve. A quarter-turn valve has a valve closure member and stem that are rotated 90° between open and closed positions. A butterfly valve may also have a wafer-type design or a flangeless design, which enables the installation of the valve between two flanges. Based on the required flow coefficient and the inlet and outlet pipe sizes, the following example shows how to size a butterfly valve.

Example 2.2 Flow coefficient values for the 70° open and fully open positions of butterfly valves are shown in Table 2.3 based on the inlet and outlet pipe sizes from 4″ to 10″.

The crude oil at 60 °F with a specific gravity of 0.8 and flow rate of 3500 US gpm is passing through the butterfly valve at the fully open position. The butterfly valve's inlet and outlet pressure values are 10 psi and 7 psi, respectively. What should be the size of the valve, inlet, and outlet pipe?

Answer

The advantage of the flow coefficient values shown in Table 2.3 is that they have been corrected to account for the effects of *piping geometry factors*. The butterfly valve C_V values in the table have been adjusted to account for pressure losses caused by pipe fittings, such as reducers and expanders, which might be connected directly to the butterfly valve's inlet or outlet connections. The second point is that in this example, the butterfly valve is used for flow isolation. Therefore, the second column from the right, which shows the flow coefficient at 70° open, is for flow control or throttling, and it would not be applicable here. The next section calculates the required valve flow coefficient for a fully open position.

Table 2.3 Flow coefficient values for butterfly valves based on the inlet and outlet pipe sizes from 4″ to 10″.

Inlet pipe size	Butterfly valve size	Outlet pipe size	Flow coefficient in 70° open position	Flow coefficient in fully open position
4″	4″	4″	496	841
6″	4″	4″	348	442
6″	4″	6″	405	539
6″	6″	6″	1025	1850
8″	6″	6″	784	1016
8″	6″	8″	913	1359
8″	8″	8″	1862	3316
10″	8″	8″	1462	1922
10″	8″	10″	1711	2633
10″	10″	10″	2948	5430

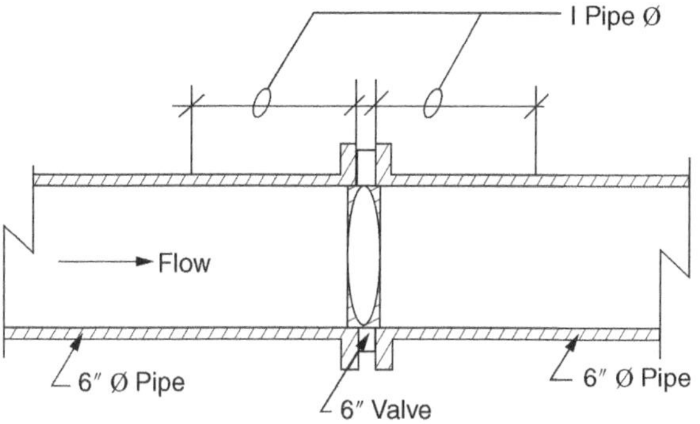

Figure 2.1 A 6″ butterfly valve connected to a 6″ pipe.

$$C_v = Q\sqrt{\frac{SG}{\Delta P}} \rightarrow C_v = 3500\sqrt{\frac{0.8}{3}} = 1807\ \text{gpm/psi}$$

According to Table 1.3, a 4″ butterfly valve can provide a maximum of 841 gpm/psi; that is not sufficient. Furthermore, a 6″ butterfly valve with an 8″ inlet pipe and 6″ or 8″ outlet pipe sizes only can give 1016 or 1359 gpm/psi flow coefficient values that are less than the required flow coefficient. An 8″ butterfly valve with both 8″ inlet and outlet pipe sizes is counted as a valve and piping oversizing scenario since it gives a very large flow coefficient value of 3316 gpm/psi. The best solution, as shown in Figure 2.1, is to select a 6″ butterfly valve with both 6″ inlet and outlet pipe sizes to provide a C_v value of 1850 gpm/psi.

2.3 Nonreturn (Check) Valve Sizing

Check valves are automatic valves that open with the forward flow and close with the reverse flow. Among the different types of check valves are swing check valves, lift check valves, dual plate check valves, etc. Check valves are sometimes installed after mechanical equipment, such as pumps and compressors, to prevent fluid reversal. Check valves are also designed to prevent massive backflow when a pipe breaks and to prevent backflow toward lower pressures. In contrast to a ball, butterfly, or gate valve, a check valve does not require an operator, and it is operated automatically by the fluid inside the pipe. Figure 2.2 illustrates a swing check valve at the top and a piston check valve at the bottom. In this case, the closure member swings around the hinge, which is mounted in the seat. As illustrated in Figure 2.2, top right, a swing check valve is illustrated with the disk moving away from the seat by the force of the fluid. However, the reverse flow on the top left is stopped by the disk in the closed position. As shown in the figure, the piston check valves at the bottom exhibit the same concept of flow movement and stoppage.

Swing check valves have been used widely in water and wastewater piping systems. These valves are available on the market at a low cost and provide low head loss when they are open. The disk of the valve swings between 60 and 90°

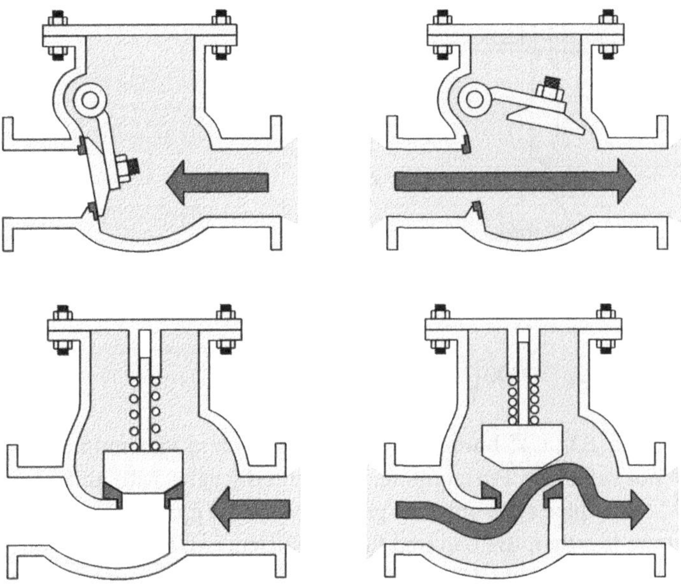

Figure 2.2 A swing check valve at the top and a piston check valve at the bottom. *Source:* Sergey Merkulov/Shutterstock.

(fully open) during the passage of fluid. The main problem with this valve during operation is that the disk slams against the seat during the closing of the valve. The long stroke of the disk and the sudden closure of the valve because of the weight of the disk exacerbate the slamming effect in swing check valves. Thus, it can conclude that a swing check valve is an economical option for a check valve with a high slamming effect. The alternative solution to minimize the slamming impact is to use a dual plate (see Figure 2.3) or an axial flow nozzle check valve (see Figure 2.4). There are double disks instead of one for the dual plate check valve. So each half-disk applies less weight force and slamming effect in its closing action. In addition, the disks are closed by spring force rather than just relying on their weight. The advantages of a dual plate check valve compared to a swing check valve are not limited to space-saving and lower slamming rate. The total cost of a dual plate check valve including the sum of initial, maintenance, and energy costs are less than a swing check valve.

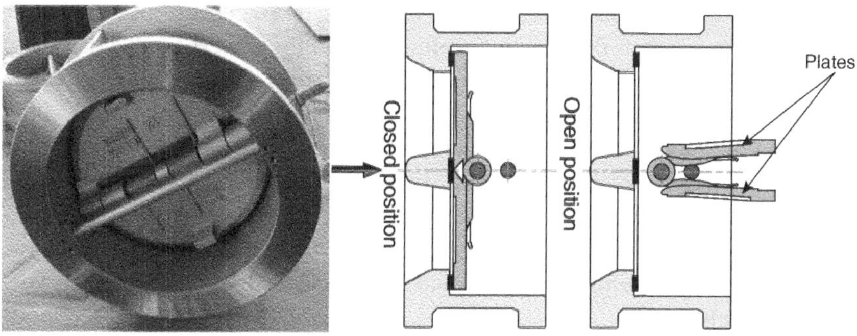

Figure 2.3 Dual plate check valve highlighting double plates.

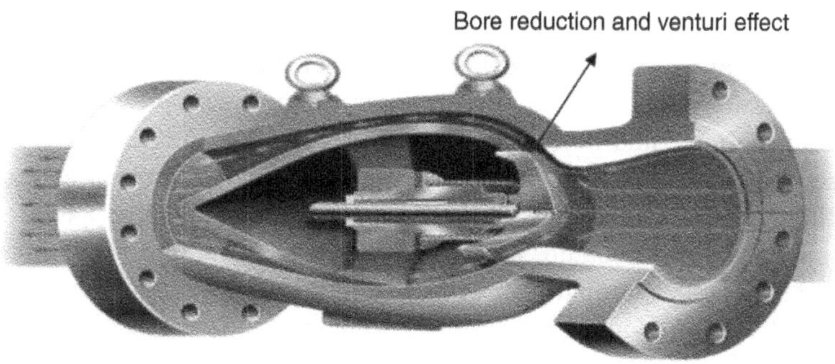

Figure 2.4 Axial flow nozzle check valve.

There are different pieces of literature in which nozzle check valves are proposed to be selected after rotating equipment such as pumps and compressors to protect this expensive equipment from backflow damage. These highly engineered check valves (see Figure 2.4) are specially designed to eliminate rotating equipment failure from backflow. The unique design of nozzle check valves minimizes the effect of water hammering and chattering. In addition, this type of valve provides very low-pressure drop as well as fast cycling (opening and closing).

It is a common practice to size a check valve in accordance with the pipe diameter, as discussed for sizing isolation valves. In order to prevent high-pressure drops and an increase in energy consumption inside the valve, it is necessary to ensure that the minimum flow capacity maintains the valve disk or closure member in the fully open position. During periods of sufficient flow, the valve's disk is continuously pushed upward and can be completely opened. As the flow capacity decreases to a minimum and the valve cannot remain fully open, the disk begins to vibrate. This causes repeated slams on the body and wears out the valve. Therefore, unscheduled maintenance work may need to be performed, or the valve may need to be replaced. The occurrence of chatter or flutter occurs when the flow is insufficient to fully open the valve disk (e.g. flow through the valve is less than the critical velocity). A valve that chatters or flutters will eventually fail prematurely. Another disadvantage of a low flow rate is that the flow cannot achieve the flow capacity and C_v value that the valve can deliver because of high pressure. In the event that a check valve is too small, it results in excessive pressure drop, reduced flow, and increased pump energy requirement. On the other hand, a check valve sized too large may have an unstable disk, resulting in valve wear, increased maintenance, or even failure of the valve.

In order to determine the check valve size, there are two key terms: the first is the cracking pressure of the valve, and the second is the critical velocity. During the initial opening of a check valve, for example, during the start-up of the piping system, the pressure applied by the flow upstream of the valve to the valve's disk or closure member shall overcome the force exerted by the spring and any downstream backpressure acting on the back part of the disk. Cracking pressure is the pressure differential that occurs when the upstream pressure overcomes the spring force and other applicable forces and the valve disk cracks open. A second significant factor is critical velocity, which is defined as the required fluid velocity to keep the plates or disks of the valves fully open. In addition, all check valves should be in a fully open position, and for spring-loaded check valves such as dual-plate or axial-flow check valves, the force provided by the flowing fluid must be greater than the spring force. Equation (2.1) is used to calculate the minimum flow necessary to fully open a swing check valve.

Minimum Flow Velocity to Fully Open a Swing Check Valve

$$V_{\min} = 55 \times \sqrt{\frac{1}{\rho}} \tag{2.1}$$

where:

$V_{\min}$: Minimum flow velocity (ft/s);
ρ : Fluid density (lb/ft^3).

Example 2.3 A dual-plate check valve in 12″ and Class 300 has been selected for the gas service with a design temperature of 100 °C and density of 5.87 kg/m^3 in a piping system with the same size as the valve. The pipe minimum and maximum flow capacities are 954.4 m^3/h and 7818.45 m^3/h, respectively. Is the valve size correct?

Answer

Calculation of the minimum and maximum velocities in the pipe is the first step. The fluid velocity in the pipe is not uniform throughout the pipe. As a result, a mean velocity is used, calculated using Eq. (2.2). The velocity of the fluid depends on the cross-section area of the pipe and the liquid flow rate, as shown in Eq. (2.2) and Figure 2.5.

Relationship Between the Flow Rate, Flow Velocity, and Cross-Section Area of the Pipe

$$Q = A \times V \tag{2.2}$$

where:

Q = liquid flow rate in pipe (m^3/s);
A = Cross-section area of the pipe or channel (m^2);
V = Mean fluid velocity in pipe (m/s).

The area of the pipe (A) is calculated using Eq. (2.3).

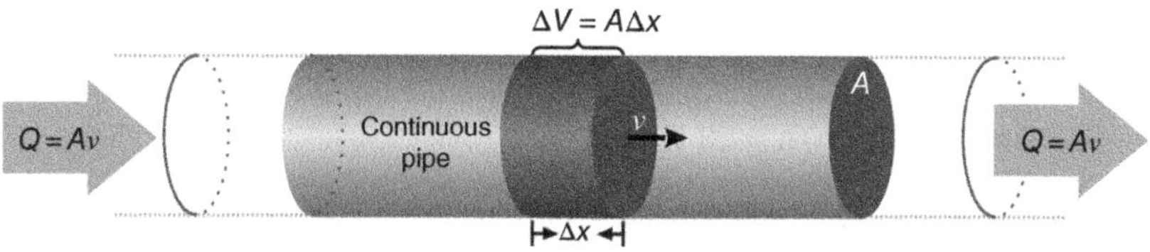

Figure 2.5 Relationship between the flow rate, flow velocity, and cross-section area of the pipe. *Source:* ScientificStock/Adobe Stock.

Pipe Cross-Section Area Calculation

$$A = \pi \left(\frac{\text{ID}}{2}\right)^2 \tag{2.3}$$

where:

ID = internal diameter of the pipe (mm).

The internal diameter of the pipe is calculated using Eq. (2.4).

Pipe Internal Diameter Calculation

$$\text{ID} = \text{OD} - 2 \times t \tag{2.4}$$

where:

OD = outside diameter of the pipe (mm);
t = piping wall thickness (mm).

The outside diameter of a pipe depends on the NPS; in this case, the outside diameter of a 12″ pipe can be determined by referring to the American Society of Mechanical Engineers (ASME) B36.10M or ASME B36.19M standard. Generally, the outside diameter of a pipe equals the NPS if the pipe is 14″ and over. According to ASME B36.19M standards, the outside diameter of a 12″ pipe is 12.750″. As a general rule, pipe thickness is calculated based on the relevant ASME code, such as ASME B31.3, process piping code. In the example, the pipe thickness is 6.35 mm, so the pipe ID will be calculated as follows:

$$\text{ID} = 12.750\text{ in.} \times \frac{25.4\text{ mm}}{1\text{ in.}} - 2 \times 6.35 = 323.85 - 12.7 = 311.15\text{ mm}$$

The next step is to calculate the pipe cross-section area as follows:

$$A = \pi \left(\frac{\text{ID}}{2}\right)^2 = \pi \left(\frac{311.15}{2}\right)^2 = 76{,}035.5\text{ mm}^2 = 0.07603\text{ m}^2$$

The next step is to calculate the minimum and maximum flow velocities in the pipe using Eq. (2.2).

$$\begin{aligned} Q_{\max} &= A \times V_{\max} \rightarrow 7818.45\text{ m}^3/\text{h} = 0.07603\text{ m}^2 \times V_{\max} \rightarrow V_{\max} \\ &= 1{,}02{,}834\text{ m/h} = 28.56\text{ m/s} \end{aligned}$$

$$\begin{aligned} Q_{\min} &= A \times V_{\min} \rightarrow 954.4\text{ m}^3/\text{h} = 0.07603\text{ m}^2 \times V_{\min} \rightarrow V_{\min} \\ &= 12{,}553\text{ m/h} = 3.49\text{ m/s} \end{aligned}$$

The next step is to calculate the flow velocity in the valve using Eq. (2.5).

Flow Velocity Calculation Inside the Nozzle Check Valves

$$V_{\text{valve}} = V_{\text{pipe}} \times \frac{A_{\text{pipe}}}{A_{\text{valve}}} \tag{2.5}$$

where:

V_{valve}: Fluid velocity inside the valve;
V_{pipe} : Fluid velocity inside the pipe;
A_{pipe}: Fluid area inside the pipe; and
A_{valve}: Fluid area inside the valve.

Assuming that the valve flow area is 70% of the pipe cross-section area, the minimum and maximum flow velocity values in the valve are calculated as follows:

$$V_{\text{valve (min.)}} = V_{\text{pipe (min.)}} \times \frac{A_{\text{pipe}}}{A_{\text{valve}}} = \times 3.49 \times \frac{1}{0.7} = 4.99\ \text{m/s}$$

$$V_{\text{valve (max.)}} = V_{\text{pipe (max.)}} \times \frac{A_{\text{pipe}}}{A_{\text{valve}}} = 28.56 \times \frac{1}{0.7} = 40.8\ \text{m/s}$$

A critical velocity is equal to the minimum flow velocity required to completely open the valve. The critical velocity of spring-loaded check valves, such as dual plate and axial flow check valves, is dependent on the spring torque. A valve manufacturer, for example, uses four categories of spring torque: Mini-torque, Low Torque, Standard Torque, and Super Torque. Table 2.4 shows the critical velocity associated with each spring torque. The critical velocity values are affected by various factors, such as the valve design, the fluid type, the spring torque value, or the valve installation direction. Therefore, they differ from one manufacturer to another.

In this example, the valve manufacturer has selected a low torque spring giving the critical water flow velocity of 2 m/s. If a high or super torque spring is used, the gas flows inside the piping cannot fully open the valve. The valve engineer evaluates the critical velocity values given in the table for a valve on the horizontal

Table 2.4 Minimum flow or critical velocity for the dual plate check valve based on the spring torque in this case.

Spring torque	Critical velocity (m/s)
Mini-torque	1.5
Low torque	2
High torque (standard)	3
Super torque	4.4

pipe. In general, valve manufacturers do not recommend installing the check valves on the vertical lines with a downward flow because the combination of valve disk weights plus the downward fluid can keep the valve always in the fully open position. However, for the valve installed in a vertical line with the upward flow, the critical velocity should be adjusted to be sufficient to overcome the weight of the valve's disks and the spring torque. The other important consideration is that the values provided in Table 2.4 are relevant to the water flow. Equation (2.6) shows the relationship between the critical water velocity from Table 2.4 and the valve medium critical velocity.

Critical Velocity Conversion Between Water and the Medium

$$V_{\text{medium}} = V_{\text{water}}\sqrt{\frac{\rho_{\text{water}}}{\rho_{\text{medium}}}} \tag{2.6}$$

where:

V_{medium}: Critical medium velocity or minimum medium velocity to fully open the dual plate check valve (m/s);
V_{water}: Critical water velocity or minimum water velocity to fully open the dual plate check valve (m/s);
ρ_{water}: Water density equal to 1000 kg/m^3;
ρ_{medium}: Medium density in kg/m^3.

$$V_{\text{medium}} = V_{\text{water}}\sqrt{\frac{\rho_{\text{water}}}{\rho_{\text{medium}}}} = 2 \times \sqrt{\frac{1000}{5.87}} = 26.10\ \text{m/s}$$

The minimum velocity to keep the valve in the fully open position is 26.10 m/s, while the fluid velocity in the valve varies between 4.99 and 40.8 m/s. Thus, the fluid inside the valve cannot always keep the valve in the fully open position. The valve is probably oversized. So the solution is to reduce the valve size. This solution can be coupled with selecting a lower spring torque to reduce the valve flow restriction. The next example reviews the minimum flow requirement for an axial flow check valve.

Example 2.4 An axial flow or nozzle check valve in 10″ and Class 150 has been selected for a 10″ pipe containing the gas service with a design temperature of 97.5 °C and density of 5.87 kg/m^3. The valve's flow rate is constant and equal to 31,263 kg/h, and the valve has a low torque spring (see Table 2.5). Assuming that the flow area inside the valve is 75% of the pipe cross-section area, is the valve sized correctly? (*Note*: Pipe ID is 264 mm.)

Table 2.5 Minimum flow or critical velocity for the axial flow check valve in this case.

Spring torque	Critical velocity (m/s)
Mini-torque	1.5
Low torque	2
High torque (standard)	2.5
Super torque	3

Answer

The critical water velocity for the axial valve considering the low torque spring is equal to 2 m/s, according to Table 2.5. However, the fluid service, in this case, is a gas. So the critical water velocity shall be converted to the critical gas velocity by Eq. (2.6) as follows:

$$V_{\text{medium}} = V_{\text{water}}\sqrt{\frac{\rho_{\text{water}}}{\rho_{\text{medium}}}} = 2 \times \sqrt{\frac{1000}{5.87}} = 26.10\ \text{m/s}$$

The minimum flow velocity to keep the check valve open is 26.10 m/s. The next step is to calculate the minimum flow rate in kg/h, and compare it with the valve normal flow rate, which is 25,263 kg/h.

The cross-section area of the pipe is calculated by Eq. (2.3) as follows:

$$A = \pi\left(\frac{\text{ID}}{2}\right)^2 = 3.14\left(\frac{0.264}{2}\right)^2 = 0.055\ \text{m}^2$$

The area of the valve is 75% of the pipe cross-section area equal to $75\% \times 0.055\% = 0.041\ \text{m}^2$.

Now it is possible to calculate the minimum flow rate to keep the check valve fully open by Eq. (2.2) as follows:

$$\begin{aligned} Q\left(\text{m}^3/\text{s}\right) &= V\,(\text{m/s}) \times A\left(\text{m}^2\right) \rightarrow Q = 26.10\left(\text{m}^2/\text{s}\right) \times 0.055\left(\text{m}^2\right) \\ &= 1.4355\left(\text{m}^3/\text{s}\right) \times 3600\,(\text{s/h}) = 5168\left(\text{m}^3/\text{h}\right) \times 5.87\left(\text{kg/m}^3\right) \\ &= 30{,}335\,(\text{kg/h}) \end{aligned}$$

Any flow rate less than 30,335 kg/h cannot keep the axial flow check valve fully open. The normal flow rate in the valve is constant and equal to 31,263 kg/h. So the valve can always remain in the fully open position, and the valve is sized correctly considering the minimum flow requirement.

2.4 Control Valve Sizing

Modern processing plants use many control loops (see Figure 2.6) to produce and deliver a control product to the market. These control loops, including control valves, are designed to keep process variables such as pressure, temperature, level, flow, etc., within the required operating range and ensure that a quality product has been produced. To avoid the effect of process variations on the desired set point, sensors collect information from the process variable, and transmitters transform the information to the controller (e.g. control room). A controller processes the information and decides on the action to return the process variable to the desired point. The most common final control element in process control is the control valve.

A control valve (see Figure 2.7) is a type of instrument valve selected to control the fluid flow by changing the fluid passage size and regulating the process variable as close as possible to the desired set point. This section intends to provide a new control valve size selection method coupled with real industrial examples. Control valve size selection is divided into two categories for liquids and gases. The next section explains the sizing of valves for liquids.

2.4.1 Control Valve Sizing for Liquids

There are six steps for a control valve size selection in liquids:

1) Specify the variables required to size the valve;
2) Determine the equation constant (N);
3) Determine the piping geometry factor (F_P);
4) Determine the maximum flow rate upstream of the valve (q_{max}) and the allowable pressure drop (ΔP_{max});
5) Solve for required flow coefficient (C_v) value using an appropriate equation;
6) Select the appropriate flow coefficient table and the calculated flow coefficient valve to select the suitable valve size.

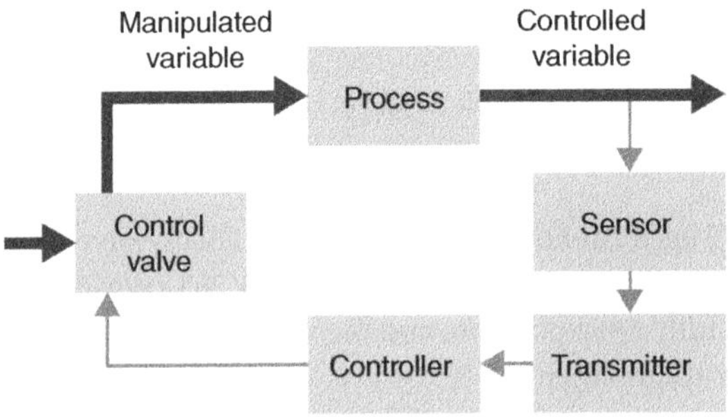

Figure 2.6 A control loop.

Figure 2.7 A control valve in the plant. *Source:* Dale/Adobe Stock.

2.4.1.1 Specify the Variables Required to Size the Valve

The first step is to specify the variables required to size the valves, such as valve pressure class, desired valve size, and design type. The pressure classes, which ASME B16.34 covers, are 150 (PN20), 300 (PN50), 600 (PN100), 900 (PN150), 1500 (PN250), 2500 (PN420), and 4500 (PN720). "PN" stands for pressure nominal. In addition to size and pressure class, the following process conditions shall be determined by the process or valve engineer:

- Process fluid (e.g. oil, water);
- Service conditions such as volume rate of flow (q), mass rate of flow (w), upstream and downstream pressure values (P_1 and P_2), pressure drop across the valve ($\Delta P = P_1 - P_2$), upstream temperature (T_1), liquid specific gravity that is a dimensionless number defined as the ratio of the liquid density at service temperature to the density of water at 60 °F (SG_l), vapor pressure absolute of liquid at inlet temperature (P_v), absolute thermodynamic critical pressure (P_c), and fluid or gas velocity (v). All parameters for control valve size selection are summarized in Table 2.6.

Table 2.6 Control valve sizing parameters and definitions.

Symbol	Definition	Symbol	Definition
C_v	Valve sizing coefficient	P_1	Upstream or inlet pressure
d	Normal valve size	P_2	Downstream or outlet pressure
D	Inlet diameter of the piping	P_c	Absolute thermodynamic critical pressure
F_D	Valve style modifier, dimensionless	P_v	Absolute vapor pressure of the liquid at inlet temperature
F_F	Liquid critical pressure ratio factor, dimensionless	ΔP	Pressure drop across the valve ($\Delta P = P_1 - P_2$)
F_k	Ratio of specific heat factor, dimensionless	$\Delta P_{max(L)}$	Maximum allowable liquid sizing pressure drop
F_L	Rated liquid pressure recovery factor, dimensionless	$\Delta P_{max(LP)}$	Maximum allowable sizing pressure drop with attached fittings
F_{LP}	Combined liquid pressure recovery factor and piping geometry factor of a valve with attached fittings, dimensionless (*Note*: if no fitting attached, $F_{LP} = F_L$)	q	Volume flow rate
F_P	Piping geometry factor, dimensionless	q_{max}	Maximum flow rate at given valve upstream condition
SG_l	Liquid-specific gravity that is a dimensionless number defined as the ratio of the liquid density at service temperature to the density of water at 60 °F	T_1	Absolute upstream temperature
SG_g	Gas-specific gravity that is defined as a dimensionless number equal to the ratio of the flowing gas density to the density of air with both at standard conditions	W	Mass rate of flow
k	Ratio of specific heats, dimensionless	X	Ratio of pressure drop to the upstream or inlet pressure ($\Delta P/P_1$)
K	Head loss coefficient of a valve due to installed fittings attached to the valve, dimensionless	X_T	Rated pressure drop ratio factor, dimensionless
M	Molecular weight, dimensionless	Y	Expansion factor (ratio of flow coefficient for a gas to that for a liquid at the same Reynolds number), dimensionless
N	Numerical constant	Z	Gas compressibility factor, dimensionless
γ	Specific weight at the valve inlet condition (*Note*: The standard condition is defined as 60 °F and 14.7 psia)	μ	Viscosity

2.4.1.2 Determine the Equation Constant (*N*)

In fact, many equations used in control valve sizing contain a numerical constant known as parameter *N*. *N* is a numerical constant in each flow equation to provide numeric values depending on the unit system used for calculations. Values for the various constants and the applicable units are given in Table 2.7. For example, use N_1 if sizing the valve for a flow rate in volumetric units such as gallon per minute (gpm) or NM^3/h. Alternatively, use N_6 if the valve is sized based on mass flow rate units such as lb/h or kg/h.

2.4.1.3 Determine Piping Geometry Factor (F_P)

F_P is a correction factor that accounts for the pressure losses due to piping fittings that are used to change the direction or size of the piping or take branches from the headers such as tees, elbows, and reducers. If fittings are used before or after the valve, the effect of piping geometry must be considered for the valve sizing. However, if no fitting is attached to the valve, F_P is equal to one and does not impact the valve-sizing process. Equation (2.7) shows how to calculate the piping geometry.

Piping Geometry Calculation

$$F_P = \left[1 + \frac{\sum K}{N_2}\left(\frac{C_V}{d^2}\right)^2\right]^{-1/2} \tag{2.7}$$

where:

N_2: Numerical constant found in Table 2.7;
d: Assumed nominal valve size;
C_V: Valve-sizing coefficient.

In Eq. (2.7), the $\sum K$ is the sum of the velocity head loss coefficients of all the fittings connected to the control valve from both upstream and downstream. The value of $\sum K$ is calculated from Eq. (2.8).

Valve Head Loss Coefficient Calculation Due to Attached Fittings

$$\sum K = K_1 + K_2 + K_{B1} - K_{B2} \tag{2.8}$$

where:

K_1: Head loss or resistance coefficient due to upstream fittings;
K_2: Head loss or resistance coefficient due to downstream fittings;
K_{B1}: Inlet Bernoulli coefficient;
K_{B2}: Outlet Bernoulli coefficient.

Table 2.7 Equation constants.

		N	w	q	p	γ	T	d, D
N_1		0.0865	—	NM^3/h	kPa	—	—	—
		0.865	—	NM^3/h	bar	—	—	—
		1.00	—	gpm	psia	—	—	—
N_2		0.00214	—	—	—	—	—	mm
		890	—	—	—	—	—	in.
N_5		0.00241	—	—	—	—	—	mm
		1000	—	—	—	—	—	in.
N_6		2.73	kg/h	—	kPa	kg/m^3	—	—
		27.3	kg/h	—	bar	kg/m^3	—	—
		63.3	lb/h	—	psia	lb/ft^3	—	—
N_7	Normal condition	3.94	—	NM^3/h	kPa	—	°K	—
	($T = 0\,°C$)	394	—	NM^3/h	bar	—	°K	—
	Standard condition	4.17	—	NM^3/h	Kpa	—	°K	—
	($T = 16\,°C$)	417	—	NM^3/h	bar	—	°K	—
	Standard condition ($T = 60\,°F$)	1360	—	SCFH	psia	—	°R	—
N_8		0.948	kg/h	—	kPa	—	°K	—
		94.8	kg/h	—	bar	—	°K	—
		19.3	lb/h	—	psia	—	°R	—
N_9	Normal condition	21.2	—	NM^3/h	Kpa	—	°K	—
	($T = 0\,°C$)	2120	—	NM^3/h	bar	—	°K	—
	Standard condition	22.4	—	NM^3/h	Kpa	—	°K	—
	($T = 16\,°C$)	2240	—	NM^3/h	bar	—	°K	—
	Standard condition ($T = 60\,°F$)	7320	—	SCFH	psia	—	°R	—

The Bernoulli coefficients are used only when the piping diameter in the inlet and outlet of the valve are not identical. In fact, if the inlet and outlet piping from the valve are equal sizes, then the Bernoulli coefficients are equal ($K_{B1} = K_{B2}$), and therefore they are dropped from the equation. Inlet and outlet Bernoulli coefficients are calculated by using Eq. (2.9).

Bernoulli Coefficients Calculation

$$K_{B1} = K_{B2} = 1 - \left(\frac{d}{D}\right)^4 \tag{2.9}$$

where:

d: Nominal valve size;
D: Internal piping diameter.

The most commonly used fitting before and after a control valve is a reducer (expander). The head loss or resistance coefficient due to reducers/expanders are calculated by using Eqs. (2.10) and (2.11).

Head Loss Coefficient for an Inlet Reducer

$$K_1 = 0.5 \times \left(1 - \frac{d^2}{D^2}\right)^2 \tag{2.10}$$

Head Loss Coefficient for an Outlet Reducer

$$K_2 = 1.0 \times \left(1 - \frac{d^2}{D^2}\right)^2 \tag{2.11}$$

Then for a control valve installed between identical reducers/expanders, the value of the head loss coefficient is calculated as follows:

$$K = K_1 + K_2 = 0.5 \times \left(1 - \frac{d^2}{D^2}\right)^2 + 1.0 \times \left(1 - \frac{d^2}{D^2}\right)^2 = 1.5 \times \left(1 - \frac{d^2}{D^2}\right)^2$$

2.4.1.4 Determine the Maximum Flow Rate (q_{max}) and Maximum Pressure Drop (ΔP_{max})

As its name implies, the maximum flow rate, also called limiting flow rate (q_{max}) or choked flow calculated from Eq. (2.12), cannot be increased even by increasing the pressure drop across the valve. For example, flashing the vapor from liquids when the static valve pressure drops below the liquid vapor pressure can cause a maximum flow rate. It is important to calculate the maximum allowable pressure drop (ΔP_{max}) for selecting the correct size of the valve if it is possible for the choked flow to develop. The calculated ΔP_{max} value is compared with the actual pressure drop across the valve specified by the process engineer, and the lesser of these two is applied to the valve sizing process. If $\Delta P_{max} > \Delta P$, the flow will not be choked and there is no need to determine the maximum flow rate as per step 4. However, if a choked flow condition exists ($\Delta P_{max} > \Delta P$), then step 5 for sizing valves for liquids shall be changed by replacing the actual service pressure differential ($\Delta P = P_1 - P_2$) with ΔP_{max}.

Maximum Flow Rate (q_{max})

$$q_{max} = N_1 F_L C_v \sqrt{\frac{P_1 - F_F P_v}{SG_l}} \tag{2.12}$$

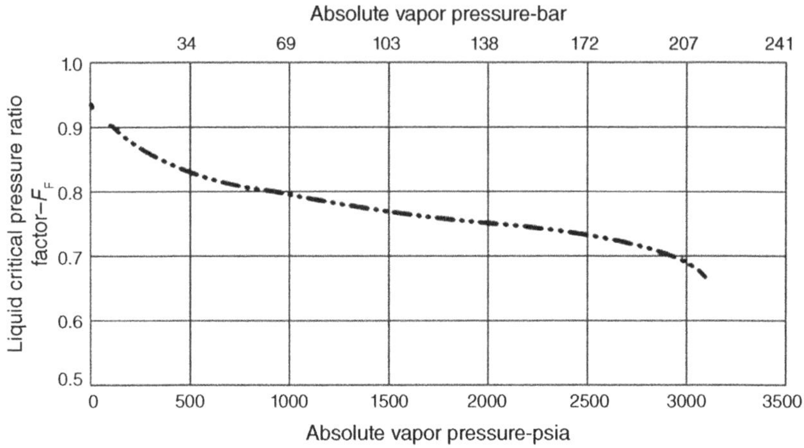

Figure 2.8 Water critical pressure ratio factor (F_F) determination chart.

The liquid critical pressure ratio factor (F_F) can be calculated by using Eq. (2.13), or Figure 2.8 for water. When a liquid pressure drops below its vapor pressure, the fluid state alters from a liquid to gas, that is called choked flow or vena contracta pressure condition. F_F is the ratio of the vena contracta pressure at choked flow condition to the vapor pressure of the liquid at inlet temperature.

Liquid Critical Pressure Ratio Factor (q_{max})

$$F_F = 0.96 - 0.28\sqrt{\frac{P_v}{P_c}} \tag{2.13}$$

The values of F_L for the globe, v-notch ball, and butterfly valves can be extracted from Tables 2.8, 2.9, and 2.10. Ball valves with standard design are not suitable for flow control; however, a v-notch ball valve with a unique v-shaped hole in the ball allows precise flow control. Furthermore, butterfly valves can be used for fluid control or throttling in addition to fluid isolation.

If the valve is installed with fittings such as reducers attached to it, the combined liquid pressure recovery factor and piping geometry factor of a valve with attached fittings known as parameter F_{LP} is applied instead of F_L in Eq. (2.12) for maximum flow rate (q_{max}) calculation. F_{LP} is calculated as per Eq. (2.14).

Combined Liquid Pressure Recovery and Piping Geometry Factor Calculation

$$F_{LP} = \left[\frac{K_1}{N_2}\left(\frac{C_v}{d^2}\right)^2 + \frac{1}{F_L^2}\right]^{-1/2} \tag{2.14}$$

Table 2.8 Representative sizing coefficients for globe control valves.

Body size, in. (DN)	Linear cage[a]						
	Line and valve size are equal		Line size is double the valve size				
	Flow coefficient (C_v)		Flow coefficient (C_v)				
	Regulating	Full open	Regulating	Full open	X_T	F_D	F_L
1 (25)	16.8	17.7	17.2	18.1	0.806	0.43	0.84
2 (50)	63.3	66.7	59.6	62.8	0.820	0.35	
3 (80)	132	139	128	135	0.779	0.30	
4 (100)	202	213	198	209	0.829	0.28	
6 (150)	397	418	381	404	0.668	0.28	

Body size, in. (DN)	Whisper™ trim cage[b]						
	Line and valve size are equal		Line size is double the valve size				
	Flow coefficient (C_v)		Flow coefficient (C_v)				
	Regulating	Regulating	Regulating	Regulating	X_T	F_D	F_L
1 (25)	16.7	17.6	15.6	16.4	0.753	0.10	0.89
2 (50)	54	57	52	55	0.820	0.07	
3 (80)	107	113	106	110	0.775	0.05	
4 (100)	180	190	171	180	0.766	0.04	
6 (150)	295	310	291	306	0.684	0.03	

[a] Cage improves the control ability of a globe valve and adjusts the amount of flow inside the valve. A linear cage gives a linear flow characteristic to the valve, as explained in Chapter 1. Linear flow means that the change in the flow is equal to the opening percentage. For example, 100% of opening a valve passes 100% of the fluid through the valve.

[b] Noise is one of the control valve's major operational problems caused by the turbulent flow of vapor, gas, or steam as the fluid passes through the valve. A Whisper™ can significantly reduce the noise level in control valves if its usage is coupled with the correct valve sizing technique. An in-depth discussion of valve noise, particularly in control valves, appears in Chapter 6.

Table 2.9 Representative sizing coefficients for single port, globe-style control valve with cage guiding and balanced plug.

Valve size	Valve plug style	Flow characteristics	Port diameter in. (mm)	Rated travel inches (mm)	C_v	F_L	X_T	F_D
1/2″	Post guided	Equal percentage	0.38 (9.7)	0.50 (12.7)	2.41	0.90	0.54	0.61
3/4″	Post guided	Equal percentage	0.56 (14.2)	0.50 (12.7)	5.92	0.84	0.61	0.61
1″	Micro-form	Equal percentage	3/8″ (9.5)		3.07	0.89	0.66	0.72
			1/2″ (12.7)		4.91	0.93	0.80	0.67
			3/4″ (19.1)	3/4″ (19.1)	8.84	0.97	0.92	0.62
	Cage guided	Linear	15/16″ (33.3)		20.6	0.84	0.64	0.34
		Equal percentage	15/16″ (33.3)		17.2	0.88	0.67	0.38
1 1/2″	Micro-form	Equal percentage	3/8″ (9.5)		3.20	0.84	0.65	0.72
			1/2″ (12.7)		5.18	0.91	0.71	0.67
			3/4″ (19.1)	3/4″ (19.1)	10.2	0.92	0.80	0.62
	Cage guided	Linear	1 7/8″ (47.6)		39.2	0.82	0.66	0.34
		Equal percentage	1 5/16″ (47.6)		35.8	0.84	0.68	0.38
2″	Cage guided	Linear	2 5/16″ (58.7)	1 1/8″ (28.6)	72.9	0.77	0.64	0.33
		Equal percentage	2 5/16″ (58.7)		59.7	0.85	0.69	0.31
3″	Cage guided	Linear	3 7/16″ (87.3)	1 1/2″ (38.1)	148	0.82	0.62	0.30
		Equal percentage	—	—	136	0.82	0.68	0.32
4″	Cage guided	Linear	4 3/8″ (111)	2″ (50.8)	236	0.82	0.69	0.28
		Equal percentage	—	—	224	0.82	0.72	0.28
6″	Cage guided	Linear	7″ (178)	2″ (50.8)	433	0.84	0.74	0.28
		Equal percentage	—	—	394	0.85	0.78	0.26
8″	Cage guided	Linear	8″ (203)	3″ (76.2)	846	0.87	0.81	0.31
		Equal percentage	—	—	818	0.86	0.81	0.26

Table 2.10 Representative sizing coefficients for the v-notch ball and high-performance butterfly valves used to control the flow.

Valve size (in.)	Valve type	Degree of valve opening	Flow coefficient (C_v)	F_L	X_T	F_D
1″	V-notch ball valve	60	15.6	0.86	0.53	—
		90	34.0	0.86	0.42	—
1 1/2″	V-notch ball valve	60	28.5	0.85	0.50	—
		90	77.3	0.74	0.27	—
2″	V-notch ball valve	60	59.2	0.81	0.53	—
		90	132	0.77	0.41	—
	High-performance butterfly valve	60	58.9	0.76	0.50	0.49
		90	80.2	0.71	0.44	0.70
3″	V-notch ball valve	60	120	0.80	0.50	0.92
		90	321	0.74	0.30	0.99
	High-performance butterfly valve	60	115	0.81	0.46	0.49
		90	237	0.64	0.28	0.70
4″	V-notch ball valve	60	195	0.80	0.52	0.92
		90	596	0.62	0.22	0.99
	High-performance butterfly valve	60	270	0.69	0.32	0.49
		90	499	0.53	0.19	0.70
6″	V-notch ball valve	60	340	0.80	0.52	0.91
		90	1110	0.58	0.20	0.99
	High-performance butterfly valve	60	664	0.66	0.33	0.49
		90	1260	0.55	0.20	0.70
8″	V-notch ball valve	60	518	0.82	0.54	0.91
		90	1820	0.54	0.18	0.99
	High-performance butterfly valve	60	1160	0.66	0.31	0.49
		90	2180	0.48	0.19	0.70
10″	V-notch ball valve	60	1000	0.80	0.47	0.91
		90	3000	0.56	0.19	0.99
	High-performance butterfly valve	60	1670	0.66	0.38	0.49
		90	3600	0.48	0.17	0.70
12″	V-notch ball valve	60	1530	0.78	0.49	0.92
		90	3980	0.63	0.25	0.99
	High-performance butterfly valve	60	2500	—	—	0.49
		90	5400	—	—	0.70
16″	V-notch ball valve	60	2380	0.80	0.45	0.92
		90	8270	0.37	0.13	1.00
	High-performance butterfly valve	60	3870	0.69	0.40	—
		90	8600	0.52	0.23	—

and

$$K_1 = K_1 + K_{B1}$$

K_1: Resistance coefficient of upstream fittings;
K_{B1}: Inlet Bernoulli coefficient.

The maximum pressure drop across the valve (ΔP_{max}) is determined from Eq. (2.15) or (2.16).

Maximum Pressure Drop Calculation for Valves Without Any Attached Fittings

$$\Delta P_{max} = F_L^2 (P_1 - F_F \times P_v) \tag{2.15}$$

Maximum Pressure Drop Calculation for Valves With Attached Fittings

$$\Delta P_{max} = \left(\frac{F_{LP}}{F_P}\right)^2 (P_1 - F_F \times P_v) \tag{2.16}$$

2.4.1.5 Solve for Flow Coefficient

Equation (2.17) is used to calculate the modified flow efficient (C_v) value if $\Delta P_{max} > \Delta P$ meaning that the flow will not be choked. Equation (2.18) applies to adjusted flow coefficient calculation if $\Delta P_{max} < \Delta P$ when the flow will be choked.

Adjusted Flow Coefficient Calculation Without a Choked Flow ($\Delta P_{max} > \Delta P$)

$$C_v = \frac{q}{N_1 F_P \sqrt{P_1 - P_2 / SG_l}} \tag{2.17}$$

Adjusted Flow Coefficient Calculation with a Choked Flow ($\Delta P_{max} > \Delta P$)

$$C_v = \frac{q}{N_1 F_P \sqrt{\Delta P_{max} / SG_l}} \tag{2.18}$$

2.4.1.6 Select the Correct Valve Size

As a general rule, the smallest valve with a flow coefficient larger than that calculated based on either Eq. (2.17) or Eq.(2.18), according to the information provided in Tables 2.8, 2.9, and 2.10, shall be selected.

Example 2.5 A control globe valve in CL300 (PN of 50 bar) with an equal percentage flow characteristic and the guided cage is installed on an 8″ line. The standard concentric reducers are used to install the valve on the line. The fluid service is liquid propane with a flow rate of 800 gpm and specific gravity of 0.5. The valve's inlet pressure, outlet pressure, and inlet temperature values are 300 psig, 275 psig, and 70 °F, respectively. The values of absolute thermodynamic critical pressure and absolute vapor pressure of the liquid at inlet temperature are 616.3 psia and 124.3 psia, respectively. Select the correct size of the control valve.

Answer

The flow coefficient of the valve is calculated based on the basic C_v calculation for liquids as per Eq. (1.1) as follows:

$$C_v = Q\sqrt{\frac{SG}{\Delta P}} = 800\sqrt{\frac{0.5}{300 - 275}} = 113.13$$

Refer to Table 2.9. The smallest globe control valve size with a flow coefficient larger than 113.13 and an equal percentage flow characteristic is 3″ with a C_v value of 136. Thus, a 3″ control valve is assumed first and checked to see whether it can provide the required flow rate by considering various parameters such as the installation of reducers and the possibility of choke flow. The next step is to specify the necessary variables required to size the valve as follows:

- Desired valve design is a 3″ CL300 globe valve with an equal percentage flow characteristic
- Service condition – $q = 800$ gpm
- $P_1 = 300$ psig (20.7 bar) $= 314.7$ psia(21.7 bara)
- $P_2 = 275$ psig (19.0 bar) $= 289.7$ psia(20.0 bara)
- $\Delta P = 25$ psi (1.7 bar)
- $T_1 = 70\,°\text{F} = 21\,°\text{C}$
- $SG_l = 0.5$
- $P_v = 124.3$ psia(8.6 bara)
- $P_c = 616.3$ psia(42.5 bara)

The next step is to select an equation constant (N) from Table 2.7. By considering the pressure values (p) in psia unit and flow rate (q) value in gpm, $N_1 = 1$.

A 3″ valve is installed on an 8″ line meaning that reducers/expanders are installed before and after the valve. Thus, it is essential to determine the piping geometry factor (F_P) from Eq. (2.7) to correct the flow coefficient by flow losses caused by fittings connected to the valve.

$$F_P = \left[1 + \frac{\sum K}{N_2}\left(\frac{C_V}{d^2}\right)^2\right]^{-1/2}$$

where:

$N_2 = 890$ refer to Table 2.7, by considering the diameter values in inch units;
d: Nominal valve size = 3 in. (76.2 mm);
C_V: flow coefficient = 136 for a 3″ CL300 globe control valve with an equal percentage flow characteristic as per Table 2.9.

For a control valve installed between identical reducers/expanders, the value of the head loss coefficient is calculated as follows:

$$K = K_1 + K_2 = 0.5 \times \left(1 - \frac{d^2}{D^2}\right)^2 + 1.0 \times \left(1 - \frac{d^2}{D^2}\right)^2 = 1.5 \times \left(1 - \frac{d^2}{D^2}\right)^2$$

where:

D: Nominal pipe size = 8″;
d: Nominal valve size = 3″.

$$\rightarrow K = 1.5 \times \left(1 - (3^2/8^2)\right)^2 = 1.11$$

$$F_P = \left[1 + \frac{\sum K}{N_2}\left(\frac{C_V}{d^2}\right)^2\right]^{-1/2} = \left[1 + \frac{1.11}{890}\left(\frac{136}{3^2}\right)^2\right]^{-1/2}$$
$$= [1.28]^{-1/2} = 0.88$$

The value of the rated liquid pressure recovery factor, F_L, for a 3″ control globe valve with a guided cage and an equal percentage flow characteristic as per Table 2.9 is 0.82. Because the given valve in this example is attached to a couple of reducers (fittings) during the installation, F_{LP}, the combined liquid pressure recovery factor and piping geometry factor of the valve with attached fittings must be calculated according to Eq. (2.14).

$$F_{LP} = \left[\frac{K_1 + K_{B1}}{N_2}\left(\frac{C_V}{d^2}\right)^2 + \frac{1}{F_L^2}\right]^{-1/2} = \left[\frac{1.35}{890}\left(\frac{136}{9}\right)^2 + \frac{1}{0.82^2}\right]^{-1/2}$$
$$= [0.35 + 1.49]^{-1/2} = 0.74$$

$$K_{B1} = 1 - \left(\frac{d}{D}\right)^4 = 1 - \left(\frac{3}{8}\right)^4 = 0.98$$

$$K_1 + K_{B1} = 0.98 + 0.37 = 1.35$$

The liquid critical pressure ratio factor (F_F) can be calculated by using Eq. (2.13) as follows:

$$F_F = 0.96 - 0.28\sqrt{\frac{P_v}{P_c}} = 0.96 - 0.28\sqrt{\frac{124.3}{616.3}} = 0.834$$

Now it is possible to calculate the maximum pressure drop (ΔP_{max}) from Eq. (2.16):

$$\Delta P_{max} = \left(\frac{F_{LP}}{F_P}\right)^2 (P_1 - F_F \times P_v) = \left(\frac{0.74}{0.88}\right)^2 (300 - 0.834 \times 124.3)$$
$$= 0.71 \times 196.33 = 139.40 \text{ psi}$$

$\Delta P_{max} = 139.40$ psi > 25 psi. So choked flow will not happen. The modified flow coefficient is calculated from Eq. (2.17):

$$C_v = \frac{q}{N_1 F_P \sqrt{P_1 - P_2/SG_l}} = \frac{800}{1 \times 0.88\sqrt{25/0.5}} = 128.58$$

The required flow coefficient of 128.58 is less than the assumed valve capacity, which has C_v of 136. So 3″ is the correct size. However, if the value of the modified flow coefficient was higher than 136, then the next larger size (4 in.) would be the correct valve size most likely.

2.4.2 Control Valve Sizing for Gas and Steam

There are six steps for a control valve size selection in liquids:

1) Specify the variables required to size the valve;
2) Determine the equation constant (N);
3) Determine the piping geometry factor (F_P);
4) Determine expansion factor (Y);
5) Solve for required flow coefficient (C_v) value using an appropriate equation;
6) Select the appropriate flow coefficient table and the calculated flow coefficient valve to select the suitable valve size.

2.4.2.1 Specify the Variables Required to Size the Valve

The first step is to specify the variables required to size the valves, such as valve pressure class, desired valve design (e.g. balanced globe with a linear cage), and valve size.

- Process fluid (e.g. oil and water);
- Service conditions such as volume rate of flow (q), mass rate of flow (w), Upstream and downstream pressure values (P_1 and P_2), pressure drop across the valve ($\Delta P = P_1 - P_2$), upstream temperature (T_1), gas-specific gravity that is defined as a dimensionless number equal to the ratio of the flowing gas density to the density of air with both at standard conditions (SG_g), molecular weight (M),

the ratio of specific heats (k), gas compressibility factor (Z), and specific weight at the valve inlet condition (γ). If any of the aforementioned parameters are new or unfamiliar, refer to Table 2.6, where all parameters for control valve size selection are summarized.

2.4.2.2 Determine the Equation Constant (*N*)

N, as explained before, is a numerical constant in each of the flow or sizing procedures to adjust the relevant equation for using different sizing units. Values of equation constant as well as their applicable units are given in Table 2.7 in the liquid valve sizing section. If sizing the valve, use either N_7 or N_9 for a flow rate in volumetric units such as SCF per hour (SCFH) or cubic meter per hour m^3/h. N_7 can be used if the gas-specific gravity, SG_g, is specified, and N_9 can be used only if the gas's molecular weight, M, is known. Use either N_6 or N_8 if sizing the valve for a mass flow rate such as lb/h or kg/h. N_6 can be used if the specific weight at the inlet valve condition, γ, is known, whereas N_8 can be selected if parameter M, molecular weight has been specified.

2.4.2.3 Determine Piping Geometry Factor (F_P)

F_P is a correction factor that accounts for the pressure losses due to piping fittings that are used to change the direction or size of the piping or take branches from the headers such as tees, elbows, and reducers. If fittings are used before or after the valve, the effect of piping geometry must be considered for the valve sizing. However, if no fitting is attached to the valve, F_P is equal to one and does not impact the valve sizing process. Equation (2.7) shows how to calculate the piping geometry.

2.4.2.4 Determine the Expansion Factor (*Y*)

When sizing components such as valves in the piping system handling compressible gas, it is essential to account for all factors, including the expansion factor that would affect the flow rate and pressure drop in the piping system. As compressible gas flows through each component in the piping system, the head loss causes expansion as well as change in density, temperature, and velocity. The gas expansion process adds resistance to the flow rate, resulting in a flow rate reduction for a given pressure drop. Thus, the gas expansion factor, Y, shall be taken into account by using Eq. (2.19) as follows.

Gas Expansion Factor Calculation

$$Y = 1 - \frac{X}{3F_k X_T} \tag{2.19}$$

where,

F_k: The ratio of specific heat factor $= k/1.4$;
k: Ratio of specific heats;

X: The pressure drop ratio = $\Delta P/P_1$; If $X < 0.2$, the change in gas density and expansion is small, the fluid can be assumed incompressible, and the parameter Y can be neglected. For $X > 0.4$, the expansion factor shall be considered as the fluid can be assumed compressible;

X_T: The pressure drop ratio factor for valves installed without attached fittings can be extracted from Tables 2.8, 2.9, and 2.10. More precisely, X_T is the pressure drop ratio required to produce critical or maximum flow through the valve when $F_k = 1$. It should be noted that the value of Y cannot be less than 0.667. When the calculated Y value is less than 0.0667, the Y value shall be assumed to be 0.667.

Suppose the control valve is installed with fittings such as reducers or elbows attached to it. In that case, the effect of fittings shall be taken into account by replacing X_T by a factor of X_{TP} that can be calculated from Eq. (2.20) as follows.

Pressure Drop Ratio Factor Calculation with Fittings Attached to the Control Valve (X_{TP})

$$X_{TP} = \frac{X_T}{F_P^2}\left[1 + \frac{X_T K_i}{N_5}\left(\frac{C_v}{d^2}\right)^2\right]^{-1} \tag{2.20}$$

where:

N_5: Numerical constant is taken from Table 2.7;

d: Assumed nominal valve size;

C_v: Valve flow coefficient for a fully open valve taken from either Table 2.8 or 2.9 or 2.10;

F_P: Piping geometry factor;

X_T: Pressure drop ratio for the valve without any fitting attached can be taken from Tables 2.8, 2.9, and 2.10;

K_i: Inlet head loss coefficient, which is calculated as follows:

$$K_i = K_1 + K_{B1} = 0.5 \times \left(1 - \frac{d^2}{D^2}\right)^2 + \left[1 - \left(\frac{d}{D}\right)^4\right]$$

where:

K_1: Resistance coefficient of upstream fittings is calculated from Eq. (2.10);

K_{B1}: Inlet Bernoulli coefficient is calculated from Eq. (2.9).

2.4.2.5 Solve for the Required Flow Coefficient (C_v)

Flow coefficient value is different from volumetric flow rate than mass flow rate units. When the gas-specific gravity, SG_g, is known for volumetric flow rate units, the flow coefficient is calculated as per Eq. (2.21).

Flow Coefficient (C_v) Calculation for Volumetric Flow Rate and Knowing the Gas-Specific Gravity:

$$C_v = \frac{q}{N_7 F_P P_1 Y \sqrt{X/(SG_g\, T_1 Z)}} \tag{2.21}$$

If the molecular weight, M, is known for volumetric flow rate units, the flow coefficient is calculated as per Eq. (2.22).

Flow Coefficient (C_v) Calculation for Volumetric Flow Rate and Knowing the Molecular Weight

$$C_v = \frac{q}{N_7 F_P P_1 Y \sqrt{X/(M\, T_1 Z)}} \tag{2.22}$$

If the specific weight at the valve inlet, γ, is known for mass flow rate units, the flow coefficient is calculated as per Eq. (2.23).

Flow Coefficient (C_v) Calculation for Mass Flow Rate and Knowing the Specific Weight at the Valve Inlet

$$C_v = \frac{W}{N_6 F_P Y \sqrt{X P_1 \gamma}} \tag{2.23}$$

If the molecular weight, M, is known for mass flow rate units, the flow coefficient is calculated as per Eq. (2.24).

Flow Coefficient (C_v) Calculation for Mass Flow Rate and Knowing the Molecular Weight

$$C_v = \frac{W}{N_8 F_P P_1 Y \sqrt{XM/T_1 Z}} \tag{2.24}$$

Example 2.6 A control globe valve in CL300 (PN of 50 bar) with a linear cage is installed on a 6″ line with standard concentric reducers on both sides. The fluid service is superheated steam with a flow rate of 125,000 lb/h. The valve's inlet pressure, outlet pressure, and inlet temperature values are 500 psig, 250 psig, and 500 °F, respectively. The values of specific weight at the valve inlet condition (γ) and ratio of specific heats (k) are 1.0434 lb/ft^3 and 1.28 lb/ft^3, respectively. Select the correct size of the control valve.

Answer

The first step is to summarize the data associated with the valve and process conditions.

1) Desired valve design: CL300 control globe valve with a linear cage;
2) Process fluid: superheated steam;
3) Service conditions:

$$W = q = 125{,}000\ \text{lb/h}$$

$$P_1 = 500\ \text{psig}(34.5\ \text{bar}) = 514.7\ \text{psia}(35.5\ \text{bar})$$

$$P_2 = 250\ \text{psig}(17\ \text{bar}) = 264.7\ \text{psia}(18.3\ \text{bar})$$

$$\Delta P = 500 - 250 = 250\ \text{psig}\ (17\ \text{bar})$$

$$X: \text{The pressure drop ratio} = \frac{\Delta P}{P_1} = \frac{250}{514.7} = 0.49$$

$$T_1 = 500°\text{F}\ (260°\text{C})$$

$$\gamma = 1.0434\ \text{lb/ft}^3$$

$$k = 1.28$$

The second step is to calculate the valve's flow coefficient for superheated steam. There are different formulas to calculate the valve flow coefficient handling the steam. First, the valve flow coefficient for saturated steam shall be determined, which depends on critical or noncritical pressure drop. In this case, the outlet pressure is less than 50% of the inlet pressure, which is considered a critical pressure drop. A noncritical pressure drop is if the outlet pressure exceeds 58% of the inlet pressure. The flow coefficient for saturated steam and critical pressure drop is calculated from Eq. (2.25).

Flow Coefficient (C_v) Calculation for Saturated Steam and Critical Pressure Drop

$$C_v = \frac{W}{1.61 \times P_1} \tag{2.25}$$

where:

W: Steam flow rate (lb/h);
P_1: Inlet steam absolute pressure (psia).

$$C_v = \frac{w}{1.61 \times P_1} = \frac{125,000}{1.61 \times 514.7} = 150.84$$

The flow coefficient for saturated steam and noncritical pressure drop is calculated from Eq. (2.26).

Flow Coefficient (C_v) Calculation for Saturated Steam and Noncritical Pressure Drop

$$C_v = \frac{m}{2.1\sqrt{(P_1 + P_2)\Delta P}} \tag{2.26}$$

The temperature of saturated steam, which is pure steam, is 298 °F. The flow coefficient of superheated steam is calculated according to Eq. (2.27).

Flow Coefficient (C_v) Calculation for Superheated Steam

$$C_v(\text{Superheated steam}) = C_v\,(\text{Saturated steam})(1 + 0.00065\text{d}t) \tag{2.27}$$

where:

dt: Steam temperature above saturation temperature at the actual pressure (°F).

$$C_v\,(\text{Superheated steam}) = C_v\,(\text{Saturated steam})(1 + 0.00065\text{d}t)$$
$$= 150.84\,[1 + 0.00065\,(500 - 298)] = 150.84 \times 1.1313 = 170.65$$

Table 2.11 extracted from Table 2.10 includes flow coefficient values for control globe valves with a guided cage in size ranges from 3″ to 6″.

Table 2.11 Flow coefficient values for cage-guided control globe valves in size range from 3″ to 6″.

Valve size	Valve plug style	Flow characteristics	Port diameter inches (mm)	Rated travel inches (mm)	C_v
3″	Cage guided	Linear	3 7/16″ (87.3)	1 1/2″ (38.1)	148
		Equal percentage	—	—	136
4″	Cage guided	Linear	4 3/8″ (111)	2″ (50.8)	236
		Equal percentage	—	—	224
6″	Cage guided	Linear	7″ (178)	2″ (50.8)	433
		Equal percentage	—	—	394

A 3″ valve is undersized because it can produce a maximum flow coefficient of 148, which is lower than 170.65 and insufficient. Conversely, a 6″ valve is oversized. So a 4″ valve with a linear flow characteristic and a flow coefficient value of 236 can be a suitable choice based on the preliminary flow coefficient calculation without considering the corrective parameters.

The next step is to find an appropriate equation constant, N, from Table 2.7. Because the specified flow rate is in the mass unit, pound per hour, and the specific weight of steam, γ, are given, N_6 constant is applied, and $N_6 = 63.3$.

The piping geometry factor is calculated as follows:

$$F_P = \left[1 + \frac{\sum K}{N_2}\left(\frac{C_V}{d^2}\right)^2\right]^{-1/2}$$

where:

$N_2 = 890$ refer to Table 2.7, by considering the diameter values in inches;
d: Nominal valve size = 4 in. (101.6 mm);
C_V: Flow coefficient = 236 for a 4″ CL300 globe control valve with a linear flow characteristic as per Tables 2.9 and 2.11.

For a control valve installed between identical reducers/expanders, the value of the head loss coefficient is calculated as follows:

$$K = K_1 + K_2 = 0.5 \times \left(1 - \frac{d^2}{D^2}\right)^2 + 1.0 \times \left(1 - \frac{d^2}{D^2}\right)^2 = 1.5 \times \left(1 - \frac{d^2}{D^2}\right)^2$$

where:

D: Nominal pipe size = 6″;
d: Nominal valve size = 4″.

$$\rightarrow K = 1.5 \times \left(1 - \frac{4^2}{6^2}\right)^2 = 0.463$$

$$F_P = \left[1 + \frac{\sum K}{N_2}\left(\frac{C_V}{d^2}\right)^2\right]^{-1/2} = \left[1 + \frac{0.463}{890}\left(\frac{236}{4^2}\right)^2\right]^{-1/2} = 0.95$$

The expansion factor, Y, is calculated as follows:

$$Y = 1 - \frac{X}{3F_k X_T}$$

$$F_k = \frac{k}{1.4} = \frac{1.28}{1.4} = 0.91 \text{ and } X = 0.49 \text{ as calculated in step 1}$$

Since the control valve size is smaller than the connected piping, fittings (reducers) must be used upstream and downstream of the valve. In that case, the effect of fittings shall be taken into account by replacing X_T by a factor of X_{TP} that can be calculated from Eq. (2.20) as follows:

$$X_{TP} = \frac{X_T}{F_P^2}\left[1 + \frac{X_T K_i}{N_5}\left(\frac{C_v}{d^2}\right)^2\right]^{-1}$$

where:

$X_T = 0.69$ extracted from Table 2.9 for a 4″ globe control valve with a guided cage and linear flow characteristic;
$F_P = 0.95$ as calculated in the previous step;
$N_5 = 1000$, from equation constants (Table 2.7);
$C_v = 236$;
$d = 4''$.

$$K_i = K_1 + K_{B1} = 0.5 \times \left(1 - \frac{d^2}{D^2}\right)^2 + \left[1 - \left(\frac{d}{D}\right)^4\right]$$

$$= 0.5 \times \left(1 - \frac{4^2}{6^2}\right)^2 + \left[1 - \left(\frac{4}{6}\right)^4\right] = 0.96$$

$$X_{TP} = \frac{0.69}{0.95^2}\left[1 + \frac{0.69 \times 0.96}{1000}\left(\frac{236}{16}\right)^2\right]^{-1} = 0.67$$

$$Y = 1 - \frac{X}{3F_k X_T} = 1 - \frac{0.49}{3 \times 0.91 \times 0.67} = 0.73$$

Because the specific weight at the valve inlet, γ, is known for mass flow rate units, the flow coefficient is calculated as per Eq. (2.23).

$$C_v = \frac{W}{N_6 F_P Y \sqrt{X P_1 \gamma}} = \frac{12{,}500}{63.3 \times 0.95 \times 0.73\sqrt{0.49 \times 514.7 \times \gamma \times 1.0434}}$$
$$= 176$$

The conclusion is that the 4″ globe control valve initially selected was correct because the assumed valve has a flow coefficient of 236, which is higher than the corrected C_v value of 176. A 3″ control globe valve with the guided cage and linear flow characteristic has a C_v value of only 148 which is insufficient.

Example 2.7 On this occasion, it has been decided to install an 8″ V-notch ball valve in CL150 (PN of 20 bar) on an 8″ pipe. Natural gas is the fluid service with a flow rate of 6×10^6 standard cubic feet (SCF). Inlet pressure, outlet pressure, and inlet temperature for the valve are 214.7 psia, 64.7 psia, and 60 °F, respectively.

Accordingly, the values of molecular weight (M), gas specific gravity (SG_g), and the ratio of specific heats (k) are 17.38, 0.60, and 1.31, respectively. Is the valve size correctly selected? (*Note*: assume an equal percentage flow characteristic for the valve.)

Answer

First, it is necessary to summarize the data pertaining to the valve and process conditions.

1) Desired valve design: CL150 V-notch ball valve;
2) Process fluid: natural gas;
3) Service conditions:

$$W = q = 6 \times 10^6 \text{ standard cubic feet (SCF).}$$

$$P_1 = 200\,\text{psig}\,(13.8\,\text{bar}) = 214.7\,\text{psia}\,(14.8\,\text{bar})$$

$$P_2 = 50\,\text{psig}\,(3.4\,\text{bar}) = 64.7\,\text{psia}\,(4.5\,\text{bar})$$

$$\Delta P = 200 - 50 = 150\,\text{psig}\,(10.3\,\text{bar})$$

$$X : \text{The pressure drop ratio} = \frac{\Delta P}{P_1} = \frac{150}{214.7} = 0.70$$

$$T_1 = 60°\text{F}\,(16°\text{C}) = 520°\text{R}$$

$$M = 17.38$$

$$SG_g = 0.6$$

$$k = 1.31$$

The second step is to determine the correct value of equation constant (N) from Table 2.7. As in the example, the flow rate value in SCF is provided, it is possible to use either N_7 or N_9 at the standard condition. Consider the use of N_7, which is equal to 1360.

The third step involves finding the value of the piping geometry factor, parameter F_P. Because the valve and the connected piping are the same size, there are no reducers required. Therefore, $F_P = 1$.

The fourth step is to find the value of Y, which is the expansion factor. It is necessary to find Y, which is the expansion factor from Eq. (2.19), in the fourth step.

$$F_k : \text{The ratio of specific heats factor} = \frac{k}{1.4} = \frac{1.31}{1.4} = 0.94$$

According to Table 2.10, the value of X_T for an 8″ V-notch ball valve at the full opening percentage is equal to 0.18. In the first step, parameter X, the pressure drop ratio, is calculated, which equals 0.70.

$$Y = 1 - \frac{X}{3F_k X_T} 1 - \frac{0.7}{3 \times 0.94 \times 0.18} = -0.38$$

Y cannot be less than 0.667. Thus, in this case, where the value of Y is calculated to be −0.038, the minimum Y value of 0.667 should be considered.

Using Eq. (2.21), the fifth step involves calculating the adjusted flow coefficient (C_v) as follows. It should be noted that the gas Z-factor is assumed to be one.

$$C_v = \frac{q}{N_7 F_P P_1 Y \sqrt{X/SG_g T_1 Z}} = \frac{6 \times 10^6}{1360 \times 1 \times 214.7 \times 0.667\sqrt{0.7/(0.6 \times 520 \times 1)}}$$
$$= 655$$

In accordance with Table 2.10, an 8-in. V-notch ball valve at the fully open position will have a C_v value of 3000. In the following step, we shall determine the opening percentage of the valve that results in a flow efficiency of 655 when we consider an equal percentage flow characteristic. Based on Eq. (1.13) taken from Chapter 1, the valve flow coefficient is calculated in terms of valve opening percentage for an equal percentage flow characteristic relevant in this case.

$$C_{vvp} = C_{vmax} \times \alpha^{vp-1}$$

where:

vp: Valve opening percentage, which is unknown in this case;
C_{vvp}: The valve flow coefficient of 655 is linked to an unknown valve opening percentage;
C_{vmax}: Maximum flow coefficient of the valve at the fully open position = 3000;
$\alpha = 50$.

$$655 = 3000 \times 50^{vp-1} \rightarrow 50^{vp-1} = 0.2183 \rightarrow (vp-1) \times 1.7 = -0.66 \rightarrow vp = 61.2\%$$

So the 8″ V-notch ball valve is capable of providing a flow coefficient value of 655 at 61.2% opening. Therefore, the 8″ valve size is appropriate.

2.5 Safety Relief Valve Sizing

It is impossible for a chemical process facility to be immune to overpressure, dictating the need for overpressure protection. It is the purpose of the pressure relief valve (PRV) installed on pressurized equipment (labeled Protected System in Figure 2.9) to release the overpressure gas or fluid from the equipment into the flare system. In API standard 520 part 1, the sizing of PRVs is clearly described. According to API 520, it is critical to understand the similarities and differences between a relief valve and a safety valve. Both types of safety valves are spring-loaded and used to protect pressure piping and facilities from overpressure. This relief valve is typically opened in proportion to the increase in pressure over the opening pressure, and it is used for noncompressible fluids. Contrary to this, a safety valve is opened rapidly and is normally used for compressible services.

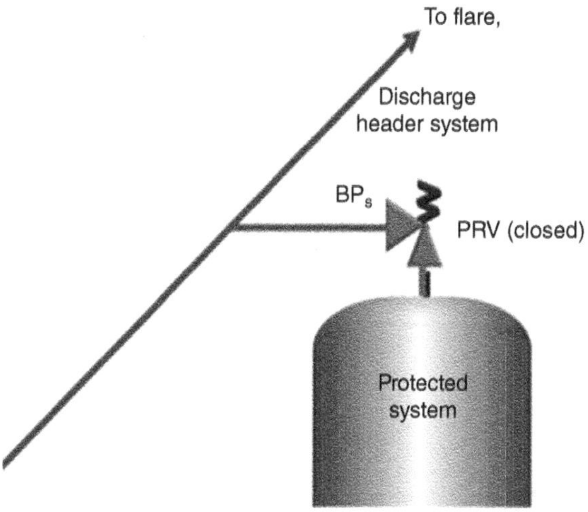

Figure 2.9 PRV installation on pressurized equipment connected to a flare system.

There is the possibility of using a safety relief valve suitable for both applications. Chapter 7 discusses PSVs or PRVs and provides relevant equations and calculations.

A basic safety valve sizing procedure involves three steps. In the first step, the set pressure of the PSV is established, which is the gauge pressure at which the pressure relief device will open under normal operating conditions. In the second step, the relief capacity must be determined. Last, but certainly not least, the pressure safety device must be sized so that it can provide a capacity equivalent to the relief pressure.

For sizing pressure safety or relief valves, it is necessary to pay attention to the valve size, which must correspond to both the inlet and the discharge pipe sizes. The inlet and outlet ports of a pressure relief or safety valve are typically different sizes, resulting in the valves being recognized in two sizes: one inlet and the other outlet. The API 526 standard, which deals with flanged steel PRVs, defines the sizes of PSVs alphabetically. ASME's Boiler and Pressure Vessel Code (BPVC) Sec. VIII refers to boiler construction rules and contains the same letters for PSV sizing corresponding to different orifice sizes to provide *effective discharge area*. In Table 2.12, the API 526 orifice sizes are compared with those of ASME Sec. VIII based on the alphabetic letters from "D" through "T". Safety valves may have several different sizes of inlet and outlet connections, even when the orifice letter is the same. A $2'' \times J \times 3''$ and $3'' \times J \times 4''$ safety valve have the same size orifice, illustrated by the letter "*J*," but have different sizes of inlet and outlet as shown before and after the orifice letter, respectively.

Table 2.12 ASME and API standard orifice sizes.

Serial number	Orifice designation	API effective area (in.^{2})	ASME effective area (in.^{2})	PSV inlet × outlet sizes (in.)
1	D	0.110	0.124	1″ × 2″ 1.5″ × 2″ 1.5″ × 2.5″
2	E	0.196	0.221	1″ × 2″ 1.5″ × 2″ 1.5″ × 2.5″
3	F	0.307	0.347	1″ × 2″ 1.5″ × 2″ 1.5″ × 2.5″
4	G	0.503	0.567	1.5″ × 2.5″ 1.5″ × 3″ 2″ × 3″
5	H	0.785	0.887	1.5″ × 3″ 2″ × 3″
6	J	1.287	1.453	2″ × 3″ 2.5″ × 4″ 3″ × 4″
7	K	1.838	2.076	3″ × 4″
8	L	2.853	3.221	3″ × 4″ 4″ × 6″
9	M	3.600	4.065	4″ × 6″
10	N	4.340	4.900	4″ × 6″
11	P	6.380	7.205	4″ × 6″
12	Q	11.05	12.47	6″ × 8″
13	R	16.00	18.06	6″ × 8″ 6″ × 10″
14	T	26.00	29.35	8″ × 10″

What is the reason behind the larger orifice size or effective discharge area in ASME than API? The required orifice area that is calculated using the equations discussed later in this chapter is inversely proportional to the *coefficient of discharge* designated with parameter K in API 520, and with parameter K_D in ASME BVPC Sec. VIII. The ASME Sec. VIII code was revised in 1962 to require that "K" be used in sizing calculations with 10% safety factor instead of K_D. ($K = K_D \times 0.9$). In some texts, the modified coefficient of discharge based on the code or regulation is referred to as the "*rated* coefficient of discharge." In the following subsection, we discuss the sizing of safety valves for gas or vapor in light of critical flow conditions.

2.5.1 Sizing for Gas or Vapor Relief

2.5.1.1 Critical Flow

When a compressible gas expands across a nozzle or orifice, its velocity, mass flow rate, and volume increase with decreasing downstream pressure. The ratio of pressure at the valve outlet to pressure at the inlet is known as the *critical pressure ratio*. As a general rule of thumb, a critical flow occurs when the outlet pressure is 40–60% of the outlet pressure. Critical flow occurs when the outlet pressure or downstream pressure reaches 6–4 bars when the inlet pressure is 10 bar. An alternative condition for critical flow occurrence is if the critical flow pressure calculated from Eq. (2.40) is higher than the backpressure. Backpressure is defined later in this chapter. When operating under critical flow conditions, pressure relief devices in gas or vapor service are sized according to API 520 using Eqs. (2.28) and (2.29) for mass flow rate, and Eqs. (2.30), (2.31), (2.32), and (2.33) for volumetric flow rate. In order to select a PRV, it must have a discharge area as large as or wider than the calculated value of "*A*" as shown in the equations later.

Sizing Relief Devices for Gas or Vapor in a Critical Flow Condition According to Mass Flow Rate

$$A = \frac{W}{C\,K_d P_1 K_b K_c}\sqrt{\frac{TZ}{M}}\ \text{(US Customary units)} \tag{2.28a}$$

$$A = \frac{13,160\,W}{C\,K_d P_1 K_b K_c}\sqrt{\frac{TZ}{M}}\ \text{(SI units)} \tag{2.29a}$$

Sizing Relief Devices for Gas or Vapor in a Critical Flow Condition According to Volumetric Flow Rate

$$A = \frac{V\sqrt{TZM}}{6.32CK_d P_1 K_b K_c}\ \text{(US Customary units)} \tag{2.30}$$

$$A = \frac{V\sqrt{TZG}}{1.175\,CK_d P_1 K_b K_c}\ \text{(US Customary units)} \tag{2.31}$$

$$A = \frac{32{,}250\,W\sqrt{TZM}}{6.32\,C\,K_d P_1 K_b K_c}\ \text{(SI units)} \tag{2.32}$$

$$A = \frac{189{,}750\,V\,\sqrt{TZG}}{C\,K_d P_1 K_b K_c}\ \text{(SI units)} \tag{2.33}$$

where:

A: Required effective discharge area of the safety relief device, in.^{2} or mm^2;
W: Mass flow rate through the safety device, lb/h or kg/h;

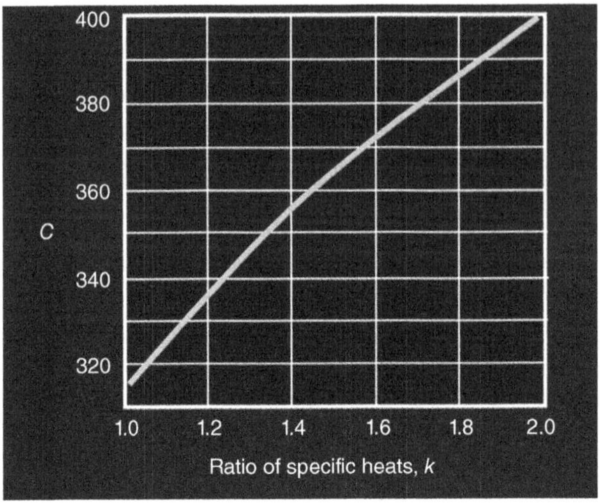

Figure 2.10 Curve for assessing the *C* factor from the specific heat ratio in the assumption of ideal gas behavior.

C: An ideal gas constant or coefficient is derived from the ratio of specific heats (k) as indicated in Eq. (2.34), Figure 2.10, and Table 2.13, respectively, assuming ideal gas behavior. (If not known, it can be assumed $C = 315$.) Table 2.14 presents the *C* factor and "k" ratio of specific heats for different gases.

Calculation of the Gas Constant from the Specific Heat Ratio for an Ideal Gas

$$C = 520\sqrt{k\left(\frac{2}{k+1}\right)^{(k+1)/(k-1)}} \tag{2.34}$$

K_d: Effective coefficient of discharge and the following values are required to be considered according to API 520 for primary sizing; the coefficient is equal to 0.975 when the PRV that handles gas, vapors, or steams is installed with a rupture disk or without a rupture disk. When a PRV is not installed and the sizing is performed for a rupture disk, 0.62 is used. A rupture disk, also called a pressure safety disk or a burst disk, is a one-time-use device to protect pressure equipment from overpressure situations.

P_1: Inlet flowing pressure or relieving pressure, psig or KPa as per Eq. (2.35). If this pressure is expressed as an absolute pressure. Based on Eq. (2.36), it is equal to the set pressure plus the allowable overpressure plus atmospheric pressure minutes loss pressure. Allowable overpressure, which is typically 10% percent of the set pressure or design pressure, is used to overcome the spring force acting against lifting the disk in order to open the valve. A safety relief valve is

Table 2.13 Values of the gas constant (*C*) based on the ratio of specific heat (*k*) according to API 520.

k	*C*	*k*	*C*	*k*	*C*	*k*	*C*
1.00	315	1.31	347	1.60	372	1.90	394
1.01	317	1.31	348	1.61	373	1.91	395
1.02	318	1.32	349	1.62	374	1.92	395
1.03	319	1.33	350	1.63	375	1.93	396
1.04	320	1.34	351	1.64	376	1.94	397
1.05	321	1.35	352	1.65	376	1.95	397
1.06	322	1.36	353	1.66	377	1.96	398
1.07	323	1.37	353	1.67	378	1.97	398
1.08	325	1.38	354	1.68	379	1.98	399
1.09	326	1.39	355	1.69	379	1.99	400
1.10	327	1.41	356	1.70	380	2.00	400
1.11	328	1.41	357	1.71	381	—	—
1.12	329	1.42	358	1.72	382	—	—
1.13	330	1.43	359	1.73	382	—	—
1.14	331	1.44	360	1.74	383	—	—
1.15	332	1.45	360	1.75	384	—	—
1.16	333	1.46	361	1.76	384	—	—
1.17	334	1.47	362	1.77	385	—	—
1.18	335	1.48	363	1.78	386	—	—
1.19	336	1.49	364	1.79	386	—	—
1.20	337	1.50	365	1.80	387	—	—
1.21	338	1.51	365	1.81	388	—	—
1.22	339	1.52	366	1.82	389	—	—
1.23	340	1.53	367	1.83	389	—	—
1.24	341	1.54	368	1.84	390	—	—
1.25	342	1.55	369	1.85	391	—	—
1.26	343	1.56	369	1.86	391	—	—
1.27	344	1.57	370	1.87	392	—	—
1.28	345	1.58	371	1.88	393	—	—
1.29	346	1.59	372	1.89	393	—	—
1.30	347	1.60	373	1.90	394	—	—

Table 2.14 *C* factor and "*k*" ratio of specific heats for different gases.

Gas	Molecular weight	"*C*" factor	Ratio of specific heat (*k*)
Air	29.96	356	1.40
Ammonia	17.03	356	1.40
Argon	40	348	1.31
Carbon dioxide	44.01	356	1.40
Carbon monoxide	28	356	1.40
Ethane	30.07	336	1.19
Ethylene	28.03	341	1.24
Helium	4	377	1.66
Hexane	86.18	322	1.06
Hydrogen	2.02	357	1.41
Hydrogen sulfide	34	349	1.32
Methane	16.4	348	1.31
N-Butane	58.12	326	1.09
Natural gas (specific gravity = 0.60)	18.9	344	1.27
Nitrogen	28	356	1.40
Oxygen	32	356	1.40
Pentane	72.15	323	1.07
Propane	44.09	330	1.13
Propylene	47.08	332	1.15
Steam	18.01	348	1.31
Sulfur dioxide	64.04	346	1.29

illustrated in Figure 2.11, including its inlet and outlet, as well as some of its main components, such as the spring and disk.

Calculation of Safety Device Inlet, Upstream, or Relieving Pressure

$$P_1 = P_{\text{set}} + P_{\text{over}} - P_{\text{loss}} \tag{2.35}$$

Calculation of Safety Device Inlet, Upstream, or Absolute Relieving Pressure

$$P_1 = P_{\text{set}} + P_{\text{over}} + P_{\text{atmosphere}} - P_{\text{loss}} \tag{2.36}$$

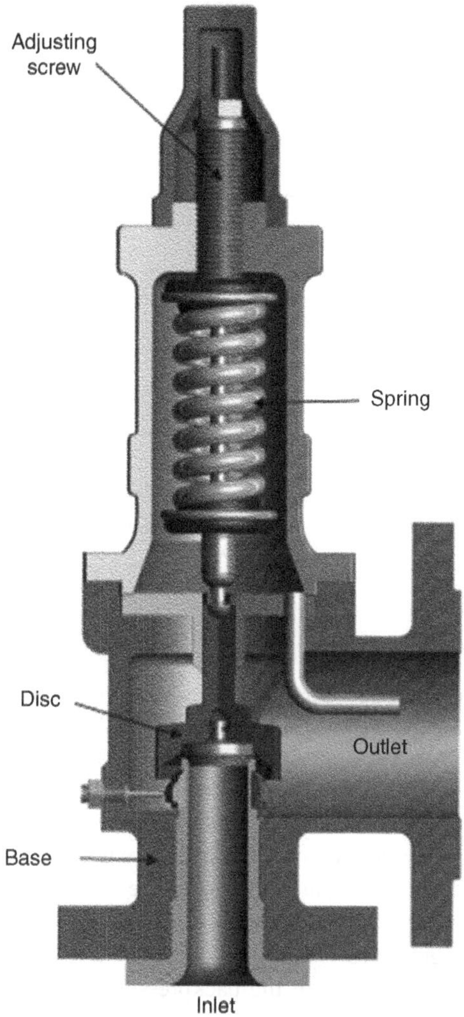

Figure 2.11 Safety relief valve.

K_b: Capacity correction factor also called backpressure correction factor due to backpressure that applies only to conventional or balanced bellows safety valves. Capacity correction factor is the ratio of the safety valve capacity with backpressure included C_1 to the rated valve capacity without backpressure C_2 that is typically applicable for valves handling vapors and gases, based on Eq. (2.37).

Calculation of Backpressure Correction Factor

$$K_b = \frac{C_1}{C_2} = \frac{\text{Safety valve capacity with backpressure included}}{\text{Safety valve capacity without backpressure}} \tag{2.37}$$

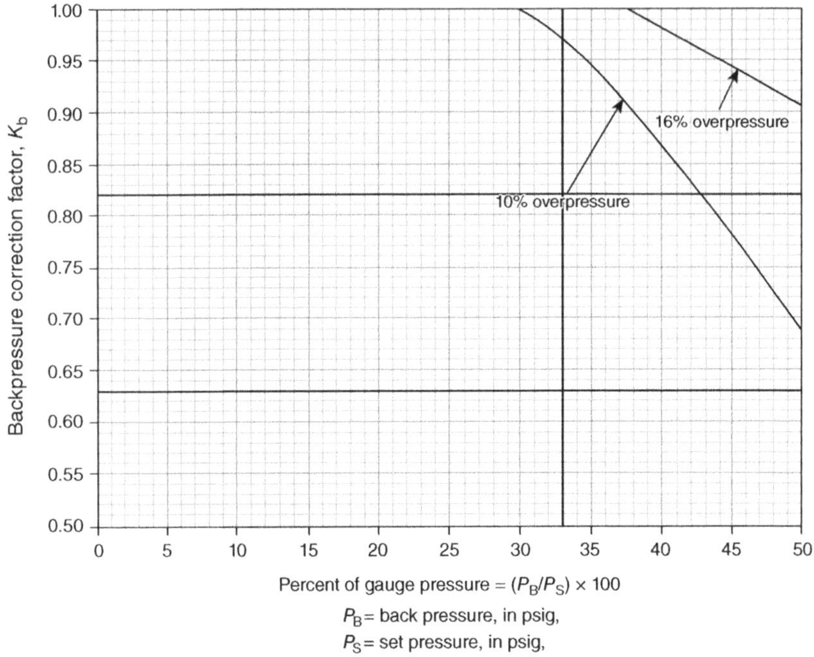

Figure 2.12 Backpressure correction factor (K_b) for balanced bellows PSVs in vapors and gases.

The balanced bellows neutralize the effect of backpressure on the safety valve's operation. The backpressure correction factor for conventional and pilot-operated safety valves is one. Backpressure is the pressure at the outlet of the pressure relief device as a result of the pressure in the discharge system. The value of K_b for balanced bellows safety valves can be obtained from the valve manufacturer or from Figure 2.12.

Calculation of the backpressure correction factor for balanced bellows safety valves is based on the *percentage of gauge backpressure* calculated according to Eq. (2.38).

Calculation of Gauge Backpressure for Bellows Safety Valves

$$\text{Percentage of gauge backpressure} = \frac{P_b}{P_s} \times 100 = \frac{\text{Backpressure}}{\text{Set pressure}} \times 100 \tag{2.38}$$

It is possible to determine the value of K_b using the diagram in Figure 2.13 for conventional safety valves. In Figure 2.13, it is shown that the percentage of

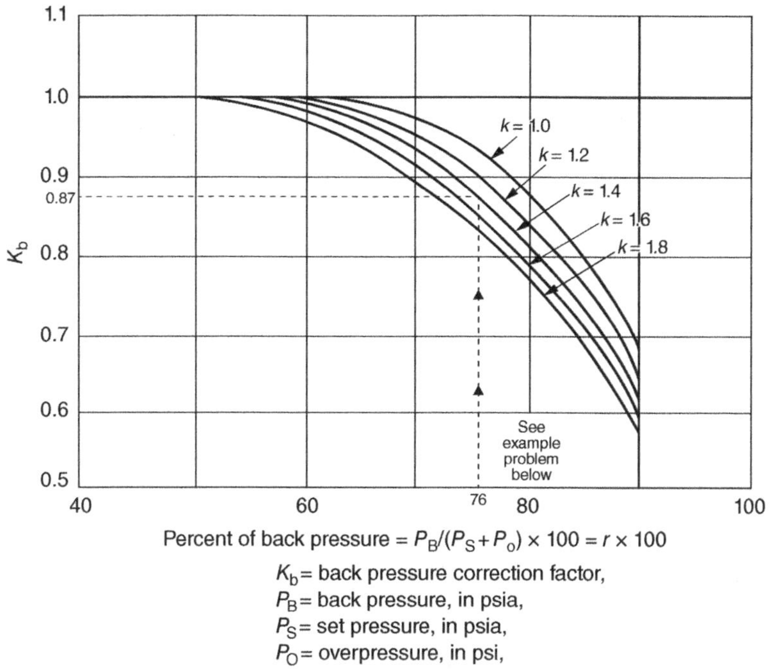

Figure 2.13 Backpressure correction factor for gas and vapors based on API 520 for conventional relief valves.

backpressure and the ratio of specific heat k are used to calculate the value of K_b for conventional relief valves handling gases or vapors. Equation (2.39) can be used to calculate the *percentage of backpressure* for conventional safety valves. A ratio of specific heat is defined as the ratio of heat capacity at constant pressure to heat capacity at constant volume. An object's thermal capacity, or heat, is defined as the amount of heat required to change its temperature by 1°. The ratio of specific heat k for a variety of gases can be determined from Table 2.14.

Calculation of Backpressure Percentage for Conventional Safety Valves

$$\text{Backpressure percentage} = \frac{P_b}{P_s + P_o} \times 100$$
$$= \frac{\text{Absolute backpressure}}{\text{Absolute set pressure} + \text{Overpressure}} \times 100 \tag{2.39}$$

K_c: The combination correction factor for installation with a rupture disk upstream of the PRV. This value is one if there is no rupture disk installed and is 0.9 if a rupture disk is installed in conjunction with a PRV.

T: Relieving temperature of the inlet gas or vapor, Rankine or Fahrenheit;
Z: Gas compressibility factor or z-factor that is equal to one for an ideal gas;
M: Molecular weight of the gas or vapor at inlet relieving conditions. The molecular weight for some gases can be found in Table 2.14.
V: The flow velocity through the device in SCF per minute (SCFM) at 14.7 psia and 60 °F, or normal cubic meter per minute at 0 °C and 101.325 KPa.
G: Gas-specific gravity at standard conditions referred to air at standard conditions. Accordingly, the specific gravity of air is one at 14.7 psia and 60 °F or at 0 °C and 101.325 KPa.

Example 2.8 When the valve handles air service, it discharges the pressure at 6 bars into the atmosphere. Can it be determined whether the flow is critical or not?

Answer

It should be noted that according to Table 2.14, the specific heat k of air is 1.40. The discharge pressure P_1 is 6 bars, and the backpressure is equal to the atmospheric pressure, which is 1.01325 bars. The critical flow pressure is calculated according to Eq. (2.40) as follows:

Critical Flow Pressure Calculation

$$P_{cf} = P_1 \left[\frac{2}{k+1}\right]^{k/(k-1)} \tag{2.40a}$$

where:

P_{cf}: Critical flow nozzle pressure in psia;
P_1: Upstream or relieving pressure in psia;
k: Ratio of specific heats for the gas.

$$P_{cf} = 6\left[\frac{2}{1.4+1}\right]^{1.4/(1.4-1)} = 6 \times 0.52828 = 3.17 \text{ bar}$$

As the critical pressure of 3.17 bar is larger than the backpressure of 1.1325 bar, the flow is critical.

Gas Z-factor or compressibility factor indicates how much the real gas deviates from an ideal gas at the given pressure and temperature. It should be noted that gases are composed of complex compounds and structures. They are composed of billions of energetic gas molecules that interact with one another. Due to this reason, it is very difficult and complex to model a real gas. As a result, the concept of an ideal gas was developed in order to simplify the behavior of gases. Scientists and engineers use the theory of ideal gases in order to model and predict the behavior of gases. However, the behavior of real gases matches that of ideal gases only in some cases, such as when the pressure is atmospheric and the temperature is in the range of room temperature. Generally, the temperature in a room is the temperature at which the majority of people would feel comfortable. The room temperature is typically in the range of 20 – 22 °C. At lower and higher temperatures, the characteristics of real gases diverge significantly from those of ideal gases. There are two main rules associated with ideal gases: the first rule states that all ideal gases have molecules that are in constant motion, rather than atoms that attract or repel each other. It is believed that intermolecular forces are negligible in ideal gases. There is no type of interaction between the molecules in an ideal gas model, except for elastic collisions, which are produced when molecules come into contact with each other and exchange kinetic energy. However, the total molecular kinetic energy inside the gas is constant, and there is no change in kinetic energy before and after a collision. The second important rule is that the molecules occupy negligible volume. Helium is known as a type of gas that behaves in a very similar way to ideal gases. Unlike an ideal gas, real gas does not behave according to the laws of kinetic molecular theory. Therefore, real gases are also called nonideal gases. As a matter of fact, gases such as carbon dioxide, carbon monoxide, helium, oxygen, and nitrogen are all real gases. Generally, a real gas has a smaller volume than an ideal gas at the same pressure and temperature. Thus, real gas can be considered supercompressible. The ratio of real volume to ideal volume is called the *Z-factor* or *compressibility factor*, represented by parameter Z and calculated according to Eq. (2.41).

Gas Z-Factor or Compressibility Factor Calculation

$$Z = \frac{\text{Actual or real volume of the gas at specified pressure and temperature}}{\text{Ideal volume of the gas at specified pressure and temperature}} \tag{2.41}$$

Different methods can be used to calculate the Z-factor or compressibility factor of a gas. The following section presents the theorem of corresponding states for the calculation of a Z-factor. The theorem of corresponding states that the Z-factor of any gas mixture is determined solely by the reduced temperature and pressure, as indicated by Eq. (2.42).

Relationship Between Z-Factor and Reduced Pressure and Temperature

$$Z = f(p_r T_r) \tag{2.42}$$

where:

Z: Z-factor of the gas;
p_r: Reduced pressure (MPa, psi);
T_r: Reduced temperature (°K, °R).

The reduced temperature, pressure, and volume are calculated according to Eqs. (2.43) and (2.44) by dividing the actual temperature, pressure, and volume by the *critical temperature*, *critical pressure*, and *critical volume*, respectively.

Reduced Temperature Calculation

$$T_r = \frac{T}{T_C} \tag{2.43}$$

where:

T: Actual temperature (°K, °R);
T_C: Critical temperature (°K, °R);
T_r: Reduced temperature (dimensionless).

Critical temperature is defined as the maximum temperature of a gas, above which the gas cannot be liquefied through the application of pressure alone, regardless of the amount of pressure.

Reduced Pressure Calculation

$$P_r = \frac{P}{P_C} \tag{2.44}$$

where:

P: Actual pressure (psi, MPa);
P_C: Critical pressure (psi, MPa);
P_r: Reduced pressure (dimensionless).

Critical pressure is defined as the pressure a gas exerts when it is in equilibrium with a liquid phase and at the critical temperature. In other words, the required pressure to liquefy the gas at a critical temperature.

Table 2.15 shows critical temperature and pressure for some common substances. In addition, the value of the compressibility factor can be determined from Figure 2.14 as a function of reduced temperature and reduced pressure.

Table 2.15 The critical pressure and temperature for some common substances.

Substance	Critical temperature		Critical temperature	
	°F	°C	psig	barg
Air	−220.94	−140.52	549.08	37.858
Ammonia (NH_3)	270	132.4	1636	112.8
Argon	−188	−122	705.6	48.7
Butane	305.6	152	550.4	38
Carbon dioxide (CO_2)	87.8	31.2	1071.6	73.8
Carbon monoxide (CO)	−220.5	−140.3	507.5	35
Chlorine	291	144	118.7	77.1
Decane	653	345	301.7	20.8
Ethane	90.0	32.2	708	48.9
Ethanol (alcohol)	467	242	914	63
Ethylether	381	194	522	36
Ethylene	48.9	9.4	735	50.7
Fluorine	−200	−129	808.5	55.8
Helium	−456	−271	33.2	2.3
Hydrogen	−400	−240	188.2	13.0
Hydrogen chloride	125	51.6	1198	82.7
Isobutane	275	135	592.2	36.5
Isobutylene	293	145	580	40
Isononane	590	310	335.1	23.1
Methane	−117	−82.6	673.3	46.5
Nitrogen (N)	−232.6	−147	492.4	34
Nitrous oxide (N_2O)	97.4	36.4	1047.6	72.3
Oxygen (O_2)	−181.5	−118.6	732	50.5
Propane	206.1	96.7	617.4	42.6
Propylbenzene	689	365	464.2	32
Propylene	198	92.4	670.3	46.3
Undecane	691	366	287.2	19.8
Water	705	374	3206.2	220.5

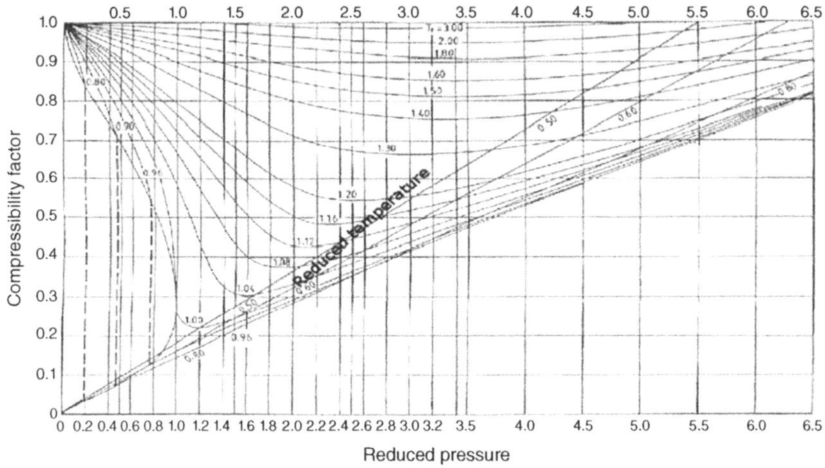

Figure 2.14 Compressibility factor determination chart.

Example 2.9 If the relieving condition is 55 °C and 62 bara, what is the compressibility factor of ethylene?

Answer

According to Table 2.15, the critical temperature and pressure for ethylene are 9.4 °C and 50.7 barg (51.7 bara), respectively. The next step is to find the reduced temperature and pressure according to Eqs. (2.43) and (2.44), respectively. The temperature values in Eq. (2.43) must be expressed in Kelvin (0 °C + 273.15 = 273.15 °K)

$$T_r = \frac{T}{T_C} = \frac{55 + 273.15}{9.04 + 273.15} = \frac{328.15}{282.19} = 1.16$$

$$P_r = \frac{P}{P_C} = \frac{62}{51.7} = 1.20$$

The compressibility factor of the gas is approximately equal to $Z = 0.71$ according to Figure 2.14.

Example 2.10 A conventional PSV is designed without its combination with a rupture disk to handle hydrocarbon gas, which is a mixture of butane and pentane with a molecular weight of 65, Z-factor = 0.84, and k = 1.09. The required hydrocarbon flow rate is 24,260 kg/h. The operating temperature of the valve is 167 °F (627 °R or 348 °K). Meanwhile, the set pressure and the backpressure of the safety valve are 75 psig and 14.7 psia, respectively. If an overpressure of 10% is tolerable, select the nozzle size for the PSV based on API 526.

Answer

By using Eq. (2.36) as a starting point, the inlet pressure can be calculated as follows:

$$P_1 = P_{set} + P_{over} + P_{atmosphere} - P_{loss} = 75 + 7.5 + 14.7 - 0 = 97.2\ \text{psia}$$

The second step is to determine whether the flow rate is critical or subcritical. An easy way to understand this is to compare the outlet pressure or backpressure with the critical flow pressure. It can be concluded that this is a critical flow condition if the backpressure is less than the critical flow pressure. The critical flow pressure can be obtained from Eq. (2.40).

$$P_{cf} = P_1 \left[\frac{2}{k+1}\right]^{k/(k-1)} \rightarrow P_{cf} = 97.2 \times \left[\frac{2}{1.09+1}\right]^{1.09/(1.09-1)}$$

$$= 97.2 \times 0.957^{12.11} = 97.2 \times 0.587 = 57.08\ \text{psia}$$

Due to the fact that the critical flow pressure (57.08 psia) is greater than the backpressure (14.7 psia), the critical flow condition is met.

In this case, the effective coefficient of discharge, K_d, should be considered equal to 0.975. From Table 2.13, $C = 326$. Due to the use of a conventional safety valve, the K_b factor, or capacity correction factor due to backpressure, is equal to one. Furthermore, the combination correction factor for installation K_c is also equal to one, since no rupture disk is used in conjunction with the safety valve. The other parameters are in US Customary units except for the mass flow rate. So by converting the mass flow rate from kilograms per hour to pounds per hour, Eq. (2.28) can be applied in this case for the sizing of a safety valve.

$$W = 24{,}260\ \text{kg/h} \times 2.20462\ \text{lb/kg} = 53{,}484\ \text{lb/h}$$

$$A = \frac{W}{C K_d P_1 K_b K_c}\sqrt{\frac{TZ}{M}} = \frac{53{,}484}{326 \times 0.975 \times 97.2 \times 1 \times 1}\sqrt{\frac{627 \times 0.84}{65}}$$

$$= 1.7311 \times 2.8465 = 4.93\ \text{in.}^2 = 3179\ \text{mm}^2 \tag{2.28b}$$

In accordance with API 526, the size of the standard nozzle must be determined using Table 2.12. In this example, nozzle designation N is 4.34 in.^{2} which is small in size. Therefore, the smallest nozzle with a flow area greater than 4.93 in.^{2}, nozzle designation "P," should be selected with the effective discharge area of 6.38 in.^{2} (4116 mm^2).

Example 2.11 Currently, a PRV with a set pressure of 30 psig (207 kPag) is handling the air service with an airflow capacity of 227 kg/h discharge to the atmosphere. The maximum overpressure that can be allowed is 3 psi and the temperature at the inlet is 21 °C. What type of API orifice should be used and what would be the nozzle discharge area? (*Note*: the compressibility factor of air in this example is equal to one.)

Answer

To begin with, it is necessary to identify the type of flow. According to Table 2.15, the critical pressure for air is 37.858 barg (3785.8 kPag). According to Table 2.14, the specific heat of air is 1.40. The discharge pressure P_1 is calculated as follows:

$$\begin{aligned} P_1 &= \text{Set pressure} + \text{Overpressure} + \text{Atmospheric pressure} \\ &= 207\ \text{kPag} + 20.7\ \text{kPag} + 101.325\ \text{kPa} = 329.025\ \text{kPag} \end{aligned}$$

The critical flow pressure is calculated according to Eq. (2.40) as follows:

Critical Flow Pressure Calculation

$$P_{\text{cf}} = P_1\left[\frac{2}{k+1}\right]^{k/(k-1)} = 329.025 \times \left[\frac{2}{1.4+1}\right]^{1.4/0.4} = 208.58\ \text{kPag} \tag{2.40b}$$

The backpressure is equal to atmospheric pressure, which is 1.01325 bars (101.325 kPag). The flow is considered critical since the backpressure is less than the critical flow pressure.

The capacity correction factor due to backpressure is one since backpressure equals atmospheric pressure. The value of gas constant C associated with the ratio of specific heat $k = 1.4$ is 356 refer to Table 2.14. The same table indicates that the molecular weight of air M is 29.96. Because the safety valve is designed for use with air, the flow is critical, and SI units are in use, Eq. (2.29) is applicable in this case.

$$W = 227\ \text{kg/h}$$

$$T = 21 + 273 = 294°\text{K}$$

$$A = \frac{13{,}160\ W}{C\,K_{\text{d}}P_1K_{\text{b}}K_{\text{c}}}\sqrt{\frac{TZ}{M}}\ \text{(SI units)}$$

$$\begin{aligned} \rightarrow A &= \frac{13{,}160 \times 227}{356 \times 0.975 \times 329.025 \times 1 \times 0.975 \times 1}\sqrt{\frac{294}{29.96}} = 26.828 \times 3.1326 \\ &= 84.04\ \text{mm}^2 = 0.130{,}26\ \text{in.}^2 \end{aligned}$$

As a result, orifice E with an area of 0.196 in.^{2} should be selected according to Table 2.12.

2.5.1.2 Subcritical Flow

Subcritical flows are defined as those in which the backpressure exceeds the critical pressure. As explained earlier, the critical pressure is calculated according to Eq. (2.40). In the event the system operates under subcritical flow conditions, the pressure relief device should be sized according to API 520 using Eqs. (2.45) and (2.46) for mass flow rates, and Eqs. (2.47)–(2.50) for volumetric flow rates. It is necessary to select a PRV that has a discharge area that is as large as, or larger than, the calculated value of "A," as shown in the following equations.

Sizing Relief Devices for Gas or Vapor in a Subcritical Flow Condition According to Mass Flow Rate

$$A = \frac{W}{735\,F_2K_dK_c}\sqrt{\frac{ZT}{MP_1(P_1-P_2)}}\ \text{(US Customary units)} \tag{2.45}$$

$$A = \frac{17.9\times W}{F_2K_dK_c}\sqrt{\frac{ZT}{MP_1(P_1-P_2)}}\ \text{(SI units)} \tag{2.46}$$

Sizing Relief Devices for Gas or Vapor in a Subcritical Flow Condition According to Volumetric Flow Rate

$$A = \frac{V}{4645\,F_2K_dK_c}\sqrt{\frac{ZTM}{P_1(P_1-P_2)}}\ \text{(US Customary units)} \tag{2.47}$$

$$A = \frac{V}{864\,F_2K_dK_c}\sqrt{\frac{ZTG}{MP_1(P_1-P_2)}}\ \text{(US Customary units)} \tag{2.48}$$

$$A = \frac{47.95\times V}{F_2K_dK_c}\sqrt{\frac{ZTM}{P_1(P_1-P_2)}}\ \text{(SI units)} \tag{2.49}$$

$$A = \frac{258\times V}{F_2K_dK_c}\sqrt{\frac{ZTG}{P_1(P_1-P_2)}}\ \text{(SI units)} \tag{2.50}$$

where:

G: The specific gravity of a gas is compared to the specific gravity of air at standard or normal conditions. It is crucial to know that $G = 1$ for air at 14.7 psia (101.325 KPa and 0 °C);

F_2: The coefficient of subcritical flow can either be derived from Figure 2.15 or calculated from Eq. (2.51).

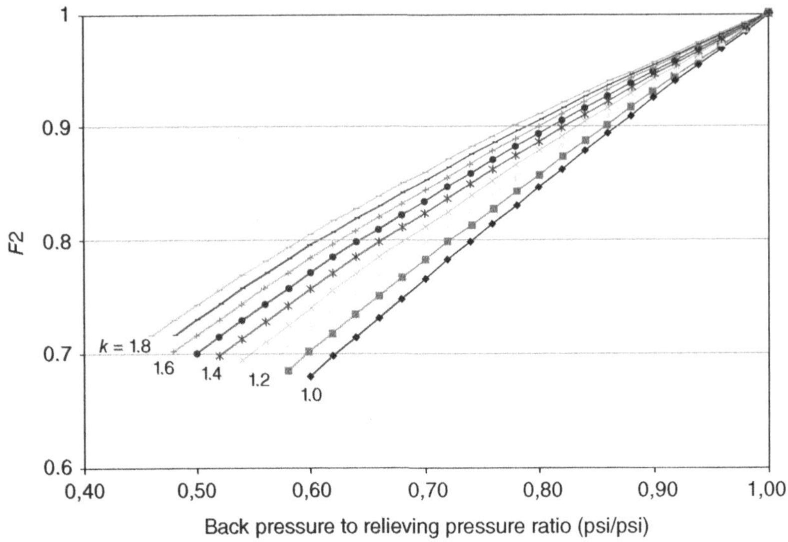

Figure 2.15 Values of subcritical flow coefficient (F_2).

Calculation of Coefficient of Subcritical Flow (F_2)

$$F_2 = \sqrt{\left(\frac{k}{k-1}\right)(r)^{2/k}\left[\frac{1-r^{(k-1)/k}}{1-r}\right]} \tag{2.51}$$

where:

k: Ratio of specific heats;
r: Ratio of backpressure to upstream pressure, P_2/P_1.

Other parameters and their values are the same as those discussed previously for the critical condition.

Example 2.12 A PSV that is installed on the top of a pressure equipment must have a flow capacity of 53,500 lb/h. In the pressure equipment, the hydraulic fluid is a mixture of butane and pentane containing a molecular weight of 65, a Z-factor of 0.84, and a specific heat ratio of 1.09. The set pressure of the safety valve is equal to the design pressure of the pressure equipment, which is equivalent to 75 psig with 10% overpressure allowed. The relieving temperature is 627 °R, corresponding to 167 °F. In this conventional safety valve, a constant backpressure of 55 psi gauge as well as 7.5 psig buildup backpressure are applied. According to API standards, what size orifice is required for this valve?

Answer

The first step is to determine whether the fluid is critical or subcritical.

$$P_1 = \text{Set pressure} + \text{Overpressure} + \text{Atmospheric pressure}$$
$$= 75 + 7.5 + 14.7 = 97.2\ \text{psia}$$

$$P_{cf} = P_1 \left[\frac{2}{k+1}\right]^{k/(k-1)} = 97.2 \times \left[\frac{2}{1.09+1}\right]^{1.09/0.09} = 57.03\ \text{psia}$$

$$\text{Total back pressure} = 55\ \text{psig} + 7.5\ \text{psig} = 62.5 + 14.7 = 77.2\ \text{psia}$$

As the backpressure (77.2 psia) is higher than the critical pressure (57.03 psia), the flow is subcritical. Because the flow in this case is subcritical and given as mass flow rate, and since the units of measurement are US Customary, Eq. (2.45) applies. Obtaining the coefficient of subcritical flow F_2 is the next step. To do so, first the ratio of backpressure to relieving pressure must be calculated as follows:

$$\text{Ratio of backpressure} = \frac{P_2}{P_1} = \frac{77.2\ \text{psia}}{97.2\ \text{psia}} = 0.794$$

It should also be noted that the coefficient of subcritical flow F_2 which is equal to 0.86 in Figure 2.15 is a consequence of combining the ratio of specific heat ($k = 1.09$) with the ratio of backpressure (0.794). The value of coefficient of discharge $K_d = 0.975$. In addition, K_c is the combination correction factor for installation with a rupture disk upstream of the PRV. This value is one in this case because there is no rupture disk installed. Inlet relieving temperature should be in Rankine unit, which is 627°.

$$A = \frac{W}{735F_2K_dK_c}\sqrt{\frac{ZT}{MP_1(P_1 - P_2)}}$$

$$= \frac{53{,}500}{735 \times 0.86 \times 0.975 \times 1}\sqrt{\frac{0.84 \times 627}{65 \times 97.2 \times (97.2 - 77.2)}}$$

$$= 86.81 \times 0.6456 = 5.60\ \text{in.}^2 = 3613\ \text{mm}^2$$

In accordance with Table 2.12, the orifice size of "*P*" with a discharge area of 6.38 in.^{2} = 4116 mm^2 should be selected.

2.5.2 Sizing for Steam Relief

Equations (2.52) and (2.53) are used to size safety valves in the steam service in US Customary and SI units, respectively.

Sizing Relief Devices for Steam

$$A = \frac{W}{51.5 \times P_1 K_d K_b K_c K_N K_{SH}} \text{ (US Customary units)} \quad (2.52)$$

$$A = \frac{190.4 \times W}{P_1 K_d K_b K_c K_N K_{SH}} \text{ (SI units)} \quad (2.53)$$

where:

A: Minimum required effective discharge area, in.^{2}(mm^2);
W: Required flow rate or capacity lb/h (kg/h);
P_1: The upstream relieving pressure, psia (kPaa). Upstream or inlet relieving pressure equals the sum of the set pressure, allowable overpressure and atmospheric pressure;
K_d: Effective coefficient of discharge that is equal to 0.975 when a pressure safety or relief valve is installed with or without a rupture disk in combination. Alternatively, this value is equal to 0.62 when the PRV is not installed and the sizing is done for a rupture disk;
K_b: The backpressure correction factor was discussed earlier in the chapter as to how it can be calculated;
K_C: Combination correction factor due to the installation of a rupture disk upstream of a PSV. The value is one if no rupture disk is installed, and 0.9 if a rupture disk is installed in conjunction with a safety valve;
K_N: Correction factor for the Napier equation. The dimensionless correction factor is only applicable to the steam flow as per API 520 and is calculated according to Eqs. (2.54) and (2.55). The factor is only used if the steam flow inlet relieving pressure P_1 is in the range of 1500–3200 psia. The correction factor is equal to one if the inlet relieving pressure is less than or equal to 1500 psia (10,339 kPa).

Correction Factor for the Napier Equation Applicable to Steam Flow

$$\begin{aligned} &K_N = 1 \text{ if } P_1 \leq 1500 \text{ psia } (10{,}339 \text{ kPa}) \\ &\text{If}: 1500 \text{ psia } (10{,}339 \text{ kPa}) < P_1 \leq 3200 \text{ psia } (22{,}057 \text{ kPa}) \end{aligned} \quad (2.54)$$

$$\begin{aligned} K_N &= \frac{0.1906 \times P_1 - 1000}{0.2292 \times P_1 - 1061} \text{ (US Customary units)} \\ K_N &= \frac{0.02764 \times P_1 - 1000}{0.03324 \times P_1 - 1061} \text{ (SI units)} \end{aligned} \quad (2.55)$$

K_{SH}: Superheat correction factor that can be obtained from Table 2.16. For saturated steam at any pressure, $K_{SH} = 1$. Steam that is saturated occurs when the liquid and gaseous phases of water are present simultaneously at a given

Table 2.16 Superheat steam correction factor, K_{SH}, according to API 520 standard.

Set pressure (psig)	Temperature (°F)									
	300	400	500	600	700	800	900	1000	1100	1200
15	1.00	0.98	0.93	0.88	0.84	0.80	0.77	0.74	0.72	0.70
20	1.00	0.98	0.93	0.88	0.84	0.80	0.77	0.74	0.72	0.70
40	1.00	0.99	0.93	0.88	0.84	0.81	0.77	0.74	0.72	0.70
60	1.00	0.99	0.93	0.88	0.84	0.81	0.77	0.75	0.72	0.70
80	1.00	0.99	0.93	0.88	0.84	0.81	0.77	0.75	0.72	0.70
100	1.00	0.99	0.94	0.89	0.84	0.81	0.77	0.75	0.72	0.70
120	1.00	0.99	0.94	0.89	0.84	0.81	0.78	0.75	0.72	0.70
140	1.00	0.99	0.94	0.89	0.85	0.81	0.78	0.75	0.72	0.70
160	1.00	0.99	0.94	0.89	0.85	0.81	0.78	0.75	0.72	0.70
180	1.00	0.99	0.94	0.89	0.85	0.81	0.78	0.75	0.72	0.70
200	1.00	0.99	0.95	0.89	0.85	0.81	0.78	0.75	0.72	0.70
220	1.00	0.99	0.95	0.89	0.85	0.81	0.78	0.75	0.72	0.70
240	1.00	1.00	0.95	0.90	0.85	0.81	0.78	0.75	0.72	0.70
260	1.00	1.00	0.95	0.90	0.85	0.81	0.78	0.75	0.72	0.70
280	1.00	1.00	0.96	0.90	0.85	0.81	0.78	0.75	0.72	0.70
300	1.00	1.00	0.96	0.90	0.85	0.82	0.78	0.75	0.72	0.70
350	—	1.00	0.96	0.90	0.86	0.82	0.78	0.75	0.72	0.70
400	—	1.00	0.96	0.91	0.86	0.82	0.78	0.75	0.72	0.70
500	—	1.00	0.96	0.92	0.86	0.82	0.78	0.75	0.73	0.70
600	—	1.00	0.97	0.92	0.87	0.82	0.79	0.75	0.73	0.70
800	—	—	1.00	0.95	0.88	0.83	0.79	0.76	0.73	0.70
1000	—	—	1.00	0.96	0.89	0.84	0.78	0.76	0.73	0.71
1250	—	—	1.00	0.97	0.91	0.85	0.80	0.77	0.74	0.71
1500	—	—	—	1.00	0.93	0.86	0.81	0.77	0.74	0.71
1750	—	—	—	1.00	0.94	0.86	0.81	0.77	0.73	0.70
2000	—	—	—	1.00	0.95	0.86	0.80	0.76	0.72	0.69
2500	—	—	—	1.00	0.95	0.85	0.78	0.73	0.69	0.66
3000	—	—	—	—	1.00	0.82	0.74	0.69	0.65	0.62

temperature and pressure. Simplified, steam is at equilibrium with the heated water. In the saturated condition, the rate at which water is vaporized is equal to the rate at which it is condensed.

Example 2.13 Choose the correct size of PSV that can handle saturated steam with a relief capacity of 153,000 lb/h. As the set pressure of the valve is 1600 psig, there is a 10% allowance for overpressure. The valve is designed to discharge into the atmosphere. A note should be made that a rupture disk is not used in the safety system in order to prevent catastrophic failure.

Answer

Since the PSV is used for steam and the data given are in US Customary units, Eq. (2.52) is applicable. Obtaining the values of the parameters in the equation is an important step before using the equation.

$$W = 153{,}000\ \text{lb/h};$$

$$P_1 = \text{Set pressure} + \text{Overpressure} + 14.7$$
$$= 1600 + 0.1 \times 1600 + 14.7 = 1774.7\ \text{psia}$$

K_d: An effective coefficient of discharge of 0.975 is achieved when a pressure safety or relief valve is installed without a rupture disk combined with a pressure safety or relief valve.

K_b: The backpressure correction factor is equal to one in this case because the valve discharge is into the atmosphere.

K_C: The combination correction factor is due to the installation of a rupture disk upstream of a PSV. It is equal to one since there is no rupture disk installed.

K_N: According to Eq. (2.54), the correction factor for the Napier equation can be calculated as follows.

$$K_N = \frac{0.1906 \times P_1 - 1000}{0.2292 \times P_1 - 1061} = \frac{0.1906 \times 1774.7 - 1000}{0.2292 \times 1774.7 - 1061} = \frac{-661.74}{-654.24} = 1.011$$

Superheat correction factor $K_{SH} = 1$ because saturated steam is handled by the valve.

$$A = \frac{W}{51.5 \times P_1 K_d K_b K_c K_N K_{SH}} = \frac{153{,}000}{51.5 \times 1774.7 \times 0.975 \times 1 \times 1.011 \times 1}$$
$$= 1.698\ \text{in.}^2$$

In accordance with Table 2.12, the selected orifice is K with an effective discharge area of 1.838 in.^{2}.

2.5.3 Sizing for Liquid Relief

In order to properly size a PSV for liquid relief, the following classifications are made; PRVs that require capacity certification and PRVs that do not require capacity certification.

2.5.3.1 Sizing for Liquid Relief with Capacity Certification

According to ASME Sec. VIII Div. 01, it is mandatory for safety valves designed for liquid services to obtain capacity certification. An application for capacity certification involves testing the valve to determine the rated coefficient of discharge for liquid valves at 10% overpressure in order to obtain certification. Thus, safety valves in liquid services that need to comply with ASME code requirements concerning the capacity certificate need to be sized according to Eqs. (2.56) and (2.57).

Sizing Relief Devices for Liquid According to ASME Code

$$A = \frac{Q}{38\,K_d K_w K_c K_v}\sqrt{\frac{G}{P_1 - P_2}}\ \text{(US Customary units)} \tag{2.56}$$

$$A = \frac{11.78 \times Q}{K_d K_w K_c K_v}\sqrt{\frac{G}{P_1 - P_2}}\ \text{(SI units)} \tag{2.57}$$

where:

A: Required discharge area, in.^{2} (mm^2);
Q: Flow rate US gallons per minute (l/min);
K_d: This is the coefficient of discharge that should be obtained from the valve manufacturer. The value of the discharge coefficient, which is used for a preliminary valve sizing calculation, can be determined as follows: It reaches a value of 0.65 when a pressure safety or relief valve is installed either with or without a rupture disk. It is also worth noting that this value is equal to 0.62 when the PRV is not installed, and the sizing is done for a rupture disk;
K_w: There is a term for this effect called backpressure correction and it is equal to one if the backpressure is discharged directly into the atmosphere. For bellows balanced type safety valves, however, the backpressure correction factor must be obtained from Figure 2.16;
K_C: This is a combination correction factor due to the installation of a rupture disk upstream of a PSV. There is a value of one if there is no rupture disk installed; however, the value becomes 0.9 if the rupture disk in conjunction with the safety valve is installed;
K_v: This is the correction factor due to viscosity that can be obtained from Eq. (2.58) or Figure 2.17. The value of the correction factor as a result of viscosity derived from the equation or figure depends on the Reynolds number, which will be

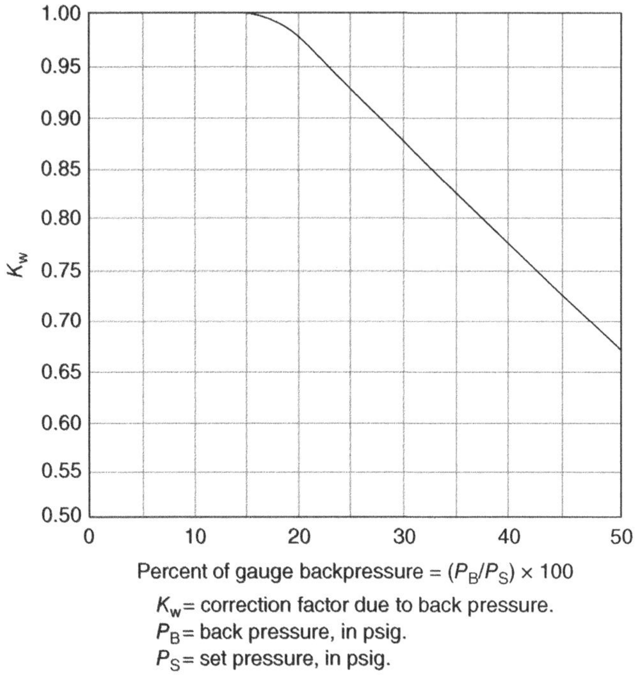

Figure 2.16 Backpressure correction factor for balanced bellows PRVs in liquid service (parameter K_w).

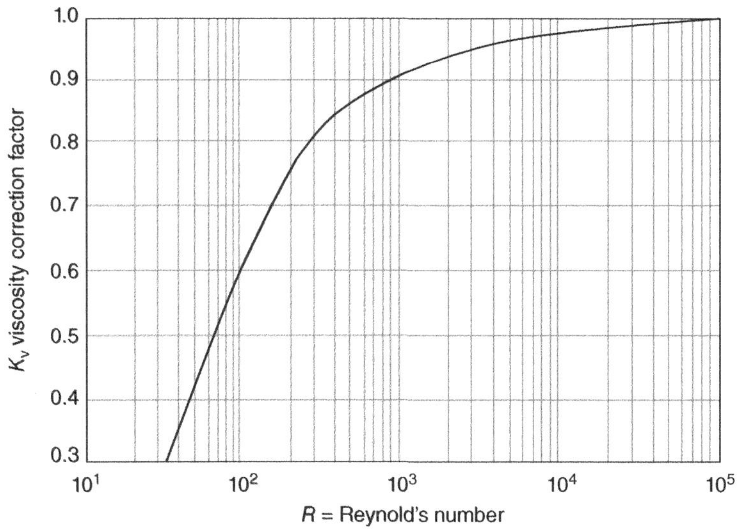

Figure 2.17 Viscosity correction factor for balanced bellows PRVs in liquid service (parameter K_v).

explained in more detail further in this chapter. The value of $K_v = 1$ for a nonviscous fluid.

Calculation of Correction Factor Due to Viscosity (K_v)

$$K_v = \left(0.9935 + \frac{2.878}{R^{0.5}} + \frac{342.75}{R^{1.5}}\right)^{-1.0} \tag{2.58a}$$

G: The specific gravity of the liquid corresponds to the density of the liquid relative to that of water at the temperature of the liquid.

P_1: The upstream relieving pressure, psia (kPaa). Upstream or inlet relieving pressure equals the sum of the set pressure, allowable overpressure and atmospheric pressure.

P_2: Backpressure psig (kPag)

First, a safety valve is sized for a nonviscous fluid where $K_v = 1$. After this, the calculated orifice discharge area is adjusted based on the Reynolds number and the viscosity correction factor. It is important to note that after the Reynolds number is determined from either Figure 2.17 or any other equation explained in the next section, the viscosity correction factor will be calculated and then applied to the calculated preliminary discharge area.

Using the Reynolds number (R), it is possible to predict flow patterns in various fluid flow situations. At low Reynolds numbers, the flow pattern tends to be laminar (sheet-like), whereas at high Reynolds numbers, the flow pattern tends to be turbulent. The Reynolds number can be calculated by taking the absolute viscosity at the flowing temperature in centipoises, shown with parameter μ in Eqs. (2.58) and (2.59), or the viscosity at the flowing temperature in saybolt universal seconds (SSUs), shown with parameter U as per Eqs. (2.60) and (2.61). Viscosity is a measure of the resistance that a fluid exhibits when it is being deformed by either a shear or a tensile force. The term viscosity is derived from the Latin word viscum, which is a reference to anything that sticks. A fluid with a low viscosity is referred to as thin, while a fluid with a high viscosity is referred to as thick. Viscosity describes the internal friction of moving fluids, and a fluid with a higher amount of viscosity is resistant to movement since the molecules of the fluid create a lot of internal friction. Low-viscosity fluids, on the other hand, flow better. In Figure 2.18, there is a comparison between the viscosity of water and honey as a function of their gravitational flow rates.

Calculation of Reynolds Number Based on Parameter μ

$$R = \frac{Q \times (2800 \times G)}{\mu \times \sqrt{A}} \quad \text{(US Customary units)} \tag{2.58b}$$

Figure 2.18 Viscosity comparison between water and honey by studios guy. *Source:* https://imgur.com/gallery/O9ahk.

$$R = \frac{Q \times (18{,}800 \times G)}{\mu \times \sqrt{A}} \text{ (SI units)} \tag{2.59}$$

Calculation of Reynolds Number Based On Parameter μ

$$R = \frac{12{,}700 \times Q}{U \times \sqrt{A}} \text{ (US Customary units)} \tag{2.60}$$

$$R = \frac{85{,}220 \times Q}{U \times \sqrt{A}} \text{ (SI units)} \tag{2.61}$$

where:

R: Reynolds number;

Q: Flow rate at the given temperature, US gallons per minute (gpm) (l/min);

G: A liquid's specific gravity is determined by comparing its density to water at the liquid's temperature;

μ: Absolute viscosity at the flowing temperature, centipoise;

A: Effective discharge area of the safety valve in.^{2} (mm^2);

U: Viscosity at the flowing temperature in saybolt universal seconds (SSUs).

Example 2.14 It is essential to size the safety relief valve properly so that it can handle crude oil with a specific gravity of 0.85 and a viscosity of 2000 saybolt universal seconds (SSUs) at the flow temperature. The valve has a flow capacity of

2000 gpm, a set pressure of 200 psig, and a backpressure of 40 psig. For valve sizing, what are the discharge flow area and orifice letter with a 10% allowable overpressure scenario?

Answer

The following data can be obtained from this example:

$$Q = 2000\,\text{gpm}, \quad G = 0.85, \quad \text{and} \quad U = 2000\,\text{SSUs};$$

$$P_1 = \text{Set pressure} + \text{Overpressure} + 14.7 = 200 + 0.1 \times 200 + 14.7$$
$$= 234.7\,\text{psia} = 220\,\text{psig}$$

$$P_2 = \text{Backpressure} = 40\,\text{psig}$$

K_d: When a pressure safety or relief valve used for liquid services is installed without a rupture disk together with a pressure safety or relief valve, an effective coefficient of discharge of 0.65 can be achieved.

K_w: The term "backpressure correction" is commonly used for this effect and the backpressure correction factor can be obtained from Figure 2.16 in the case of bellows balanced type safety valves. The value of K_w depends on the ratio of backpressure to set pressure P_b/P_S.

$$\frac{P_b}{P_S} = \frac{40}{200} \times 100 = 20\% \rightarrow K_w = 0.97$$

K_C: The combination correction factor is due to the installation of a rupture disk upstream of a PSV. It is equal to one since there is no rupture disk installed.

Due to the fact that the values given are in US Customary units, Eq. (2.56) is used for sizing the safety valves. But one should be remembered that the valve should be sized first for a viscid fluid with $K_v = 1$.

$$A_R = \frac{Q}{38\,K_d K_w K_c K_v}\sqrt{\frac{G}{P_1 - P_2}} = \frac{2000}{38 \times 0.65 \times 0.97 \times 1 \times 1}\sqrt{\frac{0.85}{220 - 40}}$$
$$= 83.4759 \times 0.0687 = 5.73\,\text{in.}^2$$

Using the formula above, the orifice area calculated above equals 5.73 in.^{2} without considering viscosity correction and is shown as A_R. The next step is to determine the appropriate orifice discharge area that corresponds to the designated letter as per Table 2.12. The orifice size that is suitable at this stage to calculate Reynolds number would be "*P*" with a discharge area of 6.38 in.^{2}. It is then necessary to calculate the Reynolds number by applying Eq. (2.60) because all units are in US Customary and the given viscosity is expressed as a parameter *U*.

$$R = \frac{12{,}700 \times Q}{U \times \sqrt{A}} = \frac{12{,}700 \times 2000}{2000 \times \sqrt{6.38}} = \frac{12{,}700}{2.53} = 5020$$

We can now calculate the correction factor due to viscosity by using Eq. (2.58). This calculation can be done by the following formula:

$$K_v = \left(0.9935 + \frac{2.878}{R^{0.5}} + \frac{342.75}{R^{1.5}}\right)^{-1.0} = \left(0.9935 + \frac{2.878}{5020^{0.5}} + \frac{342.75}{5020^{1.5}}\right)^{-1.0}$$
$$= \left(0.9935 + \frac{2.878}{70.85} + \frac{342.75}{3{,}55{,}677}\right)^{-1.0} = (0.9935 + 0.0406 + 0.00096)^{-1.0}$$
$$= (1.03506)^{-1.0} = 0.966$$

As a result, the final orifice discharge area is calculated in the following manner:

$$A = \frac{A_R}{K_v} = \frac{5.73}{0.966} = 5.93\ \text{in.}^2$$

Upon determining the appropriate discharge area for the orifice in accordance with the designated letter, as shown in Table 2.12, the next step would be to select the appropriate orifice. For this purpose, the optimum orifice size that would be suitable at this point would be "*P*" with a discharge area of 6.38 in.^{2}.

2.5.3.2 Sizing for Liquid Relief Without Capacity Certification

According to ASME Code requirements pertaining to the capacity certificate, safety valves in liquid services that do need to comply with the ASME code requirements need to be sized according to Eqs. (2.62) and (2.63).

Sizing Relief Devices for Liquid Without Capacity Certification

$$A = \frac{Q}{38K_dK_wK_cK_vK_p}\sqrt{\frac{G}{1.25\,P - P_b}}\ \text{(US Customary units)} \tag{2.62}$$

$$A = \frac{11.78 \times Q}{K_dK_wK_cK_vK_p}\sqrt{\frac{G}{1.25\,P - P_b}}\ \text{(SI units)} \tag{2.63a}$$

K_d: The coefficient of discharge should be obtained from the manufacturer of the valve in question. To determine the preliminary size in this case, 0.62 should be taken into account.

K_p: It is the correction factor due to overpressure. If the overpressure is considered 25%, the value of the overpressure correction factor is equal to one. Otherwise, the value of this parameter must be obtained from Figure 2.19. A glance at the curve in Figure 2.19 indicates that capacity is affected by a change in lift, orifice discharge coefficient, and change in overpressure up to and including 25%

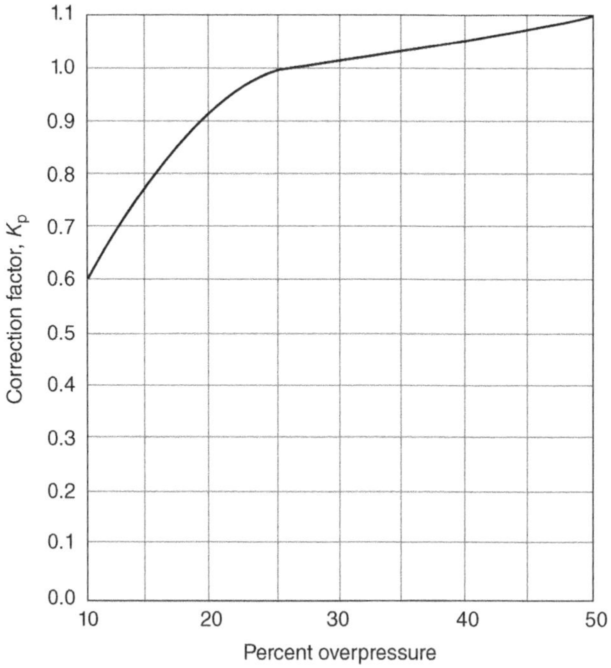

Figure 2.19 Capacity correction factor due to overpressure for noncertified pressure safety valve (PSV) in liquid services.

overpressure. As capacity rises above 25%, the change in overpressure is the only factor that affects it. The use of noncertified valves under low overpressure can cause chattering and, therefore, scenarios with more than 10% overpressure should be avoided.

P: Set pressure psig (kPag);
P_b: Total backpressure psig (kPag);

The other parameters and their values are the same for both certified and noncertified safety valves in liquid services.

2.5.4 Sizing for Two-Phase Liquid/Vapor Relief

There are various reasons why two-phase flows occur in safety valves. Some of them are listed here:

- During the safety valve's operation, a liquid vaporizes;
- From the piping or equipment connected to the valve, a two-phase mixture of liquid and vapor enters the valve;

Table 2.17 Two-phase flow scenarios.

Two-phase liquid/vapor relief scenario	Section
The valve is filled with saturated liquid and saturated vapor and the liquid flashes. There is no noncondensable gas present (flashing flow).	2.5.4.1
Condensation of supercritical fluid in the safety valve.	2.5.4.1
High subcooled liquids and either noncondensable gas or condensable vapors, or both, enter the valve, but the liquid does not flash (frozen flow).	2.5.4.2
The subcooled liquid enters the valve and flashes. At the inlet, there is no vapor or gas present.	2.5.4.2
Generic two-phase flow involving a subcooled or saturated liquid and a noncondensable gas with or without condensable vapor.	Not covered in this chapter

- Vapor condenses within the safety valve;
- A supercritical fluid is introduced into the safety valve and condenses.

The two-phase flow mixture is likely to discharge from the safety valve in all four of the cases mentioned earlier. If a relief valve handles a liquid when the liquid is at equilibrium between vapor and liquid, this may result in the production of mixed flow as a result of vapor generation. As a result of the vapor generation, the mass flow capacity of the valve decreases. According to Table 2.17, there are five possible scenarios that could occur regarding the two-phase liquid/vapor flow through the safety valves. We will cover four out of five scenarios in the following sections.

The sizing procedure for two-phase flow is given in Appendix D of API 520 based on a method called the "Omega method of Leung." For this method of sizing, the Omega parameter ω is used, which is a measure of the compressibility of the two-phase mixture. There are the following steps that must be followed to successfully complete this method:

- Calculation of the Omega parameter;
- The determination of whether the flow is critical or subcritical;
- Calculation of the mass flux, which refers to the mass flow per unit of area;
- Calculation of the required orifice discharge area and selection of the final orifice size.

The parameters used to determine the size of two-phase safety valves and their units are presented as follows:

C_p: Specific heat at constant pressure of the liquid at the inlet of the safety valve (Btu/lb°R);

G: Mass flux (lb/s ft^2);

h_{vl0}: Latent heat of vaporization at the safety valve inlet. In multicomponent systems, it is the difference between the vapor and liquid-specific enthalpies at the safety valve inlet (Btu/lb);

h_{vls}: Latent heat of vaporization at P_S; For multicomponent systems, it is the difference between the vapor and liquid specific enthalpies at P_S (Btu/lb);

k: Ratio of specific heats of the vapor. If its value is unknown, it should be assumed to be equal to one (dimensionless);

P_1 or P_0: Pressure at safety valve inlet (psia) that is equal to set pressure plus the allowable overpressure plus atmospheric pressure;

P_a: Downstream backpressure (psia);

P_c: Critical pressure (psia);

P_r: Reduced pressure (psia);

P_s: Saturation pressure (single-component flows) or bubble point pressure (multicomponent flows) at the relieving temperature T_0 (psi);

Q: Volumetric flow rate (gpm);

T_0: Temperature at safety valve inlet (°R);

T_r: Reduced temperature (°R);

v_{v0}: Specific volume of the vapor at safety valve inlet (ft^3/lb);

v_0: Specific volume of the two-phase mixture at safety valve inlet (ft^3/lb);

v_{g0}: Specific volume of the vapor, gas or combined vapor and gas at the safety valve inlet (ft^3/lb);

v_{vl0}: Difference between the vapor and the liquid specific volumes at the safety valve inlet (ft^3/lb);

v_{vls}: Difference between the vapor and the liquid specific volumes at P_s (ft^3/lb);

v_9: Specific volume evaluated at 90% of the safety valve inlet pressure (= relieving pressure), assuming isentropic flashing (ft^3/lb);

x_0: Vapor (or gas or combined vapor and gas) mass fraction (quality) at safety valve inlet (dimensionless);

η_a: Ratio between ambient pressure and relieving pressure $= P_a/P_1$ (dimensionless);

η_c: Ratio between critical pressure and relieving pressure (dimensionless);

η_s: Ratio between saturation pressure at relieving temperature and relieving pressure (dimensionless);

ρ_{l0}: Density of the liquid at the inlet of the safety valve (lb/ft^3);

ρ_9: Density evaluated at 90% of the saturation pressure (single-component flows) or bubble point pressure (multicomponent flows) P_s at T_0. The flash calculation shall be done isentropically (lb/ft^3);

ω: Omega parameter (dimensionless);

ω_s: Omega parameter for subcooled liquid flows at safety valve inlet (dimensionless).

2.5.4.1 Sizing for Saturated Liquid and Saturated Vapor, Liquid Flashes

Step 1 Calculate the Omega Parameter (ω)

The definitions of the Omega-Parameter in Eqs. (2.64)–(2.66) can be employed for multicomponent systems, whose nominal boiling range, that is the difference in the atmospheric boiling points of the heaviest and the lightest components, is less than 150 °F. For single-component systems with reduced temperature $T_r \leq 0.9$ (see Eq. 2.43) and pressure (see Eq. 2.44) $P_r \leq 0.5$, either Eq. (2.64) or Eq.(2.65) can be used.

Calculation of Omega Parameter (ω)

$$\omega = \frac{x_0 v_{vo}}{v_0} \times \left(1 - 0.37 \frac{P_1 \times v_{vl0}}{h_{vl0}}\right) + 0.185 \times \frac{C_p T_0 P_1}{v_0} \left(\frac{v_{vl0}}{h_{vl0}}\right)^2 \tag{2.64}$$

$$\omega = \frac{x_0 v_{vo}}{v_0 k} + 0.185 \times \frac{C_p T_0 P_1}{v_0} \left(\frac{v_{vl0}}{h_{vl0}}\right)^2 \tag{2.65}$$

Whenever a multicomponent system, whose nominal boiling range is greater than 150 °F, or whenever a single component system is close to its thermodynamic critical point or supercritical fluids are involved in condensing two-phase flows, Eq. (2.66) must be used.

$$\omega = 9\left(\frac{v_9}{v_0} - 1\right) \tag{2.66}$$

Step 2 Determine If the Fluid is Critical or Subcritical

The two-phase flow is critical if the critical pressure is larger than the backpressure.

If $P_c > P_a \rightarrow$ Critical flow

If $P_c < P_a \rightarrow$ Subcritical flow

The critical pressure can be calculated using Eq. (2.67) as follows:

$$P_c = \eta_c \times P_1 \tag{2.67a}$$

In order to determine the ratio of critical pressure η_c either from Eq. (2.68) or from Figure 2.20 can be used.

$$\eta_c^{\,2} + \left(\omega^2 - 2\omega\right)\left(1 - \eta_c\right)^2 + 2\omega^2 \ln(\eta_c) + 2\omega^2(1 - \eta_c) = 0 \tag{2.68a}$$

Step 3 Calculate the Mass Flux

Calculation of the mass flux G for critical and subcritical flow is based on Eqs. (2.68) and (2.69), respectively.

Calculation of Mass Flux for Critical Flow

$$G = 68.09 \times \eta_c \times \sqrt{\frac{P_1}{\omega v_0}} \tag{2.68b}$$

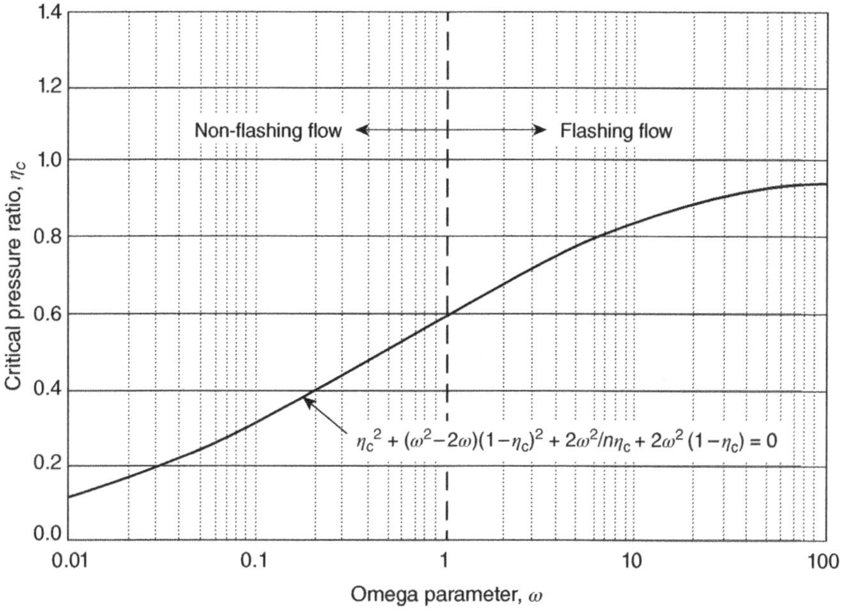

Figure 2.20 Critical flow ratio determination based on Omega parameter.

Calculation of Mass Flux for Subcritical Flow

$$G = 68.09 \times \sqrt{\frac{P_1}{v_0}} \frac{\sqrt{-2 \times [\omega \ln(P_a/P_1) + (\omega - 1)(1 - (P_a/P_1))}}{\omega((P_1/P_a) - 1) + 1} \quad (2.69)$$

Step 4 Calculate the Required Orifice Area of the Safety Valve

Finally, the required area of the safety valve can be computed from Eq. (2.70).

Calculation of the Required Orifice Area of the Safety Valve in Sizing for Saturated Liquids and Saturated Vapors, Liquid Flashes

$$A = 0.04 \times \frac{1}{K_b K_c K_d} \times \frac{W}{G} \quad (2.70)$$

where:

A: Required discharge area, in.^{2};

W: Mass flow rate (lb/h);

K_d: This is the coefficient of discharge that should be obtained from the valve manufacturer. The value of the discharge coefficient, which is used for a preliminary valve sizing calculation, can be considered 0.85 for this case;

K_b: A backpressure correction factor may also be referred to as a capacity correction factor. It should be noted that backpressure applies only to balanced bellows valves, and it can be obtained from Figure 2.12;

K_C: This is a combination correction factor due to the installation of a rupture disk upstream of a PSV. There is a value of one if there is no rupture disk installed; however, the value becomes 0.9 if the rupture disk in conjunction with the safety valve is installed.

Example 2.15 A safety valve must be sized for a two-phase flow of water and its saturated steam. The set pressure of the safety valve is 145 psig (10% overpressure) and the valve is expected to deliver a flow rate of 275,600 lb/h. In addition, the discharge of the safety valve is to the atmosphere. The saturated temperature is 830 ° R, which is the temperature at the inlet of the safety valve. At that temperature, the physical properties of saturated water and steam are provided as indicated in Table 2.18. What orifice area and designation must be chosen for this case? (*Note*: Vapor or gas or combined vapor and gas mass fraction (quality) at safety valve inlet is equal to zero.)

Answer

The relieving pressure P is calculated as follows:

$$P_1 = \text{Set pressure} + \text{Overpressure} + \text{Atmospheric pressure}$$
$$= 145 + 14.5 + 14.7 = 174.2 \text{ psia}$$

Since the mass fraction of vapor or gas as well as the combined mass fraction of gas and vapor at the safety valve inlet is equal to zero, it can be concluded that $x_0 = 0$. The temperature at the valve inlet T_0 is 830 °R.

The next step is to calculate parameter Omega by using Eq. (2.65) in the following manner:

$$\omega = \frac{x_0 v_{v0}}{v_0 k} + 0.185 \times \frac{C_p T_0 P_1}{v_0}\left(\frac{v_{vl0}}{h_{vl0}}\right)^2 = 0 + 0.185$$
$$\times \frac{1.06116 \times 830 \times 174.2}{0.01824}\left(\frac{2.6142 - 0.01824}{853.667}\right)^2$$
$$= 1{,}556{,}159.14 \times 9.25 \times 10^{-6} = 14.39$$

Table 2.18 Water and stem properties at 830 °R – the saturated temperature for water and stem.

v_{v0}	2.6142 ft^3/lb
v_{l0} and v_0	0.01824 $ft^3/lb \frac{ft^3}{lb}$
h_{vl0}	853.667 Btu/lb
C_p	1.06116 Btu/lbR

Following this step, it is necessary to determine whether the fluid is critical or subcritical. In order to do so, the ratio between critical pressure and relieving pressure $\eta_c = 0.877$ is obtained from Figure 2.20.

$$P_c = \eta_c \times P_1 = 0.877 \times 174.2 = 152.77 \text{ psia} \tag{2.67b}$$

Therefore, since the safety valve discharges into the atmosphere, the backpressure value is 14.7 psia, which is less than the critical pressure. So the flow in this example is considered critical. As a result, in the following step, the mass flux is calculated in accordance with Eq. (2.68).

$$G = 68.09 \times \eta_c \times \sqrt{\frac{P_1}{\omega v_0}} = 68.09 \times 0.877 \times \sqrt{\frac{174.2}{14.39 \times 0.01824}} = 1538.06 \text{ lb/sft}^2$$

Finally, the required area of the safety valve can be computed from Eq. (2.70). Backpressure correction factor is equal to one because the fluid in the safety valve is discharged to the atmosphere. The combination correction factor is also equal to one because to rupture disk is installed in combination with the safety valve.

$$A = 0.04 \times \frac{1}{K_b K_c K_d} \times \frac{W}{G} = 0.04 \times \frac{1}{1 \times 1 \times 0.85} \times \frac{275,600}{1538.06} = 8.43 \text{ in.}^2$$

These calculations result in the selection of orifice Q with a discharge area of 11.05 in.^{2}.

2.5.4.2 Sizing for Subcooled at the Pressure Relief Valve Inlet

As a result of the method presented in this section, a PRV can be sized to handle a subcooled fluid with a saturated liquid at the inlet using the method described in this section. Neither noncondensable vapor nor noncondensable gases should be present at the inlet. If the flow is subcooled, the subcooled liquid flashes upstream or downstream of the valve, depending on which subcooling region it falls into.

Step 1 Calculate the Omega Parameter (ω)

The sizing starts with obtaining Omega parameter same as the previous section. For subcooled liquid flows the Omega-Parameter is generally referred with ω_s. For multicomponent systems with nominal boiling range less than 150 °F ω_s can be calculated either from Eq. (2.71) or from Eq. (2.72). For single component systems with a relative temperature and pressure within the limits $T_r \leq 0.9$ and $P_r \leq 0.5$, ω_s is given by Eq. (2.71). For multicomponent systems, whose nominal boiling range is greater than 150 °F or for single-component systems close to the thermodynamic critical point, ω_s is given by Eq. (2.72).

Calculation of Omega Parameter (ω_s)

$$\omega_s = 0.185\, \rho_{l0} C_P T_0 P_s \left(\frac{v_{vls}}{h_{vls}}\right)^2 \tag{2.71}$$

$$\omega_s = 9\left(\frac{\rho_{10}}{\rho_9} - 1\right) \tag{2.72}$$

Step 2 Determine the Subcooling Region

In the event that a liquid enters the safety valve in a subcooled state, it is necessary to determine whether or not it saturates and the extent of the subcooling region on the basis of the following equations:

$$\text{If } P_s > P_0 \times \frac{2\omega_s}{1 + 2\omega_s} \tag{2.73}$$

→Low subcooling region where flashing occurs before the valve throat or orifice area

$$\text{If } P_s < P_0 \times \frac{2\omega_s}{1 + 2\omega_s} \tag{2.74}$$

→High subcooling region where flashing occurs at the valve throat or orifice area

Step 3 Determine If the Flow is Critical or Subcritical

The condition for the existence of critical and subcritical flow is provided in Table 2.19. In particular, it should be noted that the value of critical pressure is calculated from Eq. (2.67).

Step 4 Calculation of Mass Flux (*G*)

Equations (2.75) and (2.76) are used to calculate the mass flux under both low and high subcooling conditions.

Calculation of Mass Flux for Sizing the PSV in Low-Subcooling Systems

$$G = 68.09 \times \frac{\{2(1-\eta_s) + 2[\omega_S \eta_s \ln(\eta_s/\eta) - (\omega_S - 1)(\eta_s - \eta)]\}^{0.5}}{\omega_S((\eta_s/\eta) - 1) + 1} \sqrt{P_1 \rho_{l0}} \tag{2.75}$$

$\eta = \eta_c$ for critical flow and $\eta = \eta_a$ for subcritical flow

Table 2.19 Choosing critical or subcritical flow for PSV sizing in a subcooling system.

	Critical flow	Subcritical flow
Low subcooling region	$P_c > P_a$	$P_c < P_a$
High subcooling region	$P_s > P_a$	$P_s < P_a$

Calculation of Mass Flux for Sizing the PSV in High-Subcooling Systems

$$G = 96.3[\rho_{l0} \times (P_1 - P)]^{0.5} \tag{2.76}$$

For critical flow condition, $P = P_s$.
For subcritical flow condition, $P = P_a$.

Step 5 Calculate the Required Orifice Area of the Safety Valve
Finally, the required area of the safety valve can be computed from Eq. (2.77).

Calculation of the Required Orifice Area for Sizing a Subcooled Two-Phase Flow Safety Valve

$$A = 0.3208 \times \frac{1}{K_b K_c K_d} \times \frac{Q.\rho_{l0}}{G} \tag{2.77}$$

K_d: It is important to obtain the discharge coefficient from the valve manufacturer. In order to calculate a preliminary valve sizing calculation, the discharge coefficient can be considered to be 0.65 for subcooled liquids and 0.85 for saturated liquids.

K_b: The backpressure correction factor is also known as the capacity correction factor. Backpressure only applies to balanced bellows valves, and it can be calculated from Figure 2.12.

K_C: This is a combination correction factor resulting from the installation of a rupture disk upstream of a PSV. When there is no rupture disk installed, the value is one. However, if the rupture disk in conjunction with the safety valve is installed, the value becomes 0.9.

2.5.5 Sizing for Fire Case and Hydraulic Expansion

The cause of a fire may be due to the leakage of flammable material from equipment and pipelines, or an accident may occur during the operation. When accidentally ignited, this material will immediately pose a danger to adjacent vessels and equipment. In liquids and gases, burning materials can become an open, free-burning fire very quickly and can be carried some distance from the source of the leak by the slope of the ground and by air currents. In operating plants that handle or process flammable liquids or gases, fire hazards must be taken into account when selecting the size of safety relief valves. Under normal operating conditions, any pressure vessel protected by a PRV should be fire-sized in the event that the equipment may be exposed to fire (even if the contents are not flammable). When an open fire occurs surrounding equipment or vessels, heat will naturally be absorbed by anything that comes in contact with the flames and/or hot gases. In the event of prolonged heat absorption, the contents of the vessel will be heated and the pressure will rise until the safety relief valve opens.

API 521 and ISO 23251 are two standards dealing with safety requirements for systems that relieve and depressurize pressure. As shown in this section, we provide engineering and required equations for determining the size of a safety valve in two cases of hydraulic expansion and external fire. Thermal expansion or hydraulic expansion refers to the increase in the volume of liquid caused by an increase in temperature. A typical instance of this occurs for liquids trapped in vessels, pipes, heat exchangers and exposed to heat, for example, from electrical coils, ambient heat, fire, etc. The formulas in both standards are identical, except for the units. The API 521 formulas should be applied using US units, whereas the ISO 23251 formulas should be applied using SI units. Table 2.20 presents the required parameters and units based on these two standards. With respect to sizing external fire cases, API 521 distinguishes between wetted and

Table 2.20 Symbols for sizing safety valves in fire cases.

Symbol	Description	Units (US)	Units (SI)
A	Effective discharge area of valve	in.^{2}	[a]
A'	Exposed surface area of the vessel	ft^2	[a]
A_{WS}	Total wetted surface	ft^2	m^2
α_v	Cubical expansion coefficient of the liquid at the expected temperature	1/°F	1/°C
C	Specific heat capacity of the trapped liquid	Btu/(lb · °F)	J/(kg · k)
F	Environment factor	—	—
d	Relative density referred to water at 60 °F (15.6 °C)	—	—
h_{vl0}	Latent heat of vaporization	Btu/lb	J/kg
K_D	Coefficient of discharge	—	—
$\emptyset$	Total heat transfer rate	Btu/h	W
M	Molecular mass of the gas	lb/lb$_{mol}$	kg/k$_{mol}$
P_1	Upstream relieving absolute pressure	psi	[a]
Q	Total absorbed (input) heat to the wetted surface	Btu/h	J
q	Volume flow rate at the flowing temperature	gpm	m^3/s
q_m or Q	Relief load/mass flow rate	lb/h	[a]
T_1	Gas temperature at upstream relieving pressure	°R	[a]
T_w	Recommended maximum vessel wall temperature	°R	[a]

[a] ISO 23251 specifies sizing formulas that are identical to those in API 521, which are expressed in US units. The conversion factors to specified SI units have not yet been provided. It is therefore recommended that the formula be applied using US units.

unwetted vessels based on the following definitions, and presents a sizing method for each.

Wetted vessels contain liquids in equilibrium with their vapors or gases. These vessels contain temperate systems. Due to the heat transfer from the external fire, partial evaporation of the liquid occurs. Only the portion of the vessel that is in contact with the liquid within a distance of 25 ft (in ISO 23251, 7.6 m) above the fire source should be considered for sizing, as shown in Table 2.22. It may be necessary to use alternative sizing methods if fire exposure results in vapor generation due to thermal cracking.

A vessel that is unwetted is one in which the walls are either thermally insulated or filled with gases, vapors, or supercritical fluids. Unwetted vessels contain gaseous systems. In addition, vessels with separate liquid and vapor under normal conditions which become single-phase under relieving conditions may also be included here. It should be noted, however, that vessels whose walls have become thermally insulated due to the deposition of coke or materials from the fluids inside, are still considered wetted for fire sizing purposes, but additional fire protection is required. Due to the large thermal resistance in unwetted vessels, the thermal flow from the walls to the interior of the vessel is reduced in comparison to wetted vessels. When the outside surface of a vessel is exposed to a fire source for an extended period of time, the temperature within the walls of the vessel can rise to such an extreme level that a thermal rupture can occur.

2.5.5.1 Hydraulic Expansion (Thermal Expansion)

In the case of a liquid vessel exposed to a heat source, the mass flow rate for sizing the safety valve can be obtained from Eqs. (2.78) (API 521) and (2.79) (ISO 23251), provided that the trapped liquid does not evaporate. Nevertheless, according to API 521, the mass flow rates are usually so small that a safety valve with a size of NPS 3/4″ × NPS 1 (DN 20 × DN 25) should be sufficient.

Hydraulic or Thermal Expansion of a Safety Valve Based on API 521 and ISO 23251 Standards

$$q = \frac{1}{500} \frac{\alpha_V}{d} \frac{Ø}{c} \quad \text{(API 521)} \tag{2.78}$$

$$q = \frac{1}{500} \frac{\alpha_V}{d} \frac{Ø}{c} \quad \text{(ISO 23251)} \tag{2.79}$$

The cubical expansion coefficient of a liquid should be determined from the process data; however, some reference values are given in Table 2.21 for liquids at 60 °F (15.6 °C). It should be noted that process design data will provide more precise values.

Table 2.21 Value of cubical expansion coefficient for hydrocarbon liquids at 60 °F in API 521.

Gravity of liquid (°API)	α_v (1/° F)	α_v (1/° C)
3 – 34.9	0.0004	0.00072
35 – 50.9	0.0005	0.0009
51 – 63.9	0.0006	0.00108
64 – 78.9	0.0007	0.00126
79 – 88.9	0.0008	0.00144
89 – 93.9	0.00085	0.00153
94 – 100 and lighter	0.0009	0.00162
Water	0.0001	0.00018

2.5.5.2 Sizing Safety Valve for the Fire Case

There are two types of fire scenarios discussed in this section: an external fire on a wetted vessel and an external fire on an unwetted vessel.

2.5.5.2.1 External Fire – Wetted Vessel In the fire case, Table 2.22 details the portions of wetted surfaces that should be taken into account when sizing the safety valve.

Figure 2.21, for instance, shows two vessels on the left and a sphere on the right. For those two vessels on the left, the portion of a wetted surface that must be taken

Table 2.22 Wetted surfaces that should be taken into account when sizing a safety valve in a fire situation.

Class of vessels	Portions of liquid inventory
Liquid full (e.g. treaters)	All up to the height of 25 ft (7.6 m)
Surge or knockout drums, process vessels	Normal operating level up to a height of 25 ft (7.6 m)
Fractionating columns	The normal level in the bottom of the column is increased by the liquid hold-up from all trays dumped into the normal level in the column's bottom; the total wetted area of the column reaches a height of 25 ft (7.6 m).
Storage tanks	Normally, the maximum inventory level is 25 ft (7.6 m), excluding the portions of the wetted area which are in contact with the foundations and the ground
Spheres and spheroids	As high as 25 ft or as large as the largest horizontal diameter, whichever is greater

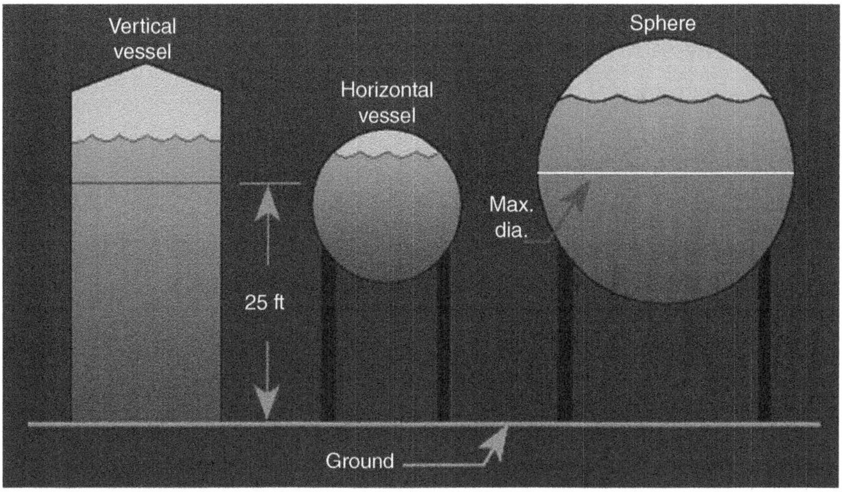

Figure 2.21 Wetted vessels.

into account is up to the level that is 25 ft above the ground. As for the sphere on the right, the maximum diameter of the sphere is greater than 25 ft. So the proportion of liquid that must be taken into consideration for the wetted area (surface) is determined by the diameter of the sphere.

The amount of heat absorbed by a noninsulated vessel filled with a liquid is dependent on the following considerations:

- The source of fuel for the fire;
- The degree to which a vessel is engulfed with flames, which depends on its size and shape; and
- The immediacy of firefighting measures and the possibility of draining flammable materials from the vessel.

The total heat absorption Q for the wetted surface can be estimated by Eq. (2.80) in case of adequate drainage and prompt firefighting measures and by Eq. (2.81) in case of insufficient drainage and/or firefighting measures. It is possible to implement an adequate drainage system for fuels that are flammable by utilizing sewers and trenches as well as the natural slope of the land.

Total Heat Absorption by the Wetted Surface by Adequate Drainage and Prompt Fire Fighting

$$\begin{aligned} Q &= 21{,}000 F A_{\mathrm{ws}}^{0.82} (\text{US units}) \\ Q &= 43{,}200 F A_{\mathrm{ws}}^{0.82} (\text{SI units}) \end{aligned} \tag{2.80}$$

Total Heat Absorption by the Wetted Surface During Insufficient Drainage and/or Prompt Fire Fighting

$$Q = 34{,}500FA_{\mathrm{ws}}^{0.82} \text{ (US units)}$$
$$Q = 70{,}900FA_{\mathrm{ws}}^{0.82} \text{ (SI units)} \quad (2.81)$$

where:

Q: Total heat absorption (input) to the wetted surface, expressed in British Thermal Units (BTU) per hour in US units or in Joules in SI units;
F: Environmental factor as per Table 2.23;
A_{WS}: Total wetted surface area ft^2, m^2 calculated by using the equations given in Table 2.24. In Figure 2.22, different dimensional designations for wetted vessels that are partially filled with liquid are shown.

The angle β in Table 2.24 is calculated from Eq. (2.82). F_{eff} is the effective liquid level up to a maximum distance of 25 ft from the flame source as determined by Eq. (2.83) or (2.84).

Table 2.23 Environmental factor.

Type of equipment	Environmental factor (F)
Bare or uninsulated vessel	1
Insulated vessel with insulation conductance values in BTUs per hour for fire exposure conditions	
4	0.3
2	0.15
1	0.075
0.67	0.05
0.5	0.0376
0.4	0.03
0.33	0.026
Water application facilities, on a bare vessel	1.0
Depressurizing and emptying facilities	1.0
Earth covered storage	0.03
Below grade storage	0.00

Table 2.24 Calculation of wetted area (surface) for different vessels.

Type of vessel	Wetted area (surface) calculation
Sphere	$A_{WS} = \pi D F_{eff}$
Horizontal cylindrical vessel with flat ends	$A_{WS} = \beta D\left[L + \frac{D}{2}\right] - D\sin\beta\left[\frac{D}{2} - F_{eff}\right]$
Horizontal cylindrical vessel with spherical ends	$A_{WS} = \pi D\left[(L - D)\frac{\beta}{\pi} + F_{eff}\right]$
Vertical cylinder with flat ends Partially filled ($F < L$) Totally filled ($F = L$)	$A_{WS} = \pi D\left[\frac{D}{4} + F_{eff}\right]$ $A_{WS} = \pi D\left[\frac{D}{2} + F_{eff}\right]$
Vertical cylinder with spherical ends	$A_{WS} = \pi D F_{eff}$

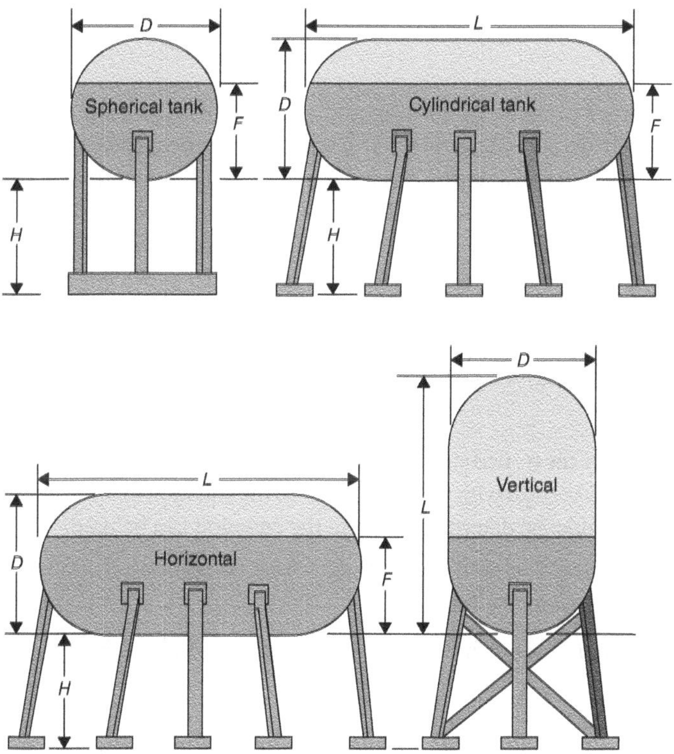

Figure 2.22 Different dimensional designations for wetted vessels that are partially filled with liquid.

Calculation of Angle β

$$\beta = \cos^{-1}\left(1 - \frac{F}{D}\right) \tag{2.82}$$

Calculation of the Effective Liquid Level

$$F_{\text{eff}} = \min(25\ \text{ft}; F) - H \quad \text{(US Customary)} \tag{2.83}$$

$$F_{\text{eff}} = \min(7.6\ \text{m}; F) - H \quad \text{(SI unit)} \tag{2.84}$$

Equation (2.85) is used to determine the mass flow rate of the safety valve. Based on the mass flow rate or relieving capacity, it is possible to size safety valves and calculate the required orifice area for vapor or gas services based on one of the methods discussed previously.

Calculation of the Mass Flow Rate for a Safety Valve in the Fire Case (Wetted Vessel)

$$W = \frac{Q}{h_{\text{vl0}}} \tag{2.85}$$

Example 2.16 At a set pressure of 200 psig and an atmospheric backpressure, benzene is contained in an uninsulated vertical vessel with spherical ends at a temperature of 100 °F (559.7 °R). The vessel has a diameter of 15 ft, a length of 40 ft, and an elevation of 15 ft. A maximum fluid level of 32 ft from the bottom of the vessel is achieved. In the event of a fire, assume that fire-fighting measures will be implemented promptly and that adequate drainage will be provided. In this case, what is the recommended type of orifice and area for the safety valve considering a critical flow? (*Note*: The latent heat of vaporization for benzene is considered 114.9 Btu/lbm, $K = 1.23$, $T = 875.5\,°\text{R}$, and $M = 875.5$.)

Answer

The first step is to calculate the wetted surface area by using one of the equations given in Table 2.24. The vessel is vertical with spherical ends. So the equation in the last row of the table applies. F_{eff} must be determined before the equation can be used.

$D = 15$ ft (Diameter), $L = 40$ ft (Length), $H = 15$ ft (Elevation), $F = 32$ ft (Liquid level from the bottom)

$$F_{\text{eff}} = \min\ (25\ \text{ft}; F) - H = \min\ (25\ \text{ft}; 32) - 15 = 25 - 15 = 10\ \text{ft}$$

$$A_{\text{WS}} = \pi D F_{\text{eff}} = 3.14 \times 15 \times 10 = 471\ \text{ft}^2$$

Using Eq. (2.80), it is now possible to calculate the thermal heat flow.

$$Q = 21{,}000FA_{ws}^{0.82} = 21{,}000 \times 1 \times 471^{0.82} = 3{,}266{,}603.6\ \text{Btu/h}$$

$$W = \frac{Q}{h_{vl0}} = \frac{3{,}266{,}603.6}{114.9} = 28{,}430\ \text{Btu/h}$$

$$K = 1.23 \rightarrow C = 340\ \text{refer to Table 2.13}$$

$$P_1 = \text{Set pressure} + \text{Overpressure} + \text{Atmospheric pressure} = 200 + (21\% \times 200) + 14.7 = 256.7\ \text{psia}$$

Note that overpressure is considered 21% due to the fire case.

Since there is no information to estimate the z factor, $Z = 1$.

Therefore, the safety valve can be sized based on the vapor or gas pressure and critical flow using Eq. (2.28) as follows:

$$A = \frac{W}{C\,K_d P_1 K_b K_c}\sqrt{\frac{TZ}{M}} = \frac{28{,}430}{340 \times 0.975 \times 256.7 \times 1 \times 1}\sqrt{\frac{875.5 \times 1}{78.11}}$$

$$= 0.3340 \times 3.35 = 1.118\ \text{in.}^2$$

For this application, Orifice J with an area of 1.287 in.^{2} should be selected.

External Fire – Unwetted Vessel A wetted vessel is one whose walls are either thermally insulated or filled with gases, vapors, or supercritical fluids. Unwetted vessels contain gaseous systems rather than liquids. In this case, the discharge area of the safety valve is determined by using Eq. (2.86).

Discharge Area of the Safety Valve in the Event of an External Fire and an Unwetted Vessel

$$A = \frac{F'A'}{\sqrt{P_1}} \tag{2.86}$$

F' refers to the bare metal temperature of the vessel at relief, which can be calculated based on Eq. (2.87). The value of 0.045 is used if F' *is* not known. In case the calculated value for F' is less than 0.01, a value equal to 0.01 must be taken, which is the recommended and minimum value. Figure 2.23 can be used as an alternative to estimate parameter F'.

Calculation of F'

$$F' = \frac{0.1406}{C \times K_d}\left[\frac{(T_W - T_1)^{1.25}}{T_1^{0.6506}}\right] \tag{2.87}$$

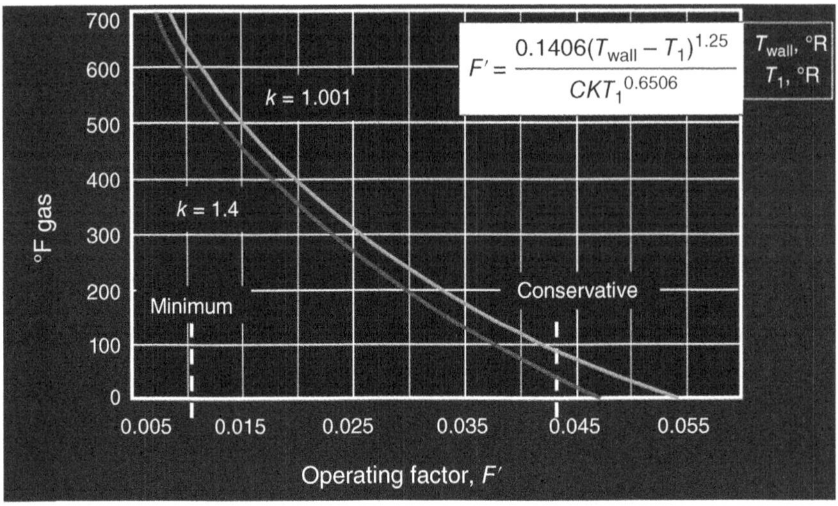

Figure 2.23 Estimation of parameter F'.

For carbon steel plate materials, the maximum vessel wall temperature T_w is recommended to be 1100 °F (593 °C). In the case of plates made of alloys, the recommended maximum wall temperature must be adjusted appropriately.

The relieving temperature T_1 is determined by Eq. (2.88) in relation to the normal operating temperature and pressure, respectively, T_n and P_n, and the relieving pressure.

Relieving Temperature Calculation

$$T_1 = T_n \frac{P_1}{P_n} \tag{2.88}$$

Carbon steel vessel ($T_w = 1560$ °R) is filled with air at a set pressure of 100 psig. This vessel has an exposed surface area of 250 ft^2. A normal temperature and pressure are 125 °F (584.70 °C) and 80 psig (94.7 psia). Which size of the orifice should be used in this situation?

Answer

Due to the fact that the vessel is filled with air rather than liquid, it must be unwetted.

$$\begin{aligned} P_1 &= \text{Set pressure} + \text{Overpressure} + \text{Atmospheric pressure} \\ &= 100 + (21\% \times 100) + 14.7 = 135.7 \text{ psia} \end{aligned}$$

$$\text{Relieving temperature calculation} = T_1 = T_n \frac{P_1}{P_n} = 584.7 \times \frac{135.7}{94.7} = 837.84°\text{R}$$

According to Table 2.14, the specific heat of air is 1.40. The value of gas constant C associated with the ratio of specific heat $k = 1.4$ is 356 (refer to Table 2.14).

We can now calculate F' using Eq. (2.87) as follows:

$$F' = \frac{0.1406}{C \times K_\text{d}} \left[\frac{(T_\text{W} - T_1)^{1.25}}{T_1^{0.6506}} \right] = \frac{0.1406}{356 \times 0.975} \left[\frac{(1560 - 837.84)^{1.25}}{837.84^{0.6506}} \right] = 0.019$$

In addition, the minimum discharge can be calculated from Eq. (2.86) as follows:

$$A = \frac{F'A'}{\sqrt{P_1}} = \frac{0.019 \times 250}{\sqrt{135.7}} = 0.40\ \text{in.}^2$$

For this application, Orifice G with an area of 0.503 in.^{2} should be selected.

Questions and Answers

2.1 Select the correct answer about valve sizing.

A Before a PSV, an 8″ × 6″ ball valve is the correct size for a subflare line.

B The water service is 60 °F with a flow rate of 1000 gpm that passes through a 3″ reduced bore ball valve with a one psi pressure drop. The valve is correctly sized in this case. (*Note*: the flow coefficient value should be taken from Table 2.1.)

C The flow rate for fully opening an axial check valve is 21,048 kg/h. The minimum flow rate in the valve is 22,234 kg/h showing that the valve is sized correctly.

D All answers are wrong.

Answer

Option A is incorrect because the valve engineer shall select a full-bore ball valve for the subflare line allowing sudden fluid or gas service release. Option B is not correct either because a 3″ reduced bore ball valve provides a flow coefficient equal to a 2″ ball valve that is 480 gpm. In contrast, the calculated flow coefficient based on process data (Temperature = 60 °F, Flow rate = 1000 gpm, Pressure drop = 1 psi, SG = 1 for water) should be 1000 gpm. It means that the valve is undersized, and it should be a 3″ full bore with a flow coefficient value equal to 1300 gpm as per Table 2.1. Option C shows that the check valve will not be at risk of chattering, and it will be fully open during the operation. However, this information does not indicate that the valve is sized correctly. Option D is the correct answer since all other options are incorrect.

2.2 Which answer is correct about sizing check valves?

A The critical velocity in a swing check valve depends on the spring torque value.

B The critical velocity to fully open a check valve is higher for water than gas.

C The critical velocity is increased for spring-loaded valves by selecting a higher torque spring.

D There is no difference between sizing a check valve installed on a horizontal or vertical line.

Answer

Option A is incorrect because a swing check valve does not have a spring. Option B is not correct either since a lower velocity of water than gas is needed to fully open a check valve due to the higher density of the water. Option C is the correct answer as per data provided in Tables 2.4 and 2.5. Option D is wrong because, for a check valve installed on a vertical line with an upward flow, the weight of disk(s) for sizing of the valve shall be considered by the valve engineer. In contrast, disk(s) weight can be neglected for sizing a check valve installed on a horizontal line.

2.3 Which statements are incorrect regarding the valve sizing in liquid services?

A The possibility of choked flow is evaluated during the valve size selection.

B Installation of fittings attached to the valve does not impact the valve size.

C Type of control valve design in terms of plug and cage type as well as flow characteristics are not taken into account during the valve sizing.

D All three options are incorrect.

Answer

Option B and D are wrong statements.

2.4 A high-performance butterfly valve in pressure class of 300 (PN of 50 bar) is installed on an 8″ line. The fluid service is water with a flow rate of 1600 gpm. The valve's inlet pressure, outlet pressure, and inlet temperature values are 300 psig, 275 psig, and 70 °F, respectively. It is assumed that the piping geometry factor (F_p) is equal to 0.90, and no choked flow can happen in the valve; which size for the butterfly valve is correct?

A 6″

B 4″

C 8″

D 10″

Answer

Required flow coefficient (C_v) by considering the effect of piping geometry factor and no choke flow occurrence is calculated by using Eq. (2.17).

$$C_v = \frac{q}{N_1 F_P \sqrt{(P_1 - P_2)/SG_l}} = \frac{1600}{1 \times 0.90\sqrt{25/1}} = 355.55$$

Referring to Table 2.10, a 3″ high performance butterfly valve at a fully open position can provide a flow coefficient value of 237, which is less than the required $C_v = 355.55$. Hence, a 3″ butterfly valve is undersized. However, a 4″ high performance butterfly valve can give a flow coefficient value of 499, which is sufficient. Selecting a valve with a larger size than 4″ is considered an oversize valve. Thus, the 4″ valve size and option B are correct.

2.5 The required hydrocarbon gas mass flow rate resulting from the operation of the safety valve is 53,500 lb/h. Fluid service consists of a mixture of butane and pentane with a molecular weight (M) of 65 and a relief temperature of 627 °R, or 348 °K. The relief valve set pressure is 75 psig, which is the design pressure of the component that may be increased by 10% to accommodate the spring force during the opening. There is a constant backpressure of 55 psig, and there is no pressure loss in the valve. The compressibility factor of the gas service is 00.84, while the specific heat is 1.09. There is a total backpressure of 62.5 psig at the valve, whereas the critical flow pressure is 42.6 psi. Which of the following statements is incorrect?

A The safety valve should be sized based on the subcritical flow condition.
B The coefficient of subcritical flow is 0.86 in this case.
C The effective orifice area in this case is 5.6 in.^{2}.
D The valve should be fitted with a "P" sized orifice.

Answer

The total backpressure of 62.5 psig is higher than the critical flow pressure of 42.6. Therefore, the safety valve should be sized according to the subcritical flow condition, and option A is correct. The subcritical flow coefficient can be calculated using Eq. (2.51) by using the parameters r and k in the following manner.

2.6 A PSV is designed for handling natural gas with a molecular weight of 19 and a flow capacity of 2675 kg/h. The set pressure for the valve is 1450 kPag, and the allowable overpressure is 10%. In the presence of atmospheric backpressure and an inlet temperature of 50 °C, and assuming a critical flow, what is the area of the orifice and its type based on API standard?

A Orifice designation F with API effective area of 0.307 in.^{2}
B Orifice designation G with API effective area of 0.503 in.^{2}

C Orifice designation H with API effective area of 0.785 in.^{2}
D Orifice designation J with API effective area of 1.287 in.^{2}

Answer

Equation (2.29), used for the sizing of safety valves for critical flow gases, is applicable here. Data that are used for safety valve sizing are summarized as follows:

$$W = 2675\ \text{kg/h}$$
$$T = 50 + 273 = 323°\text{K}$$
$$T = 50 + 273 = 323°\text{K}$$
$$P_1 = \text{Absolute relieving pressure} = \text{Set pressure} + \text{Overpressure} + \text{Atmospheric pressure}$$
$$= 1450 + 10\% \times 1450 + 101 = 1696\ \text{KPa}$$

Table 2.14 indicates that the constant C for natural gas is 344. The coefficient of discharge $K_d = 0.975$. Since the backpressure is equal to the atmospheric pressure, it cannot affect the flow capacity of the valve. So the capacity correction factor due to backpressure can be assumed as equal to 1, $K_b = 1$. K_c is the combination correction factor for installation with a rupture disk upstream of the PRV. This value is one in this example because there is no rupture disk installed. It is assumed that the Z-factor, or gas compressibility factor, is equal to one in this case since its real value is unknown.

$$A = \frac{13{,}160\ \text{W}}{C\,K_d P_1 K_b K_c}\sqrt{\frac{TZ}{M}}\quad \text{(SI units)}$$

$$\rightarrow A = \frac{13{,}160 \times 2675}{344 \times 0.975 \times 1696 \times 1 \times 1}\sqrt{\frac{323 \times 1}{19}} = 61.885765 \times 4.1231$$
$$= 255.16\ \text{mm}^2 = 0.395498\ \text{in.}^2 \tag{2.29b}$$

As a result, orifice G with an area of 0.503 in^2 should be selected and option B is the correct answer.

2.7 In the previous example, the backpressure is 725 kPag. If all other parameters are the same, which API orifice should be selected?
A Orifice designation F with API effective area of 0.307 in.^{2}
B Orifice designation G with API effective area of 0.503 in.^{2}
C Orifice designation H with API effective area of 0.785 in.^{2}
D Orifice designation J with API effective area of 1.287 in.^{2}

Answer

The main difference between this case and the previous case is that since the backpressure is not discharged into the atmosphere, the value of the backpressure correction factor is not equal to one. The ratio of backpressure to set pressure is equal to 725/1450 = 0.50 and refer to Figure 2.12, the value of $K_b = 0.69$. Because all other parameters are equal in these two examples, the orifice area is calculated as follows:

$$A = \frac{0.395498}{0.69} = 0.57 \text{ in.}^2$$

As a result, orifice *H* with an area of 0.785 in.^{2} should be selected and option C is the correct answer.

2.8 In this instance, a bellows balanced type PSV is handling sodium trisulfide, a nonviscous liquid with a specific gravity of 1.23. As a result, the valve is capable of alleviating 475 l of pressure per minute and the set pressure of the valve is 690 kPa. It should be noted that there is a 10% overpressure allowed and the backpressure is 207 kPag. If this is the case, then what type of orifice and effective area should be selected considering that the safety valve will not be certified?

A Orifice designation *G* with API effective area of 0.503 in.^{2}
B Orifice designation *H* with API effective area of 0.785 in.^{2}
C Orifice designation *J* with API effective area of 1.287 in.^{2}
D Orifice designation *K* with API effective area of 1.838 in.^{2}

Answer

Since the given values are in SI unit and the safety valve is used in liquid service, Eq. (2.63) applies. The first step is to collect the given parameters given above as follows:

$Q = 475$ l/min (Flow rate)
$K_d = 0.62$ (Correction factor due to discharge)
$P_b = 207$ kPag (Backpressure)
$P = P_s = 690$ kPag (Set pressure)

$$\frac{P_b}{P_s} = \frac{207}{690} = 0.3$$

$K_w = 0.87$ (The backpressure correction factor relates to the ratio of the backpressure to the set pressure, which is equal to 0.3 and it is derived from Figure 2.16).

$K_C = 1$ (The combination correction factor is equal to one in this case because the rupture disk does not need to be installed.).

$K_v = 1$ (The liquid is nonviscose)

$K_p = 0.6$ (Correction factor due to overpressure related to 10% overpressure and taken from Figure 2.19)
$G = 1.23$ (Specific gravity of the flowing fluid)

$$A = \frac{11.78 \times Q}{K_d K_w K_c K_v K_p} \sqrt{\frac{G}{1.25\,P - P_b}} = \frac{11.78 \times 475}{0.62 \times 0.87 \times 1 \times 1 \times 0.6} \sqrt{\frac{1.23}{1.25 \times 690 - 207}}$$
$$= 17{,}289.27 \times 0.043 = 748.93\ \text{mm}^2 = 1.160\ \text{in.}^2 \tag{2.63b}$$

In conclusion, the orifice letter "*J*" with an effective area of 1.287 in.^{2} corresponding to Table 2.12 must be selected, and option C is the correct response.

2.9 A safety valve controls the flow of subcooled propane at a volumetric rate of 100 gpm. The set pressure of the relief valve is 260 psig with an overpressure allowance of 10%. This valve has a total backpressure of 10 psig, including only build-up backpressure. It has a density of 31.92 lb/ft^3 and a specific heat of 0.6365 btu/lb R at constant pressure at the PRV inlet. In addition, the saturation pressure of propane at 60 °F is 107.6 psia, and the specific volume of propane liquid at the saturation pressure is 0.03160 ft^3/lb. The specific volume of propane vapor and latent heat of vaporization at saturation pressure are 1.001 ft^3/lb and 152.3 Btu/lb, respectively. The temperature at safety valve inlet is 519.67 °R. Which statement is incorrect?

A Omega is 8.515 for subcooled liquid flows at the safety valve inlet.
B In this example, the propane liquid corresponds to the high subcooling region.
C The mass flux value is 5670 lb/sft^2
D "*F*" is the correct size of the orifice.

Answer

Due to the fact that propane is flowing at a subcooled temperature, the calculation methods and formulas provided in Section 2.5.4.2 must be applied. A summary of the provided parameters is the first step.

$$Q = 100\ \text{gal/ min};$$

$$P_1 = P_0 = \text{Set pressure} + \text{Overpressure} + \text{Atmospheric pressure}$$
$$= 260 + 26 + 14.7 = 300.7\ \text{psia} P_b = P_a = 10\ \text{psig}$$

$$\text{Percent of gauge backpressure} = \frac{P_a}{\text{Set pressure}} = \frac{10}{260} = 0.038 = 3.8\%$$

As the downstream backpressure is less than 10% of the set pressure, a conventional relief valve may be used and the back-pressure correction factor is one. $K_b = 1$, $K_d = 0.65$, $K_c = 1$.

$$C_p = 0.6365\ \text{btu/lb}\cdot{}^\circ\text{R}$$

$$\rho_{10} = 31.92\ (\text{lb/ft}^3)$$

$$P_s = 107.6\ \text{psia}$$

v_{vls}: Difference between the vapor and the liquid specific volumes at $P_s\,(\text{ft}^3/\text{lb})$ = Specific volume of propane liquid at the saturation pressure – Specific volume of propane vapor at the saturation pressure = $1.001 - 0.03160 = 0.9694\ \text{ft}^3/\text{lb}$

$$h_{vls} = 152.3\ \text{Btu/lb}$$

$$T_0 = 519.67^\circ\text{R}$$

The next step is to calculate the saturated Omega parameter ω_s by using Eq. (2.71) as follows:

$$\omega_s = 0.185\rho_{10}C_pT_0P_s\left(\frac{v_{vls}}{h_{vls}}\right)^2 = 0.185 \times 31.92 \times 0.6365 \times 519.67 \times 107.6 \times \left(\frac{0.9694}{152.3}\right)^2 = 8.515$$

Thus, option A is correct. The next step is to determine the subcooling region.

$$P_0 \times \frac{2\omega_s}{1+2\omega_s} = 300.7 \times \frac{2 \times 8.515}{1 + 2 \times 8.515} = 284 > P_s = 107.6$$

Therefore, the liquid falls into the subcooling region, and option B is also correct. The next step is determining whether the fluid is critical or subcritical. Since $P_s = 107.6$ psia $> P_a = 10$ psig the flow of fluid is critical. The next task is to calculate mass flux according to Eq. (2.76). (For critical flow condition, $P = P_s$)

$$G = 96.3[\rho_{l0} \times (P_1 - P)]^{0.5} = 96.3[31.92 \times (300.7 - 107.06)]^{0.5} = 7571\ \text{lb/sft}^2$$

Therefore, option C is incorrect. It is now possible to calculate the orifice area using Eq. (2.77).

$$A = 0.3208 \times \frac{1}{K_bK_cK_d} \times \frac{Q.\rho_{10}}{G} = 0.3208 \times \frac{1}{1 \times 1 \times 0.65} \times \frac{100 \times 31.92}{7571} = 0.208\ \text{in.}^2$$

Accordingly, the correct orifice is F with an effective discharge area of 0.307 in.^{2}. Therefore, option D is appropriate.

Further Reading

Ahmed, T. (2007). *Equations of State and PVT Analysis*. Gulf Publishing Company.

American Petroleum Institute (API) 520 (2020). *Sizing, Selection, and Installation of Pressure-Relieving Devices Part 1 – Sizing and Selection*. Washington, DC: American Petroleum Institute (API).

American Petroleum Institute (API) 521 (2007). *Pressure-Relieving and Depressuring Systems*, 5e. Washington, DC: American Petroleum Institute (API).

American Petroleum Institute (API) 526 (2017). *Flanged Steel Pressure-Relief Valves*, 7e. Washington, DC: American Petroleum Institute (API).

American Society of Mechanical Engineers (ASME) B16.34 (2017). *Process Piping*. New York: American Society of Mechanical Engineers (ASME).

American Society of Mechanical Engineers (ASME) B36.10M (2015). *Welded and Seamless Wrought Steel Pipe*. New York: American Society of Mechanical Engineers (ASME).

American Society of Mechanical Engineers (ASME) B36.19M (2018). *Stainless Steel Pipe*. New York: American Society of Mechanical Engineers (ASME).

Ballun, J.V. (2007). *A Methodology for Predicting Check Valve Slam*. American Water Works Association (AWWA).

Crosby Valve Inc (1997). *Crosby Pressure Relief Valve Engineering Handbook*. Crosby Valve Inc. Technical document number. TP-V300.

Emerson (2019). *Control Valve Sizing*, 5e. Emerson. [online]. https://www.emerson.com/documents/automation/control-valve-handbook-en-3661206.pdf (accessed 11 February 2022).

Emerson (2022). *Valve Sizing Calculations*. Emerson. [online]. https://www.emerson.com/documents/automation/manual-valve-sizing-standardized-method-fisher-en-140724.pdf (accessed 11 February 2022).

Ford, R. (2014). Power industry applications: a valve selection overview. *Valve World Magazine* 19 (8): 96–103.

Goodwin (2016). *Check Valves Technical Catalogue*. Goodwin. [online]. https://dokumen.tips/documents/goodwin-check-valve-technical-cataloguepdf.html (accessed 8 February 2022).

Guo, B. and Ghalambor, A. (2005). *Natural Gas Engineering Handbook*, 2e. Gulf Publishing Company.

International Organization of Standardization (ISO) 23251 (2019). *Petroleum, Petrochemical and Natural Gas Industries – Pressure-Reliving and Depressurizing Systems*, 2e. Geneva: International Organization of Standardization (ISO).

International Society of Automation (ISA) 75.1 (2007). *Control Valve Sizing Equations*. Research Triangle Park, NC: International Society of Automation (ISA).

Ludtke, P.R. (1986). *Natural Gas Handbook*. National Bureau of Standards, US Department of Commerce.

Nesbitt, B. (2007). *Handbook of Valves and Actuators: Valves Manual International*, 1e. Oxford: Elsevier.

Oxler, G. (2009). Non-return valve and/or check valve for pump system – a new approach. *Valve World Magazine* 14 (4): 75–77.

Rahmeyer, W.J. (1993). Sizing swing check valves for stability and minimum velocity limits. *Transition to the ASME* 115: 406–410.

Skousen, P.L. (2011). *Valve Handbook*, 3e. New York: McGraw-Hill.

Smit, P. and Zappe, R.W. (2004). *Valve Selection Handbook*, 5e. New York: Elsevier.

Sotoodeh, K. (2015). Axial flow nozzle check valves for pumps and compressors protection. *Valve World Magazine* 20 (1): 84–87.

Sotoodeh, K. (2018). Comparing dual plate and swing check valves and the importance of minimum flow for dual plate check valves. *American Journal of Industrial Engineering* 5 (1): 31–45. https://doi.org/10.12691/ajie-5-1-5.

Sotoodeh, K. (2018). Why are butterfly valves a good alternative to ball valves for utility services in the offshore industry? *American Journal of Industrial Engineering* 5 (1): 36–40. https://doi.org/10.12691/ajie-5-1-6.

Sotoodeh, K. (2020). Challenges associated with the bypass valve of control valve in a sea water service. *Journal of Marine Science and Application* https://doi.org/10.1007/s11804-020-00132-8.

Sotoodeh, K. (2021). Analysis and failure prevention of nozzle check valves used for protection of rotating equipment due to wear and tear in the oil and gas industry. *Journal of Failure Analysis and Prevention* https://doi.org/10.1007/s11668-021-01162-2.

Sotoodeh, K. (2021). *A Practical Guide to Piping and Valves for the Oil and Gas Industry*, 1e. Austin, TX: Elsevier (Gulf Professional Publishing).

Sotoodeh, K. (2022). *Cryogenic Valves for Liquified Natural Gas Plants*, 1e. Austin, TX: Elsevier (Gulf Professional Publishing).

VALMETALIC (2018). *Design and Selection of Check Valves*. VALMETALIC. https://www.valmatic.com/Portals/0/pdfs/DesignSelectionCheckValves.pdf (accessed 7th February 2022).

3

Cavitation and Flashing

3.1 Introduction

Flow control, also known as regulation or throttling of fluids, is one application for some valves. Flow control valves adjust the amount of flow moving through the valve as well as the flow rate inside the valve; this impacts other parameters of the fluid process such as temperature, pressure, and level. Various sectors of the oil and gas industry have used globe valves extensively. Generally speaking, globe valves are covered by the American Petroleum Standard (API 602), which is titled “Gate, globe, and check valves for sizes DN 100 (4″) for the petroleum and natural gas industries.” Valves used for fluid control are susceptible to cavitation, which is a significant problem that is discussed in the following section.

3.2 Cavitation

3.2.1 What is Cavitation?

For oil and gas facilities and components such as pumps, valves, and piping, erosion or corrosion can be costly. Cavitation can be described as a type of erosion or erosion-corrosion and is the most severe operational problem for controlling globe valves. In the oil and gas industry, T-pattern (or tee-pattern) globe valves (shown in Figure 3.1) are very common. A globe valve with an actuator is referred to as a control valve. The liquid flows into the valve shown in the figure from the left side and reaches the middle portion of the valve where the plug (disk) and seats are located. As the disk is off the seat, the fluid makes two 90° turns and leaves the valve from the outlet port highlighted by pressure. These two 90° rotations of the fluid result in a substantial pressure drop inside the valve. There is a narrow area

Industrial Valves: Calculations for Design, Manufacturing, Operation, and Safety Decisions, First Edition. Karan Sotoodeh.

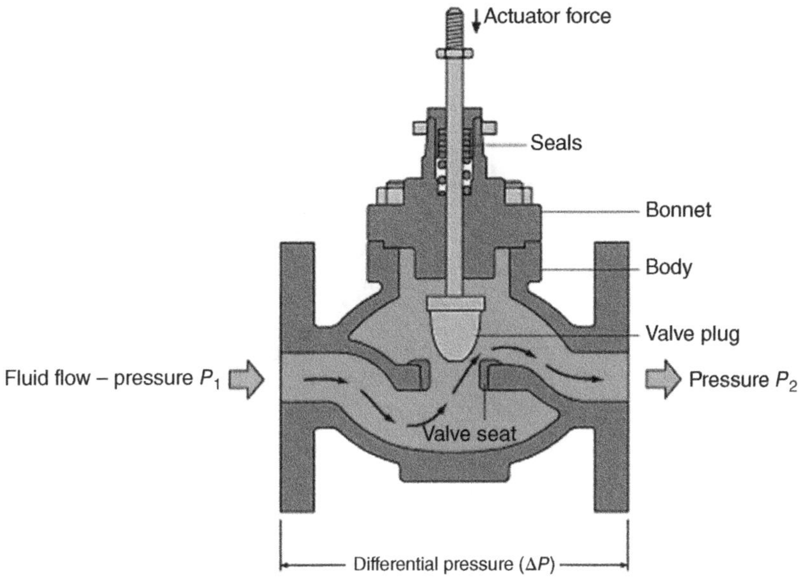

Figure 3.1 A T-pattern globe valve.

below the plug where the fluid velocity is very high and the pressure is extremely low. It is possible for the pressure to drop below the liquid's vapor pressure in this narrow area, causing some bubbles to form. The bubbles of gas that are separated from the liquid by a high-pressure drop can recover their pressure and collapse firmly, causing pressure waves. Known as cavitation, this can damage parts of the valve, such as its seats and plugs. Cavitation is a form of corrosion or erosion that causes metal loss or the creation of pits on the valve trims (internals). Cavitation also causes noise and vibration in addition to accelerated corrosion. A vapor pressure is a pressure at which a fluid is at thermodynamic equilibrium with its condensed state. Temperature affects vapor pressure. So, when the pressure of a fluid drops below its vapor pressure, it changes from liquid to gas. In Figure 3.2, irregular pitting and erosion-corrosion can be seen on a plug of a control valve due to cavitation.

3.2.2 Cavitation Essential Parameters

The following are a few of the essential parameters that affect the severity of cavitation and its consequences:

- **Pressure drop**: Cavitation is intensified by higher pressure drops in valves with higher pressure classes. As an example, a globe valve in pressure class 150 has a

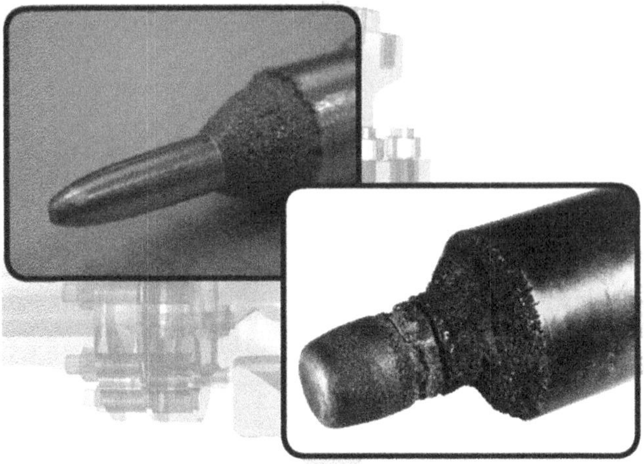

Figure 3.2 Corrosion caused by cavitation on a control valve plug.

maximum pressure drop of 20 bar, while a valve in pressure class 300 has a maximum pressure drop of 50 bar.

- **Leakage**: A leak is the movement of fluid from an area of high pressure to an area of low pressure. Cavitation is more likely to occur if the valve leaks in a closed position.
- **Material**: Harder materials are less susceptible to cavitation. The cavitation resistance of hard trim materials such as Stellite 6 (UNS R30006) or Stellite 21 in solid or overlay form, as well as 13Chromium martensitic stainless steels like UNS S41000 and UNS 415000, is higher.
- **Valve size and design**: In smaller valves, cavitation may be more severe. As a result of causing less pressure drop, Y-pattern globe and axial valves have less cavitation risk than T-pattern globe valves. API released the first edition of the new globe valve design standard (API 623) to control and avoid operation problems in globe valves such as cavitation, vibration, and leakage. The API 623 specifies applying a hardfacing alloy such as stellite to the valve internals such as the plug and seat, a larger stem diameter, and ensuring a more durable connection between the plug and the seat.
- **Trim design**: Valve trim refers to the internal components of the valve that are in contact with the fluid service. The multistep or anticavitation trim design (see Figure 3.3) minimizes cavitation risk. By creating a pressure drop in two or more stages, multistep trimming prevents high-pressure drops inside the valve and cavitation.
- **Flow regime**: Turbulent and high-velocity flows increase erosion and cavitation risks.

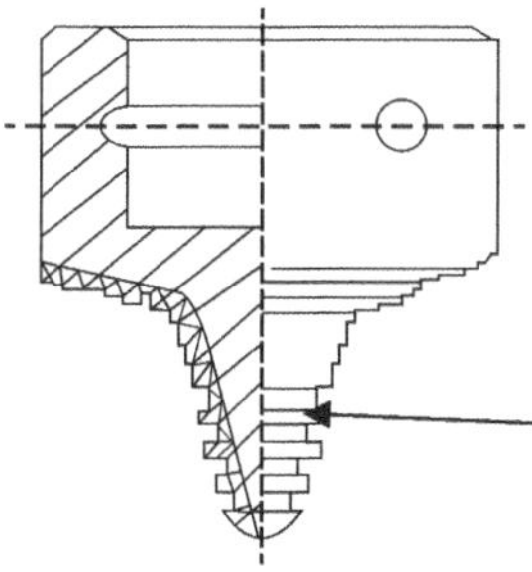

Figure 3.3 Trim design with multiple stages.

3.2.3 Cavitation Analysis

The *cavitation index* or *parameter sigma* (δ) is the most widely accepted method for estimating and measuring cavitation in control valves. The cavitation index is calculated by using Eq. (3.1) as follows:

Cavitation Index (δ) Calculation

$$\delta = \frac{(P_u - P_v)}{(P_u - P_d)} \tag{3.1}$$

where:

δ: Cavity index (dimensionless);
P_d: Downstream (outlet) pressure (psig);
P_v: Adjustment of the vapour pressure of the fluid flowing inside the valve to account for temperature and atmospheric pressure (psig);
P_u: Upstream (inlet) pressure (psig).

The severity and extension of cavitation for valves including globe valves based on cavity index values are given in Table 3.1.

Example 3.1 If the inlet and outlet pressure values are 25 bar and 20 bar, respectively, calculate the cavity index for a 6″ globe valve CL300 in seawater service at 25 °C. Water has a vapor pressure of 0.36 psig. In this case, is cavitation a possibility?

Table 3.1 Cavity severity based on cavity index values.

Criteria	Cavitation consequence
$\delta \geq 2.0$	No risk of cavitation
$1.7 < \delta < 2.0$	There is no need for cavitation control The trim is hardened for protection
$1.5 < \delta < 1.7$	There is some cavitation control required
$1.0 < \delta < 1.5$	Potential for severe cavitation
$\delta \leq 1.0$	The flashing is occurring

Answer

Inlet and outlet pressure are calculated based on psig as follows:

$$P_u = 25 \times 14.5 = 362.5 \text{ psi}$$

$$P_d = 20 \times 14.5 = 290 \text{ psi}$$

$$\delta = \frac{(P_u - P_v)}{(P_u - P_d)} = \frac{362.5 - 0.36}{362.5 - 290} = \frac{362.14}{72.5} = 4.995$$

Cavitation is not a concern due to the cavitation index exceeding 2, as shown in Table 3.1.

3.3 Flashing

Similarly, flashing is another term that is highly associated with cavitation, and it occurs when the fluid pressure falls below the vapor pressure. At this point, fluid begins to transform from a liquid to a vapor. There is a flashing in a valve (e.g. globe or control) if the valve outlet pressure is less than the vapor pressure. In this case, the pressure inside the valve will not recover, and the vapor will travel downstream (after) the valve (see Figure 3.4). Vapor pressure will eventually be restored to the pipe, and the collapsing vapor will result in mechanical damage, such as erosion and noise, similar to that which occurs during cavitation. The primary difference between cavitation and flashing is in the pressure value at the valve outlet; the cavity flow has a higher pressure than the vapor pressure at the outlet, whereas the flashing flow has a lower pressure than the vapor pressure (see Figure 3.5). There are two main conditions that must be met for cavitation and flashing to occur:

- High-pressure drop across the valve;
- The liquid pressure drops below the vapor pressure.

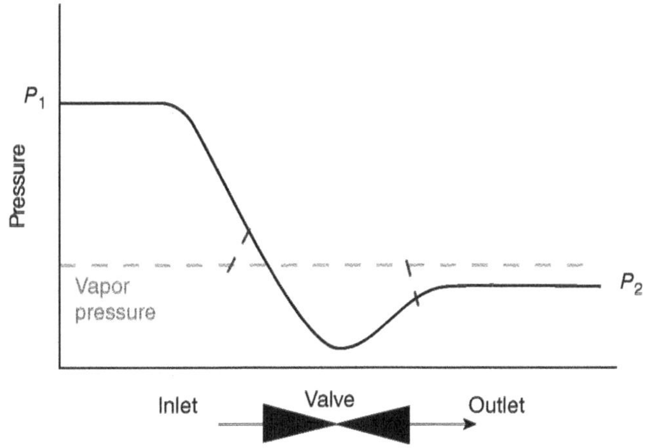

Figure 3.4 A flash occurs when the liquid pressure drops below the vapor pressure.

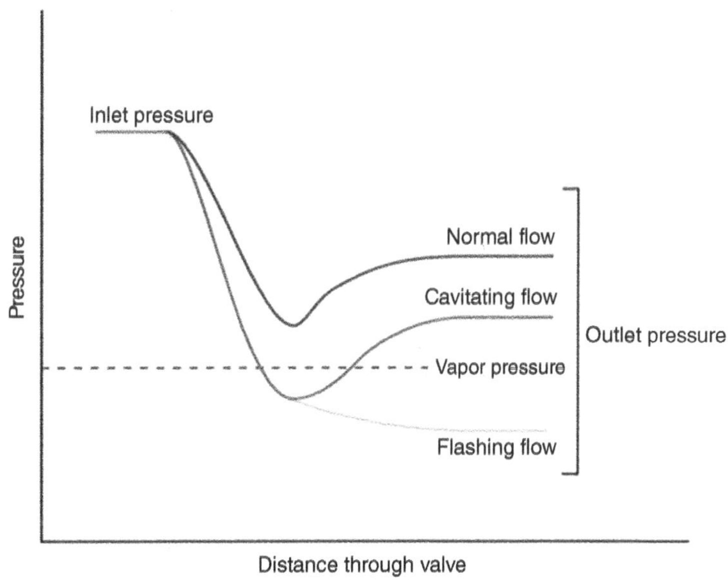

Figure 3.5 A control valve's normal, cavitation, and flashing pressure profiles.

It is pertinent to note that cavitation and flashing are the results of choked flow. A choked flow is a fluid dynamic phenomenon caused by the venturi effect. Figure 3.6 illustrates how fluid flow is compressed and choked when it passes through a restricted area. Pressure drops as velocity increases when the fluid passes through a restricted area.

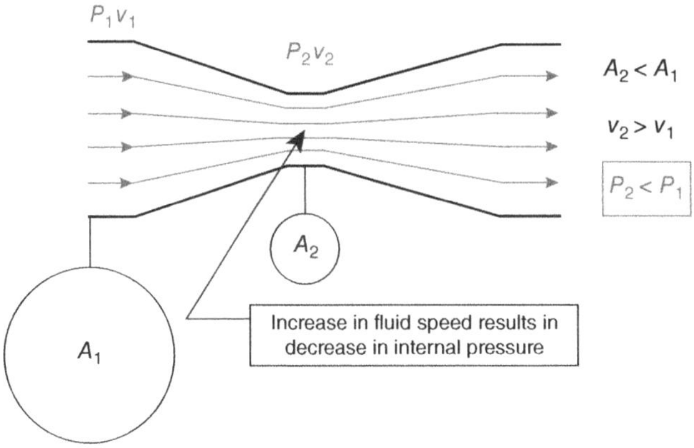

Figure 3.6 Choked flow.

Many methods can be used to prevent flashing. For example, one can install a large control valve and outlet piping. A control valve's body should not be smaller than half the diameter of the pipe it is connected to. In addition, there are other ways to minimize the risk of flashing, such as using hard materials for the trim (valve internals) and anticavitation trim, as well as incorporating a sacrificial spool into the piping system. Flashing is usually less damaging than cavitation. It is important to note, however, that a small piping outlet configuration can cause cavitation and cause severe piping damage at the valve's outlet. The flow of fluid is rotated 90° when using angle globe valves. Because 90° elbows are no longer required with angle globe valves, less piping space is required. The valve manufacturer should be consulted regarding the possibility of cavitation and flashing within valves and their outlet piping.

Questions and Answers

3.1 Which of the following statements is correct regarding cavitation?

A Cavitation only occurs in globe valves.

B Cavitation causes vibration, noise, erosion, and corrosion.

C Noise and vibration are the results of pressure drop below the vapor pressure.

D The valve's outlet pressure is less than the vapor pressure.

Answer

In fact, option A is incorrect since cavitation can occur in almost any type of valve used for flow regulation, including globe, v-notch ball, and butterfly valves. It is option B that is correct. Option C is incorrect since noise and vibration are caused by vapor pressure recovery and the collapse of the vapor. Option D is also incorrect since cavitation can only occur when the outlet pressure of a valve exceeds the vapor pressure.

3.2 What condition increases the chance and severity of cavitation?

A Applying stellite 6 or 21 to the internals of the valve
B Using an axial valve instead of a T-pattern globe valve
C Selecting a larger size and higher-pressure class valve
D Designing multistep trim

Answer

Option C is the correct answer. All other options reduce the chance and severity of cavitation.

3.3 Which cavitation remedy is illustrated on the right side of Figure 3.7?

A Selecting a smaller size and lower pressure class valve
B Designing a double-stage trim
C Applying stellite to the trim of the valve
D All these answers are incorrect.

Answer

Option B is correct since the pressure drop or reduction on the right side of the figure occurs in two stages, and the minimum pressure is still above the vapor pressure. So, cavitation will not occur.

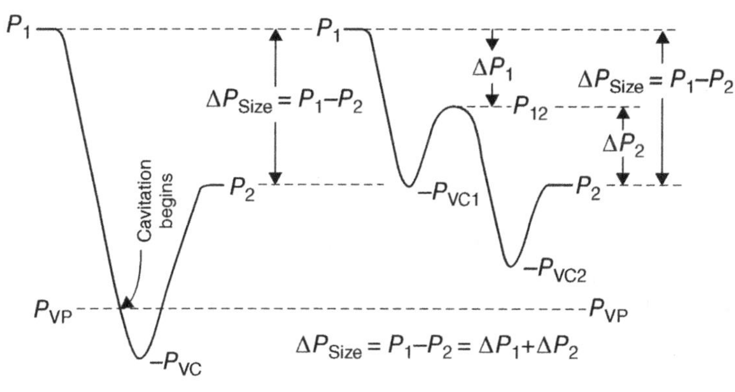

Figure 3.7 Cavitation remedy.

3.4 An 8″ butterfly valve in pressure class 150 (pressure nominal of 20 bar) is used to control flow in water service. During operation, the butterfly valve is open 50–100%. As the differential pressure across the valve varies from a minimum of 2 bar to a maximum of 4 bar, the ratio of the differential pressure to the inlet operating pressure changes from 0.4 to 0.8. The fluid temperature is 80 °F, and the water vapor pressure at this temperature is 0.5069. What is the correct option?

A The chance of cavitation is higher when the valve is half-open.
B In general, there is no risk of cavitation in this valve.
C The cavity index of the valve is higher when the valve is in a half-open position.
D All answers are wrong.

Answer

Considering that the cavitation index is greater than two at the fully open valve position, there is no risk of cavitation, as shown in Table 3.1. At 50% open, there is a risk of severe cavitation. Therefore, option A is the correct answer, and option B is wrong. When the valve is half-opened, the cavity index is lower. Therefore, option C is incorrect. Option D is incorrect because option A is correct.

$$100\% \text{ valve open} \rightarrow \Delta P = 2\,\text{bar} \rightarrow \frac{\Delta P}{P_1} = 0.4 \rightarrow \frac{2}{P_1} = 0.4 \rightarrow P_1 = 5\,\text{bar} \rightarrow P_2 = 3\,\text{bar} \rightarrow \delta = \frac{(P_u - P_v)}{(P_u - P_d)} = \frac{(5 - 0.5069)}{(5 - 3)} = 2.25$$

$$50\% \text{ valve open} \rightarrow \Delta P = 4\,\text{bar} \rightarrow \frac{\Delta P}{P_1} = 0.8 \rightarrow \frac{4}{P_1} = 0.8 \rightarrow P_1 = 5\,\text{bar} \rightarrow P_2 = 1\,\text{bar} \rightarrow \delta = \frac{(P_u - P_v)}{(P_u - P_d)} = \frac{(5 - 0.5069)}{(5 - 1)} = 1.123$$

As the cavitation index is greater than two at the fully open valve position, therefore, there is no risk of cavitation, as shown in Table 3.1. However, severe cavitation may result at 50% open position. Therefore, option A is the correct answer, and option B is false.

3.5 Find the correct statement regarding cavitation.

A Cavitation happens in gas services.
B The first stage of cavitation is the collapse of the vapor bubbles.
C Cavitation energy can cause physical damage to the valve's internal parts.
D The larger the cavitation index, the higher the cavitation risk.

Answer

Option A is incorrect because cavitation happens in liquid services. Option B is not correct either since the second stage of cavitation is the collapse of the vapor bubbles. Option C is the right answer. Option D is wrong, as a larger cavitation index implies a lower cavitation risk.

3.6 For flow throttling in the water piping system, an 8″ plug valve is used. The upstream pressure of the valve equals 11 psi, and the pressure drop is 4 psi during throttling. The flow coefficient of the valve at fully open is 1800. The valve has a flow of 600 gpm during throttling. Figure 3.8 illustrates the flow curve of the plug valve in this case. Regarding the plug valve's cavitation risk and opening percentage, which statement is true? (*Note*: In this case, the vapor pressure of water is −14.4 psi.)

A Cavitation risk is severe, and the valve is less than 20% open.

B Flashing may occur, and the valve is only at a 10% opening position.

C There is no risk of cavitation, and the valve is approximately 40% open.

D All answers are incorrect.

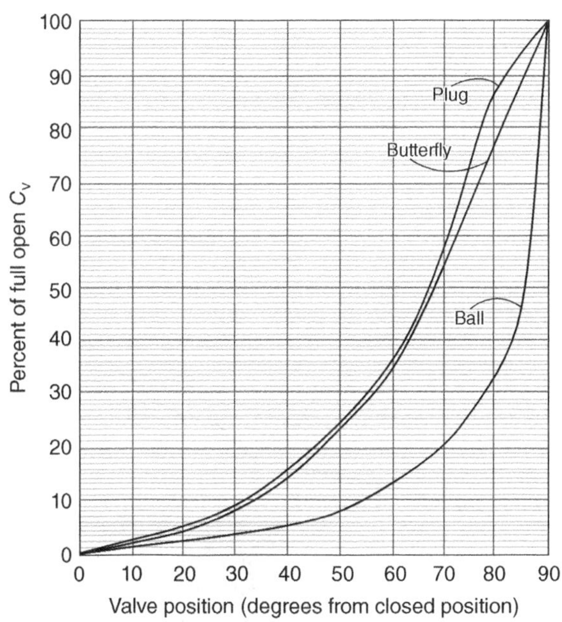

Figure 3.8 Valves flow characteristics including the plug valve.

Answer

$$\delta = \frac{(P_u - P_v)}{(P_u - P_d)} = \frac{(11 - (-14.4))}{(11 - 7)} = 6.35$$

Cavitation cannot occur since the cavity index is greater than two. It is now possible to calculate the valve flow coefficient during throttling as follows:

$$C_v = Q\sqrt{\frac{SG}{\Delta P}} = 600\sqrt{\frac{1}{4}} = 300$$

$$\text{Flow coefficient percentage during flow throttling} = \frac{300}{1800} = 16.67\%$$

According to Figure 3.8, the plug valve opening percentage corresponds to 16.67% of the maximum flow coefficient, or approximately 41%. Answer C is therefore the correct answer.

3.7 Figure 3.9 shows the cavitation characteristics of three types of valves – plug, butterfly, and ball. Select the correct statement regarding the valve's cavitation characteristics chart.

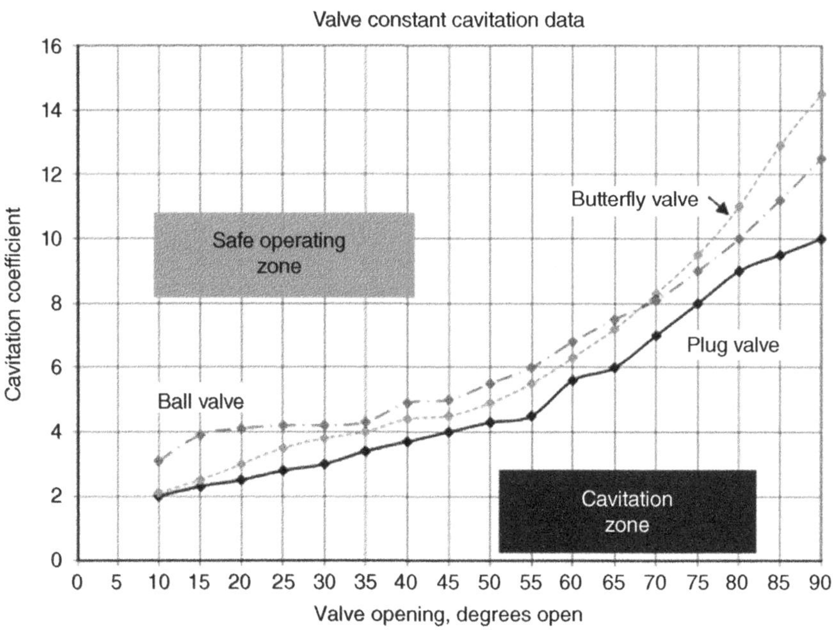

Figure 3.9 Valves cavitation characteristics.

A Increases in valve opening percentage increase the risk of cavitation.
B Compared to ball or butterfly valves, the plug valve is less prone to cavitation.
C Due to the risk of cavitation, it is not recommended to keep the flow throttling valves open less than 10%.
D A ball valve is always safer than a butterfly valve against cavitation.

Answer
Option A is incorrect because the cavitation risk is reduced by increasing the valve opening percentage. Option B is not correct either since the plug valve has the highest risk of cavitation as its curve is closer to the cavitation zone highlighted at the bottom of the chart in red. Option C is the correct answer. Option D is wrong as the ball valve is safer than the butterfly valve against cavitation for an opening percentage of less than 70%.

3.8 Which of the following statements is incorrect regarding flashing and its conditions?
A When a flow encounters a restriction, such as a reduced bore valve, its velocity increases, and flashing could occur.
B Flashing occurs when the outlet pressure is greater than the vapor pressure ($P_2 > P_v$).
C Mechanical damages such as erosion and noise are adverse consequences of flashing.
D Flashing usually causes less damage to the valve than cavitation.

Answer
Option B is incorrect since flashing occurs when the outlet pressure is less than the vapor pressure ($P_2 < P_v$).

Further Reading

American Petroleum Institute (API) 602 (2016). *Gate, Globe and Check Valves for Sizes DN100 (NPS 4) and Smaller for the Petroleum and Natural Gas Industries*, 10e. Washington, DC: American Petroleum Institute (API).

Iranian Petroleum Standard (IPS) (2015). *Engineering Standard for Control Valves. IPS-E-IN-160*. Tehran: Iranian Petroleum Standard (IPS).

Nesbitt, B. (2007). *Handbook of Valves and Actuators: Valves Manual International*, 1e. Oxford: Elsevier.

Skousen, P.L. (2011). *Valve Handbook*, 3e. New York: McGraw-Hill.

Smit, P. and Zappe, R.W. (2004). *Valve Selection Handbook*, 5e. New York: Elsevier.

Sotoodeh, K. (2016). Cavitation in globe valves and solutions. *Valve World Magazine* 21 (3): 32–36.

Sotoodeh, K. (2018). Selecting a butterfly valve instead of globe valve for fluid control in a utility service in the offshore industry (based on industrial experience). *American Journal of Mechanical Engineering* 6 (1): 27–31. https://doi.org/10.12691/ajme-6-1-4.

Sotoodeh, K. (2021). *Subsea Valves and Actuators for the Oil and Gas Industry*, 1e. Austin, TX: Elsevier (Gulf Professional Publishing).

Valve-Metallic Valve and Manufacturing (2018). *Cavitation in Valves*. Valve-Metallic Valve and Manufacturing. https://www.valmatic.com/Portals/0/pdfs/CavitationinValves_6-18.pdf (accessed 25 February 2022).

4

Wall Thickness

4.1 Introduction

The bodies and bonnets of industrial valves are pressure-containing parts that lead to leakage of valve internal fluid into the environment if they fail to function. It is essential to calculate and select the correct thickness for the body and bonnet to prevent mechanical failures. The American Society of Mechanical Engineers (ASME) B16.34, the standard for flanged, threaded, and welding end, covers various aspects of valves, including pressure-temperature rating, dimensions, tolerances, materials, nondestructive examination requirements, testing, and marking. The valve body and bonnet thickness should meet the criteria set in ASME B16.34.

4.2 ASME B16.34 Minimum Wall Thickness Calculation

4.2.1 Conservation Approach (Mandatory Appendix A)

The minimum valve body thickness according to ASME B16.34, parameter t_m, is provided in both millimeters and inches. The minimum valve body thickness in ASME B16.34 depends on two valve parameters: internal diameter (parameter d) and the pressure class of the valve. Generally, the minimum valve wall thickness increases with the internal diameter and pressure class of the valve. The pressure classes, which are covered by ASME B16.34, are 150 (PN20), 300 (PN50), 600 (PN100), 900 (PN150), 1500 (PN250), 2500 (PN420), and 4500 (PN720). "PN" stands for pressure nominal. The internal diameter of a valve is the minimum diameter of fluid passage through the valve bore. Using Table 4.1 extracted from ASME B16.34, the inside diameter of the valves can be determined in millimeters and inches. In general, ASME B16.34 specifies that the inner diameter of the valve shall not be less than 90% of the diameter of the valve end.

Industrial Valves: Calculations for Design, Manufacturing, Operation, and Safety Decisions, First Edition. Karan Sotoodeh.

Table 4.1 Valve inside diameter in millimeters and inches, according to valve size (NPS) and pressure class, as per ASME B16.34.

	CL150		CL300		CL600		CL900		
NPS	mm	in.	mm	in.	mm	in.	mm	in.	**DN**
1/2″	12.7	0.5	12.7	0.5	12.7	0.5	12.7	0.50	15
3/4″	19.1	0.75	19.1	0.75	19.1	0.75	17.5	0.69	20
1″	25.4	1.00	25.4	1.00	25.4	1.00	22.1	0.87	25
1 1/4″	31.8	1.25	31.8	1.25	31.8	1.25	28.4	1.12	32
1 1/2″	38.1	1.50	38.1	1.50	38.1	1.50	34.8	1.37	40
2″	50.8	2.00	50.8	2.00	50.8	2.00	47.5	1.87	50
2 1/2″	63.5	2.50	63.5	2.50	63.5	2.50	57.2	2.25	65
3″	76.2	3.00	76.2	3.00	76.2	3.00	72.9	2.87	80
4″	101.6	4.00	101.6	4.00	101.6	4.00	98.3	3.87	100
6″	152.4	6.00	152.4	6.00	152.4	6.00	146.1	5.75	150
8″	203.2	8.00	203.2	8.00	199.9	7.87	190.5	7.50	200
10″	254.0	10.00	254.0	10.00	247.7	9.75	238.0	9.37	250
12″	304.8	12.00	304.8	12.00	298.5	11.75	282.4	11.12	300
14″	336.6	13.25	336.6	13.25	326.9	12.87	311.2	12.25	350
16″	387.4	15.25	387.4	15.25	374.7	14.75	355.6	14	400
18″	438.2	17.25	431.8	17.00	419.1	16.50	400.1	15.75	450
20″	489.0	19.25	482.6	19.00	463.6	18.25	444.5	17.50	500
22″	539.8	21.25	533.4	21.00	511.0	20.12	489.0	19.25	550
24″	590.6	23.25	584.2	23.00	558.8	22.00	533.4	21.00	600
26″	641.4	25.25	635.0	25.00	603.3	23.75	577.9	22.75	650
28″	692.2	27.25	685.8	27.00	647.7	25.50	622.3	24.50	700
30″	743.0	29.25	736.6	29.00	695.2	27.37	666.8	26.25	750
32″	793.7	31.25	787.4	31.00	736.6	29.00	711.2	28.00	—
34″	844.5	33.25	838.2	33.00	781.0	30.75	755.6	29.75	—
36″	895.3	35.25	889.0	35.00	825.5	32.62	800.1	31.50	—
38″	946.1	37.25	939.8	37.00	872.9	34.37	844.5	33.25	—
40″	996.9	39.25	990.6	39.00	920.7	36.25	889.0	35.00	—
42″	1047.7	41.25	1041.4	41.00	965.2	38.00	933.4	36.75	—
44″	1098.5	43.25	1092.2	43.00	1012.6	39.87	977.9	38.50	—
46″	1149.3	45.25	1143.0	45.00	1057.1	41.62	1022.3	40.25	—
48″	1200.1	47.25	1193.8	47.00	1104.9	43.50	1066.8	42.00	—

Table 4.1 (Continued)

	CL150		CL300		CL600		CL900		
NPS	mm	in.	mm	in.	mm	in.	mm	in.	DN
50″	1250.9	49.25	1244.6	49.00	1149.3	45.25	1111.2	43.75	—
52″	1301.7	51.25	1295.4	51.00	1193.8	47.00	—	—	—
54″	1352.5	53.25	1346.2	53.00	1241.2	48.87	—	—	—
56″	1403.3	55.25	1397.0	55.00	1285.7	50.62	—	—	—
58″	1454.1	57.25	1447.8	57.00	1330.1	52.37	—	—	—
60″	1504.9	59.25	1498.6	59.00	1374.6	54.12	—	—	—

	CL1500		CL2500		
	mm	in.	mm	in.	DN
1/2″	12.7	0.50	11.2	0.44	15
3/4″	17.5	0.69	14.2	0.56	20
1″	22.1	0.87	19.1	0.75	25
1 1/4″	28.4	1.12	25.4	1.00	32
1 1/2″	34.8	1.37	28.4	1.12	40
2″	47.5	1.87	38.1	1.50	50
2 1/2″	57.2	2.25	47.5	1.87	65
3″	69.9	2.75	57.2	2.25	80
4″	91.9	3.62	72.9	2.87	100
6″	136.4	5.37	111.0	4.37	150
8″	177.8	7.00	146.1	5.75	200
10″	222.3	8.75	184.2	7.25	250
12″	263.4	10.37	218.9	8.62	300
14″	288.8	11.37	241.3	9.50	350
16″	330.2	13.00	276.1	10.87	400
18″	371.3	14.62	311.2	12.25	450
20″	415.8	16.37	342.9	13.50	500
22″	457.2	18.00	377.7	14.87	550
24″	498.3	19.62	412.8	16.25	600
26″	539.8	21.25	447.5	17.62	650
28″	584.2	23.00	482.6	19.00	700
30″	625.3	24.62	517.4	20.37	750

(Continued)

Table 4.1 (Continued)

	CL1500		CL2500		
	mm	in.	mm	in.	DN
32″	—	—	—	—	—
34″	—	—	—	—	—
36″	—	—	—	—	—
38″	—	—	—	—	—
40″	—	—	—	—	—
42″	—	—	—	—	—
44″	—	—	—	—	—
46″	—	—	—	—	—
48″	—	—	—	—	—
50″	—	—	—	—	—
52″	—	—	—	—	—
54″	—	—	—	—	—
56″	—	—	—	—	—
58″	—	—	—	—	—
60″	—	—	—	—	—

The ASME B16.34 standard covers pressure-temperature ratings for forgings, castings, bars, plates, and tubular products. Pressure-temperature ratings in ASME B16.34 indicate the maximum allowable pressure for a specific type of material (e.g. carbon steel) based on the product pressure class rating (e.g. CL300) and the process temperature. Bodies of valves are typically made of forging for sizes <2″ or casting for sizes ≥2″. Carbon steel body valves are in material group number 1.1 as ASME B16.34 and have forging and casting designations of A105 and A216 WCB, respectively. American Society for Testing and Materials (ASTM) covers the provided material designations consisting of a letter (A for ferrous materials) followed by a sequential number. Table 4.2 extracted from ASME B16.34 includes the maximum working pressure in bar unit for carbon steel designations mentioned earlier established on the pressure class and temperature values. As a general rule, the working pressure is reduced by increasing the temperature.

Socket-welding end or threaded-end valves may be designed for pressure classes not designated in Tables 4.1 and 4.2. In this case, it is necessary to perform interpolation to find the relevant valve inner diameter. ASME B16.34 mandatory

Table 4.2 Pressure-temperature rating values for material group 1.1 as per ASME B16.34.

	Working pressure (bar)						
Temperature °C	**CL150**	**CL300**	**CL600**	**CL900**	**CL1500**	**CL2500**	**CL4500**
−29 to 38	19.6	51.1	102.1	153.2	255.3	425.5	765.9
50	19.2	50.1	100.2	150.4	250.6	417.7	751.9
100	17.7	46.6	93.2	139.8	233.0	388.3	699.0
150	15.8	45.1	90.2	135.2	225.4	375.6	676.1
200	13.8	43.8	87.6	131.4	219.0	365.0	657.0
250	12.1	41.9	83.9	125.8	209.7	349.5	629.1
300	10.2	39.8	79.6	119.5	199.1	331.8	597.3
325	9.3	38.7	77.4	116.1	193.6	322.6	580.7
350	8.4	37.6	75.1	112.7	187.8	313.0	563.5
375	7.4	36.4	72.7	109.1	181.8	303.1	545.5
400	6.5	34.7	69.4	104.2	173.6	289.3	520.8
425	5.5	28.8	57.5	86.3	143.8	239.7	431.5
450	4.6	23.0	46.0	69.0	115.0	191.7	345.1
475	3.7	17.4	34.9	52.3	87.2	145.3	261.5
500	2.8	11.8	23.5	35.3	58.8	97.9	176.3
538	1.4	5.9	11.8	17.7	29.5	49.2	88.6

Source: ASME B16.34/American Society of Mechanical Engineers.

appendix VI provides a method based on basic equations for obtaining the minimum wall thickness of the valves according to the internal diameter and pressure class of the valve. Table 4.3 summarizes the basic equations for minimum wall thickness calculation as per ASME B16.34, mandatory appendix VI.

4.2.2 Nonconservation Method

The wall thickness values, calculated according to mandatory appendix VI, are very conservative. It means that the provided values for wall thickness are relatively high, which leads to heavier and bulkier valves. Therefore, ASME B16.34 provides another way to calculate valve wall thickness using Eq. (4.1). The minimum wall thickness values, t_m, calculated based on the equations in mandatory appendix VI are all greater than the thickness values determined by Eq. (4.1).

Table 4.3 Basic equations for minimum valve wall thickness calculation, as per ASME B16.34, mandatory appendix VI.

Valve pressure class, P_c	Inside diameter (d) (mm)	Metric equation t_m (mm)	Round
150	$3 \leq d < 50$	$t_m(150) = 0.064d + 2.34$	Off, one decimal
150	$50 \leq d \leq 100$	$t_m(150) = 0.020d + 4.5$	Off, one decimal
150	$100 < d \leq 1300$	$t_m(150) = 0.0163d + 4.70$	Off, one decimal
300	$3 \leq d < 50$	$t_m(300) = 0.080d + 2.29$	Off, one decimal
300	$50 \leq d \leq 100$	$t_m(300) = 0.030d + 4.83$	Off, one decimal
300	$100 < d \leq 1300$	$t_m(300) = 0.0334d + 4.32$	Off, one decimal
600	$3 \leq d < 25$	$t_m(600) = 0.090d + 2.54$	Off, one decimal
600	$25 \leq d \leq 50$	$t_m(600) = 0.060d + 3.30$	Off, one decimal
600	$50 < d \leq 1300$	$t_m(600) = 0.06777d + 2.54$	Off, one decimal
900	$3 \leq d < 25$	$t_m(900) = 0.160d + 2.29$	Off, one decimal
900	$25 \leq d \leq 50$	$t_m(900) = 0.060d + 4.83$	Off, one decimal
900	$50 < d \leq 1300$	$t_m(900) = 0.10449d + 2.54$	Off, one decimal
1500	$3 \leq d \leq 1300$	$t_m(1500) = 0.18443d + 2.54$	Off, one decimal
2500	$3 \leq d \leq 1300$	$t_m(2500) = 0.34091d + 2.54$	Off, one decimal
4500	$3 \leq d \leq 1300$	$t_m(4500) = 0.78488d + 2.54$	Off, one decimal
Valve pressure class, P_c	**Inside diameter (d) (in.)**	**Inch equation t_m (in.)**	**Round**
150	$0.12 \leq d < 2$	$t_m(150) = 0.064d + 0.092$	Off, two decimals
150	$2 \leq d \leq 4$	$t_m(150) = 0.020d + 0.18$	Off, two decimals
150	$4 < d < 50$	$t_m(150) = 0.0163d + 0.185$	Off, two decimals
300	$0.12 \leq d < 2$	$t_m(300) = 0.080d + 0.09$	Off, two decimals
300	$2 \leq d \leq 4$	$t_m(300) = 0.030d + 0.19$	Off, two decimals
300	$4 < d < 50$	$t_m(300) = 0.0334d + 0.17$	Off, two decimals
600	$0.12 \leq d < 1$	$t_m(600) = 0.090d + 0.10$	Off, two decimals
600	$1 \leq d \leq 2$	$t_m(600) = 0.060d + 0.13$	Off, two decimals
600	$2 < d < 50$	$t_m(600) = 0.06777d + 0.10$	Off, two decimals
900	$0.12 \leq d < 1$	$t_m(900) = 0.160d + 0.09$	Off, two decimals
900	$1 \leq d \leq 2$	$t_m(900) = 0.060d + 0.19$	Off, two decimals
900	$2 < d < 50$	$t_m(900) = 0.10449d + 0.10$	Off, two decimals
1500	$0.12 \leq d \leq 50$	$t_m(1500) = 0.18443d + 0.10$	Off, two decimals
2500	$0.12 \leq d \leq 50$	$t_m(2500) = 0.34091d + 0.10$	Off, two decimals
4500	$0.12 \leq d \leq 50$	$t_m(4500) = 0.78488d + 0.10$	Off, two decimals

Source: ASME B16.34/American Society of Mechanical Engineers.

Minimum Wall Thickness Calculation Based on ASME B16.34 (An Alternative Method to ASME B16.34 Mandatory Appendix VI)

$$t_m = 1.5 \times \left[\frac{P_c d}{2S_F - 1.2P_c}\right] \tag{4.1}$$

where:

t_m: Minimum calculated thickness (mm/in.);
P_c: Pressure class designation number (e.g. for class 150, $P_c = 150$ and for class 300, $P_c = 300$);
d: Inside diameter of the valve (mm/in.);
S_F: Stress-based constant equal to 7000.

Example 4.1 Calculate the wall thickness of a 12″ × 10″ reduced bore ball valve with a pressure class of 150 according to both methods provided in ASME B16.34 and this chapter and compare the results.

Answer
A reduced bore, port, or opening valve has a bore or opening inside the valve smaller than the end connections. The opening or bore of the ball valve is 12″ at both ends of the valve and 10″ inside the valve. Therefore, the internal diameter or bore of the valve is equal to 10 in. Using Table 4.1 to obtain the internal diameter of a valve with a 10″ bore and pressure class of 150 gives an inner diameter of 254 mm.

The next step is to use the basic calculations provided in mandatory appendix VI, as seen in Table 4.3, as follows:

$$\begin{aligned} P_c &= \text{Pressure class 150 and } 100 < d \leq 1300 \rightarrow t_m(150) \\ &= 0.0163d + 4.70 \rightarrow t_m \\ &= 0.0163 \times 254 + 4.70 = 8.84\,\text{mm} \end{aligned}$$

The next step is to calculate the valve wall thickness as per the less conservative approach provided by Eq. (4.1) as follows:

$$t_m = 1.5 \times \left[\frac{P_c d}{2S_F - 1.2P_c}\right] = 1.5 \times \left[\frac{150 \times 254}{2 \times 7000 - 1.2 \times 150}\right] = \frac{57{,}150}{13{,}820} = 4.14\,\text{mm}$$

The valve wall thickness is thinner by 4.70 mm using Eq. (4.1) compared to the basic equations provided in Table 4.3. As a general rule, the wall thickness values calculated according to basic equations in mandatory appendix VI in Table 4.3 are typically 3–5 mm thicker than the wall thickness values calculated using Eq. (4.1).

It is important to remember that the wall thickness values provided by ASME B16.34 and the methods mentioned earlier are minimums. Other parameters, such as corrosion allowance (CA), abbreviated as CA, could be factored into the minimum thickness calculated for noncorrosion-resistant alloys, such as carbon and low-alloy steels. The amount of CA could be standardized to 3 mm for carbon and low-alloy steels based on the NORSOK M-001 material selection standard and the L-001 piping and valves standard. It is essential to avoid adding any CA to exotic materials such as stainless steel and nickel alloys. CA calculations are covered in detail in Chapter 5. Equation (4.2) shows the effect of CA on the valve wall thickness.

Effect of Corrosion Allowance on the Valve Wall Thickness

$$t_{\mathrm{m,ca}} = t_{\mathrm{m}} + \mathrm{CA} \tag{4.2}$$

where:

$t_{\mathrm{m,ca}}$: Final valve wall thickness after adding a corrosion allowance (mm/in.);
t_{m}: Minimum valve wall thickness (mm/in.);
CA: Corrosion allowance (mm/in.).

Example 4.2
Find the wall thickness of a threaded-end 2″ ball valve in a pressure class of 3000 at 50 °C that is made of A216 WCB material. What is the valve's maximum pressure?

Answer
The pressure classes in Tables 4.1, 4.2, and 4.3 are applied to flanged end valves. However, welding-end and threaded-end valves are typically designed based on intermediate pressure class designations, including 3000 psi (206.9 bar), 6000 psi (413.8 bar), and 9000 psi (620.7 bar) (1 bar = 14.5 psi). It is necessary to apply multiple linear interpolations in this case. Table 4.4 shows pressure-temperature rating values for all standard pressure classes as well as 3000 psi for the maximum temperature of 100 °C. As shown in the table, 3000 psi falls between CL900 and CL1500. It is important to remember that 3000 psi pressure class indicates the maximum pressure of 206.9 bar at a temperature range from −29 to 38 °C.

But the pressure value is decreased by increasing the temperature. So the next step is to find the associated pressure value of 50 °C by using interpolation.

Pressure reduction for CL900 valve by increasing the temperature from ambient to 50 °C = 153.2 − 150.4 = 2.8 bar

Pressure reduction for CL1500 valve by increasing the temperature from ambient to 50 °C = 255.3 − 250.6 = 4.7 bar

Table 4.4 Pressure-temperature rating values for all standard pressure classes as well as 3000 psi for the maximum temperature of 100 °C.

Temperature °C	CL150	CL300	CL600	CL900	3000 psi	CL1500	CL2500	CL4500
−29 to 38	19.6	51.1	102.1	153.2	206.9	255.3	425.5	765.9
50	19.2	50.1	100.2	150.4	*X*	250.6	417.7	751.9
100	17.7	46.6	93.2	139.8	*X*	233.0	388.3	699.0

$$X = 206.9 - Y$$

$$4.7 - Y = \frac{(255.3 - 206.9) \times (4.7 - 2.8)}{(255.3 - 153.2)} = \frac{48.4 \times 1.9}{102.1} = 0.9 \rightarrow Y = 3.8$$

$$\rightarrow X = 206.9 - 3.8 = 203.1$$

Table 4.5 does not provide the internal diameters of valves in a pressure class of 3000 psi that are required to calculate the valve wall thickness. Suppose the internal diameter of the 2″ class 3000 psi valve is calculated. In that case, there is no formula in Table 4.3 to determine the wall thickness values for valves in intermediate pressure classes such as 3000 psi. Thus, the best way to calculate the wall thickness of the valve with intermediate pressure class in this example is to use linear interpolation between the thicknesses of 2″ valves in pressure classes of 900 and 1500, considering that there is a direct relationship between the pressure and thickness values of valves.

Table 4.5 Valve inside diameter for size ranges from 1/2″ to 2″ and pressure classes of 900 and 1500, as per ASME B16.34.

	CL900		3000 psi		CL1500	
Size	Mm	in.	mm	in.	mm	in.
1/2″	12.7	0.50	—	—	12.7	0.50
3/4″	17.5	0.69	—	—	17.5	0.69
1″	22.1	0.87	—	—	22.1	0.87
1 1/4″	28.4	1.12	—	—	28.4	1.12
1 1/2″	34.8	1.37	—	—	34.8	1.37
2″	47.5	1.87	*X*	*Y*	47.5	1.87

$P = 150.4$ bar @ $T = 50°C$, Valve size $= 2'' \rightarrow P_c = 900$ and $d = 47.5$ mm

$(20 \leq d \leq 50) \rightarrow t_m(900) = 0.060d + 4.83 = 7.68$ mm

$P = 250.6$ bar @ $T = 50°C$, Valve size $= 2'' \rightarrow P_c = 1500$ and $d = 47.5$ mm

$(20 \leq d \leq 50) \rightarrow t_m(1500) = 0.18443d + 0.10 = 8.86$ mm

$P = 203.1$ bar @ $T = 50°C$, Valve size $= 2'' \rightarrow t_m$?

$$t_m - 7.68 = \frac{(8.86 - 7.68)(203.1 - 150.4)}{250.6 - 150.4} = 0.62$$

$$\rightarrow t_m = 7.68 + 0.62 = 8.3 \text{ mm}$$

4.2.3 ASME Sec. VIII Div. 02 Wall Thickness Calculation

ASME BPVC (Boiler and Pressure Vessel Code) Sec. VIII Div. 02 is titled "Rules for Construction of Pressure Vessels." It covers rules for the designing, manufacturing, testing, inspection, and certification of pressure vessels and boilers. Piping and valves are also categorized as pressure vessels. The main concept of ASME BPVC Sec. VIII Div. 02 is "Design by Analysis," i.e. using stress analysis, finite element analysis (FEA), or finite element method (FEM) to demonstrate the accuracy and/or correctness of the valve design. Besides, the presented thickness calculation method by ASME Sec.VIII code is less conservative than that in ASME B16.34. Pipeline valves installed on oil and gas export pipelines are typically large and high-pressure class, making them very heavy. Even nonconservative wall thickness calculation in ASME B16.34 adds large tonnes of weight to these valves. Therefore, the main aim of this section is to propose using ASME Sec.VIII Div. 02 instead of ASME B16.34 for pipeline valves' wall thickness calculation reduces the valves' weight. According to ASME Sec. VIII Div.02, Eq. (4.3) is proposed for the valve wall thickness calculation, assuming the cylindrical valve shell.

Valve Wall Thickness Calculation as Per ASME Sec. VIII Div. 02

$$t = \frac{D}{2}\left(e^{p/s} - 1\right) \tag{4.3}$$

t = valve thickness (in.)
D = valve diameter (in.)
p = valve design pressure (psi)
s = Allowable stress (psi) taken from table A.1 of ASME B31.3
$e = 2.7182$

Example 4.3 Calculate the wall thickness of a 30″ CL1500 valve based on methods in ASME B16.34 and ASME Sec. VIII Div. 02. The valve's body is made in A216 WCB carbon steel material, and the allowable stress of carbon steel at ambient temperature is equal to 20,000 psi as per ASME B31.3, process piping code. In addition, assume that the carbon steel material grade A216 WCB can withstand a maximum of 250 bar as per ASME B16.34.

Answer

Using Table 4.1 to obtain the internal diameter of a valve with a 30″ size and pressure class of 1500 gives an inner diameter (d) of 625.3 mm.

The next step is to use the basic calculations provided in mandatory appendix VI of ASME B16.34, as seen in Table 4.3, as follows:

$$P_c = 1500 \text{ pressure class, 150 and } 3 < d \leq 1300 \rightarrow t_m(1500) = 0.18443d + 2.54$$
$$= 0.18443 \times 625.3 + 2.54 = 117.86 \text{ mm}$$

The next step is to calculate the valve wall thickness as per the less conservative approach in ASME B16.34 provided by Eq. (4.1) as follows:

$$t_m = 1.5 \times \left[\frac{P_c d}{2S_F - 1.2P_c}\right] = 1.5 \times \left[\frac{1500 \times 625.3}{2 \times 7000 - 1.2 \times 1500}\right]$$
$$= \frac{1,406,925}{12,200} = 115.32 \text{ mm}$$

The valve wall thickness is thinner by 2.54 mm using Eq. (4.1) compared to the basic equations provided in Table 4.3.

The next step is to calculate the valve wall thickness based on Eq. (4.3) from ASME Sec. VIII Div. 02:

t = valve thickness (in.)
D = valve diameter (in.) = 30″
p = valve design pressure (psi) = 250 barg = 250 × 14.5 = 3625 psi
s = Allowable stress (psi) for ASTM A216 WCB carbon steel body material of the valve = 20,000 psi at ambient temperature as per ASME B31.3
e = 2.7182

$$t = \frac{D}{2}\left(e^{p/s} - 1\right) = \frac{30}{2}\left(e^{3625/20,000} - 1\right) = 15 \times (1.1987 - 1) = 2.98 \text{ in.} = 75.71 \text{ mm}$$

The valve wall thickness saving between ASME Sec. VIII Div. 02 and nonconservative ASME B16.34 method is 39.61 mm.

4.3 Wafer Design Thickness Validation

Flange connection is a prevalent method used to attach a valve to a pipe. A flange is preferred over welding because flanges can be easily installed and dismantled for maintenance, inspection, or replacement. Valves can be designed with a flangeless body to save weight, space, and cost. In fact, a wafer design is defined as a flangeless design with a facing that permits installation between ASME and manufacturer standard (MSS SP) flanges. Figure 4.1 illustrates a drawing of the wafer body butterfly valve.

Wafer valves that are bolted between flanges (e.g. butterfly or ball valve) should be designed according to the requirements in ASME B16.34 from Sections (a) through (f):

a) The design should provide for bolting using all of the bolt holes and the bolt circle of the specified flange.
b) Bolt holes parallel to the body run, maybe either be threaded or unthreaded. Threaded holes may be blind holes suitable for use with bolt studs. When threaded, the full thread engagement, excluding champers, should be provided to a depth not less than one nominal bolt diameter.

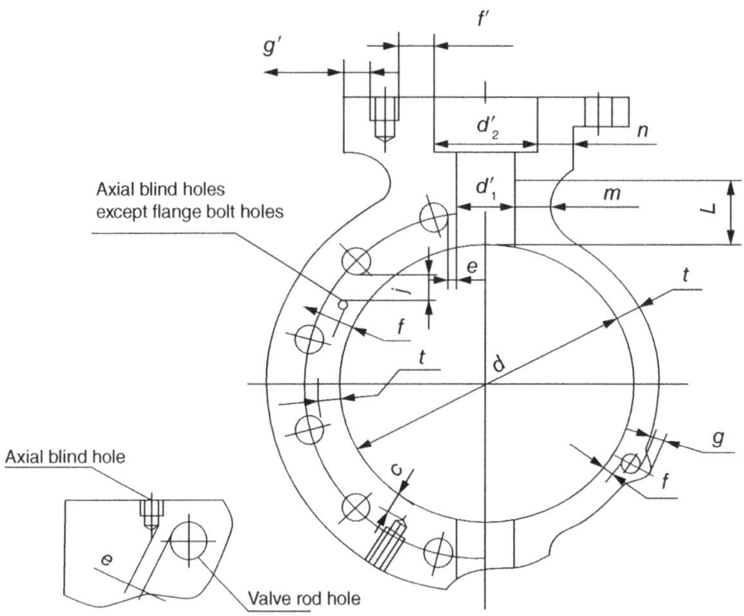

Figure 4.1 A drawing of the wafer butterfly valve.

c) The required minimum valve body wall thickness, t_m, should be measured from the valve body inside the circumference out to the lesser of the valve body outside the circumference, or the circumference of the circle inscribed through the inner tangent points to the flange bolts.

d) The inner ligament (e in Figure 4.1) of either a through hole or a blinded threaded hole in the vicinity of a stem penetration should not be less than 25% of the required wall thickness of the body neck but in no case less than 2.5 mm (0.1 in.).

$$(e \geq 0.25\,t' \text{ and } e \geq 2.5\text{ mm})$$

e) The inner ligament (f and g in Figure 4.1) for holes parallel to the body run should not be less than $0.25t_m$, but in no case less than 2.5 mm (0.1 in.). The sum of the inner and outer ligaments should not be less than t_m.

$$(f \geq 0.25t_m \text{ and } f \geq 2.5\text{ mm}, g \geq 0.25t_m \text{ and } g \geq 2.5\text{ mm, and } f+g \geq t_m)$$

f) A ligament within the minimum body wall between two adjacent holes within the minimum body valve (j in Figure 4.1) should be $0.25t_m$ or greater, but not less than 2.5 mm (1 in.).

$$(j \geq 0.25t_m \text{ and } j \geq 2.5\text{ mm})$$

a) Wall thickness of the valve body neck (t') is referred to in section (d). The neck is the top part of the body in Figure 4.1, where parameters such as d', m, n, g', and f' are shown. Within the area measured from the outer side of the valve body, along the direction of the valve neck within the distance of 1.1 $\sqrt{d \times t_m}$, as illustrated in Figure 4.2, the minimum diameter is d and t_m is the minimum wall thickness calculated by methods explained before according to ASME B16.34. Beyond the aforementioned 1.1 $\sqrt{d \times t_m}$ regions, straight, circular sections of valve body necks with inside diameter d' shall be provided with local wall thickness at least equal to t', where t' is taken from the appropriate class rating equation given in Table 4.3 using the relevant diameter value shown as d'' using either Eq. (4.4) or Eq. (4.5).

Calculation of d'', a Diameter Used to Determine the Minimum Body Wall Thickness at the Valve Neck Beyond 1.1 $\sqrt{d \times t_m}$ Regions

$$150 \leq \text{Pressure class} \leq 2500 \rightarrow d'' = \frac{2d'}{3} \tag{4.4}$$

$$1500 \leq \text{Pressure class} \leq 4500 \rightarrow d'' = \frac{d'}{48}\left(27 + \frac{P_c}{5000}\right) \tag{4.5}$$

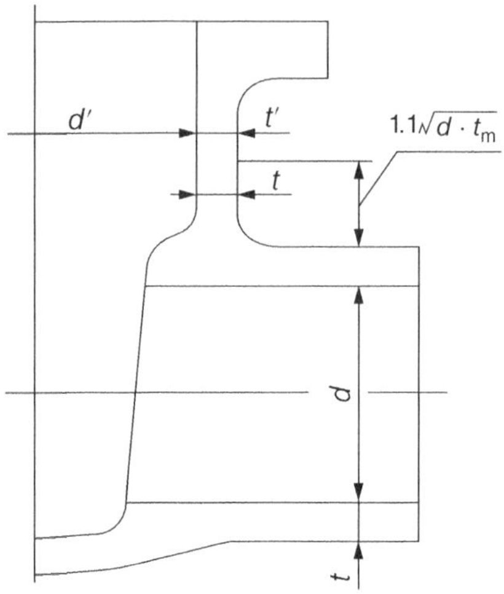

Figure 4.2 Valve neck section.

where:

P_c: Pressure class designation number as defined in Eq. (4.1);
d': Body neck inside diameter (mm/in.);
d'': The diameter used to determine body neck valve thickness value beyond 1.1 $\sqrt{d \times t_m}$ regions.

b) For the special case where $d' > 1.5d$, it is necessary that the wall thickness be equal to or greater than t' for the entire body neck length having diameter d' including the 1.1 $\sqrt{d \times t_m}$ body region.
c) If the inner diameter of the valve neck is a lot smaller than the inner diameter of the valve flow passage, i.e. $d/d' \geq 4$, the minimum body neck wall thickness over a distance of $L = t_m\left(1+\ 1.1\sqrt{d/t_m}\right)$, measured from the intersection of the valve's body inside diameter and the axis of the body neck outside diameter shall not be less than t'. t' is obtained from Table 4.3 according to the internal diameter (d') of the corresponding valve body neck. The neck thickness beyond the distance of L should be obtained from Table 4.3 according to the value of d''.
d) In cases where there is a drilling or tapping on the wall of the valve neck parallel to the direction of the valve neck axis, the sum thickness of the inner and outer sides be equal or greater than t' or t_m. Referring to Figure 4.1, $f' \geq 0.25t'_2$, $g' \geq 0.25t'_2$ and $f' + g' \geq t'_2$. (t'_2 is the valve body neck thickness associated with d'_2, also shown with parameter "n" in Figure 4.1).

Example 4.4 A ball valve in 3/4″ and CL300 made of carbon steel is supplied in the wafer design (see Figure 4.3). The valve sits between two mating flanges, and connecting bolts pass through the valve's body. It is not common to use wafer design for ball valves, unlike butterfly valves. So the concern of the valve purchaser is that supplied valves do not have sufficient wall thickness to meet the requirements in ASME B16.34. Figure 4.4 shows the machining drawing of the ball valve. Are the thickness values in the machining drawing sufficient? (*Note*: 3 mm CA is added to the valve wall thickness.)

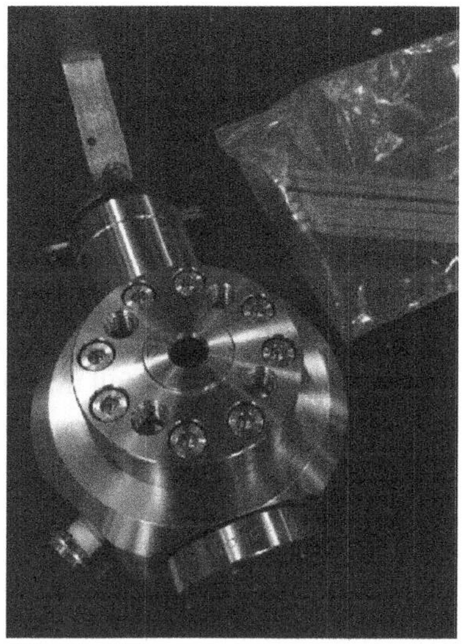

Figure 4.3 Compact 3/4″ CL300 wafer ball valve.

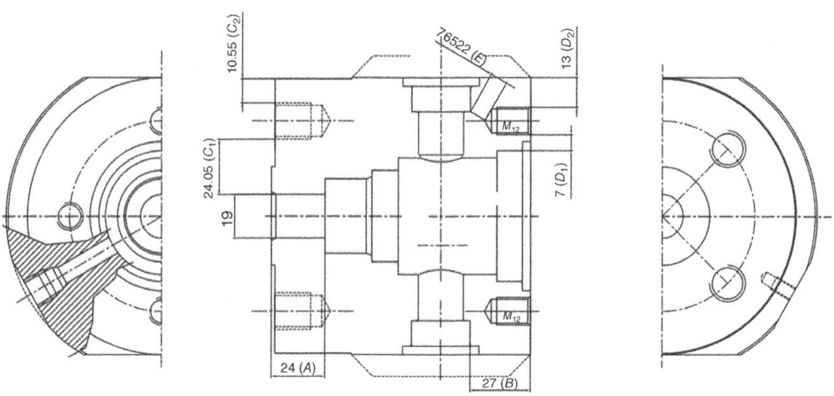

Figure 4.4 Compact wafer ball valve body machining drawing.

Answer

The internal diameter, (d), of a 3/4″ CL300 valve is 19.1 mm (refer to Table 4.1).

$$3 \leq d < 50 \rightarrow t_m(300) = 0.080d + 2.29 = 0.080 \times 19.1 + 2.29 = 3.82\ \text{mm}$$

The carbon steel valve has a CA equal to 3 mm in this case to mitigate the risk of corrosion and metal loss. The 3 mm CA should be added to the minimum valve thickness, as per Eq. (4.2).

$$t_{m,ca} = 3.82 + 3 = 6.82\ \text{mm}$$

t_m in sections (c), (d), (e), and (f) will be considered the minimum wall thicknesses, including the CA, which is equal to 6.82 mm. It is essential to check and ensure that each section of the valve body highlighted in Figure 4.4 has a higher thickness than the minimum thickness requirement and complies with ASME B16.34 sections (a)–(f).

The internal bolt holes are based on ASME B.1.1, which matches the flange bolting dimensional to satisfy the requirement in Section (a). Besides, the bolting depth in Figure 4.4 is 24 mm (equal to parameter A), and the size of the M12 bolt is equal to 12 mm, and they are threaded in the body. The bolting length is double the bolting size, which satisfies the requirement in Section (b). Section (b) states that "when threaded, the full thread engagement, excluding champers, should be provided to a depth not less than one nominal bolt diameter."

Referring to the minimum thickness in Section (c) of ASME B16.34, the minimum thickness values of the valve in Figure 4.4 are A = 24 mm, B = 27 mm, C_1 = 24.05 mm, C_2 = 10.55 mm, D_1 = 7 mm, and D_2 = 13 mm. All six parameters are more than the minimum allowable thickness equal to 6.82 mm, which qualifies the design regarding the minimum thickness and Section (c) of ASME B16.34. The provisions in paragraph (d) do not apply to this example because the picture does not include the valve body neck.

Referring to Section (e), the inner ligament (C_1 and D_1 in Figure 4.4) for holes parallel to the body run should not be less than $0.25t_m$, but in no case, less than 2.5 mm (0.1 in.). The sum of the inner and outer ligaments should not be less than t_m. The calculations here demonstrate that section (e) requirements are fulfilled.

Inside diameter (d) = 19 mm, CA = 3 mm

$$C_1 = 24.05, \quad C_2 = 10.55$$

$$t_{m(300)} = 0.08 \times 19 + 2.29 = 3.82\ \text{mm}$$

$$t_{m,ca} = 3.82 + 3 = 6.82\ \text{mm}, \quad 0.25 \times 6.82 = 1.705\ \text{mm}$$

$$C_1 = 24.05 > 0.25t_{m,ca} = 1.705 \text{ and } C_1 > 2.5\ \text{mm} \quad \text{Verified}$$

$$C_2 = 10.55 > 0.25t_{m,ca} = 1.705 \text{ and } C_2 > 2.5\ \text{mm} \quad \text{Verified}$$

$$C_1 + C_2 = 34.6 > 6.9\ \text{mm} \quad \text{Verified}$$

$$D_1 = 7\ \text{mm}, \quad D_2 = 13\ \text{mm}$$

$$D_1 = 7 > 0.25t_{\text{m,ca}} = 1.705\ \text{and}\ D_1 > 2.5\ \text{mm} \quad \text{Verified}$$

$$D_2 = 13 > 0.25t_{\text{m,ca}} = 1.705\ \text{and}\ D_2 > 2.5\ \text{mm} \quad \text{Verified}$$

$$D_1 + D_2 = 20\ \text{mm} > 6.9\ \text{mm} \quad \text{Verified}$$

Section (f) addresses ligaments within the minimum body wall between two adjacent holes within the minimum body valve. The ligament between two holes (E in Figure 4.4) should not be less than $0.25t_{\text{m,ca}}$ and greater than 0.25 mm.

$$t_{\text{m,ca}} = 6.82\ \text{mm} \rightarrow 0.25t_{\text{m,ca}} = 0.25 \times 6.82 = 1.705\ \text{mm}$$

$$E = 7.6522\ \text{mm} > 1.705\ \text{mm and}\ E = 7.6522\ \text{mm} > 0.25\ \text{mm}$$

Thus, the ligament thickness between the two holes is sufficient and complies with the requirements of Section (f) of ASME B16.34. Furthermore, all ASME B16.34 requirements regarding the minimum wall thickness are fulfilled in this case.

Example 4.5 A 4″ CL150 wafer butterfly valve in 25Cr super duplex has a drawing the same as the one illustrated in Figure 4.1. Calculate the minimum wall thickness values of "m" and "n." Consider $d_1' = 19$ mm and $d_2' = 30$ mm.

Answer

The first step is to obtain the internal diameter of the valve (d) based on the valve size and pressure class as per Table 4.1, which is 101.6 mm. Now, it is possible to obtain the minimum valve wall thickness (t_{m}) according to Table 4.3 as follows:

$$100 < d \leq 1300 \rightarrow t_{\text{m}}(150) = 0.0163d + 4.70 = 0.0163 \times 101.6 + 4.70$$
$$= 6.36\ \text{mm}$$

Since the valve body is in a super duplex, no CA is required. Refer to Section (c) of the valve body neck calculation $d/d_1' = (101.6/19) \geq 4$.

$$L = t_{\text{m}}\left(1 + 1.1\sqrt{\frac{d}{t_{\text{m}}}}\right) = 6.36\left(1 + 1.1\sqrt{\frac{101.6}{6.36}}\right) = 34.32\ \text{mm}$$

The minimum body neck thickness (m) associated with d_1' within the distance of L shall not be less than t'. t' is obtained from Table 4.3 according to the internal diameter (d_1') of the corresponding valve body neck.

$$d_1' = 19\,\text{mm}, P_c = 150, 3 \le d < 50 \rightarrow t_m'(150) = 0.064d_1' + 2.34 = 0.064 \times 19 + 2.34$$
$$= 3.556\,\text{mm} \rightarrow \text{Minimum } m = 3.556\,\text{mm}$$

The minimum body neck thickness (n) associated with d_2' beyond the distance of L should be obtained from Table 4.3 according to the value of d''.

$$150 \le \text{Pressure class} \le 2500 \rightarrow d'' = \frac{2d_2'}{3} = \frac{2 \times 30}{3} = 20$$

$$3 \le d'' < 50 \rightarrow t_m(150) = 0.064d'' + 2.34 = 0.064 \times 20 + 2.34 = 3.62$$
$$\rightarrow \text{Minimum } n = 3.62\text{ mm}$$

Questions and Answers

4.1 Which statement is correct regarding the pressure-temperature rating tables for valves?

A Pressure-temperature rating tables for valves are provided according to ASME B16.5.

B Pressure-temperature rating tables are identical for different materials such as carbon and stainless steel.

C The working pressure values are decreased by increasing the pressure class.

D The valve pressure class values are reduced by raising the temperature.

Answer

Option A is wrong because ASME B16.5 covers industrial flanges, and ASME B16.34 provides the pressure-temperature rating tables for valves. Option B is not correct either since the pressure-temperature rating tables are not identical for different materials. Option C is wrong as the working pressure values are increased by increasing the pressure class. Option D is the correct answer.

4.2 What is the thickness value for a 10″ ball valve in pressure class 600 calculated through a more conservative method?

A 17.04

B 19.32

C 20.34

D 21.38

Answer

The internal diameter of the valve as per Table 4.1 is 247.7 mm.

$$P_c = 600 \text{ and } d = 247.7\,\text{mm}(50 < d \leq 1300)$$
$$\rightarrow t_m(600) = 0.06777d + 2.54 = 19.32\,\text{mm}$$

Option B is the correct answer.

4.3 Find the correct statement about wall thickness calculations for the valves.

A The wall thickness of threaded-end and socket weld-ended valves can be calculated directly from equations in Table 4.3 according to ASME B16.34.

B The final and minimum thickness values of a valve made of carbon steel are equal.

C The thickness of a valve in the 600-pressure class is always more than a valve in the 300-pressure class.

D Inconel 625 weld overlay applied on the valve body in three-millimeter thickness to prevent corrosion shall not be counted in the valve wall thickness calculated through the method and equations provided earlier in this chapter.

Answer

Option A is wrong because threaded-end and socket weld-ended valves do not have standard pressure classes as per ASME B16.34. So it is impossible to calculate the valves' wall thickness with these two ending connections directly from the given equations provided in ASME B16.34. Linear interpolation as used in Example 4.2 to obtain the wall thickness of a socket-welded valve in an intermediate pressure class of 3000 is the correct method to calculate the wall thickness of threaded and socket-welded valves. Option B is incorrect because, typically, a 3 mm CA is required for carbon steel piping and valves based on the NORSOK L-001 standard. So the final thickness of a carbon steel valve is 3 mm thicker than the minimum wall thickness considering Eq. (4.2). Option C is not correct either since the thickness of a valve depends on the size in addition to the pressure class. For example, a 10″ CL300 valve is thicker than a 2″ CL600 valve based on the calculations as follows:

$$10''\text{CL300} \rightarrow \text{ID or } d = 254.00\,\text{mm(from Table 4.1)}\ 100 < d \leq 1300 \rightarrow t_m(300)$$
$$= 0.0334d + 4.32 = 12.80\,\text{mm}$$

$$2''\text{CL600} \rightarrow \text{ID or } d = 50.8\,\text{mm (from Table 4.1)}\ 50 < d \leq 1300 \rightarrow t_m(600)$$
$$= 0.06777d + 2.54 = 5.98\,\text{mm}$$

The correct answer is option D. For corrosion mitigation, three millimeters of Inconel 625 are overlaid on the valve's body made of carbon or low-alloy steel. However, the calculated wall thickness prevents mechanical failure of the valve body, so adding a 3 mm Inconel 625 overlay cannot be considered mechanical failure prevention, and should be included as part of the calculated wall thickness.

4.4 Which statements are not correct regarding the minimum valve wall thickness calculation according to the ASME B16.34 standard?

A The wall thickness of a ball and gate valve with an inside diameter of 370 mm and a pressure class of 600 is 27.6 mm.

B Ball valves with the same internal diameter and pressure class but are made of different materials have different body wall thicknesses according to the ASME B16.34 standard.

C A weakness of Table 4.3 extracted from ASME B16.34, which is used for wall thickness calculation, is that it does not cover all of the possible internal diameters of valves. Thus, it is impossible to have wall thickness values for internal diameters that are not covered.

D The methods and calculations provided in ASME B16.34 are always used for minimum valve wall thickness calculations.

Answer

Option A is correct. First, ASME B16.34 provides the same wall thicknesses for all valves, including ball and gate valves, as long as they have the same size and pressure class. Let's try calculating wall thickness according to the basic equations provided in mandatory appendix VI in ASME B16.34:

$$P_c = \text{Pressure class } 600 \text{ and } 50 < d \leq 1300$$
$$\rightarrow t_m\,(600) = 0.06777d + 2.54 = 27.6\ \text{mm}$$

Option B is not correct because the type of material and its mechanical strength do not change the wall thickness of the valve as per the ASME B16.34 standard. In fact, valves made of different materials will have the same wall thickness as long as they have the same internal diameter and pressure class. Option C is not completely correct: although not all of the internal diameters are not covered by ASME B16.34, it is possible to interpolate the valve wall thickness values associated with the missing internal diameters. Option D is incorrect because minimum wall thickness calculations for some valves like those installed on pipelines are typically calculated based on ASME Sec. VIII Div. 02 to save thickness and weight compared to ASME B16.34. Thus, except for option A, all other choices are wrong.

4.5 What is the calculated body thickness for a 30″ CL1500 valve installed on the oil export pipeline calculated according to the nonconservative method in ASME B 16.34? (*Note*: the minimum diameter of this valve is 671.26 mm.)

A 75.1 mm
B 90.62 mm
C 123.80 mm
D 141.9 mm

Answer

The valve has a specific (special) bore of 671.26, which does not follow the value given in Table 4.1. The wall thickness of the valve in this example is calculated as per Eq. (4.1):

$$t = 1.5 \times \left[\frac{P_c d}{2S_F - 1.2P_c}\right]$$

where:

t: Calculated thickness (mm);
P_c: Pressure class designation number (Class 1500, $P_c = 1500$);
d: Inside diameter of the valve = 671.26;
S_F: Stress-based constant equal to 7000.

$$t = 1.5 \times \left[\frac{1500 \times 671.26}{2 \times 7000 - 1.2 \times 1500}\right] = \frac{1{,}510{,}335}{12{,}200} = 123.80\ \text{mm}$$

Therefore, option C is the correct answer.

4.6 Figure 4.5 illustrates the machining drawing of a ball valve body closure that is the end section of the valve's body connected to the piping system. The ball valve is 1″ CL300 in carbon steel with three millimeters of CA. Which statements are correct?

A The minimum thickness of the valve is 4.32 mm.
B The illustrated body closure meets the minimum wall thickness requirements of ASME B16.34.
C The minimum thickness of the valve is 7.32 mm.
D The illustrated body closure does not meet the minimum wall thickness requirements of ASME B16.34.

Answer

The internal diameter, (d), of a 1″ CL300 valve is 25.4 mm (refer to Table 4.1).

$$3 \le d < 50 \rightarrow t_m(300) = 0.080d + 2.29 = 0.080 \times 25.4 + 2.29 = 4.32\ \text{mm}$$

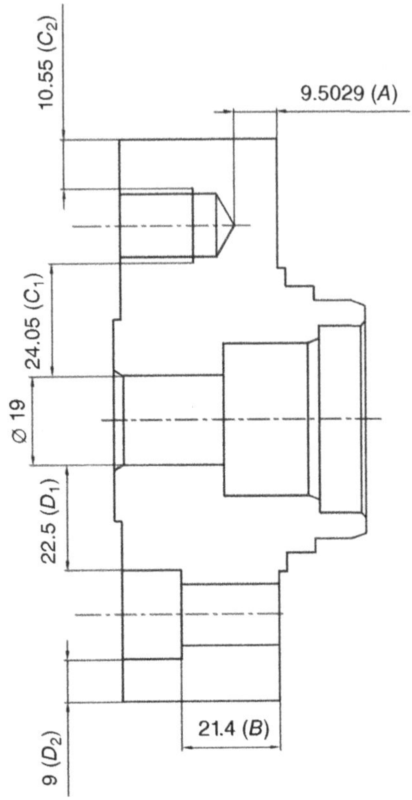

Figure 4.5 Ball valve body closure machining drawing.

The carbon steel valve has a CA equal to 3 mm in this case to mitigate the risk of corrosion and metal loss. The 3 mm CA should be added to the minimum valve thickness, as per Eq. (4.2).

$$t_{m,ca} = 4.32 + 3 = 7.32\ \text{mm}$$

Therefore, option A is incorrect and choice C is correct.

Referring to the minimum thickness in Section (c) of ASME B16.34, the minimum thickness values of the valve in Figure 4.5 are $A = 9.5$ mm, $B = 21.4$ mm, $C_1 = 24.05$ mm, $C_2 = 10.55$ mm, $D_1 = 22.5$ mm, and $D_2 = 9$ mm. All six parameters are more than the minimum allowable thickness equal to 7.32 mm, which qualifies the design regarding the minimum thickness and Section (c) of ASME B16.34. The provisions in paragraph (d) do not apply to this example because the figure does not include the valve body neck.

Referring to section (e), the inner ligament (C_1 and D_1 in Figure 4.5) for holes parallel to the body run should not be less than $0.25t_m$, but in no case

less than 2.5 mm (0.1 in.). The sum of the inner and outer ligaments should not be less than t_m. The calculations here demonstrate that section (e) requirements are fulfilled.
Inside diameter (d) = 19 mm, CA = 3 mm

$$C_1 = 24.05, \quad C_2 = 10.55$$

$$t_{m(300)} = 4.32 \text{ mm}$$

$$t_{m,ca} = 4.32 + 3 = 7.32 \text{ mm}, 0.25 \times 7.32 = 1.83 \text{ mm}$$

$$C_1 = 24.05 > 0.25t_{m,ca} = 1.83 \text{ and } C_1 > 2.5 \text{ mm} \quad \text{Verified}$$

$$C_2 = 10.55 > 0.25t_{m,ca} = 1.83 \text{ and } C_2 > 2.5 \text{ mm} \quad \text{Verified}$$

$$C_1 + C_2 = 34.6 > 7.32 \text{ mm} \quad \text{Verified}$$

$$D_1 = 22.5 \text{ mm}, \quad D_2 = 9 \text{ mm}$$

$$D_1 = 22.5 > 0.25t_{m,ca} = 1.83 \text{ and } D_1 > 2.5 \text{ mm} \quad \text{Verified}$$

$$D_2 = 9 > 0.25t_{m,ca} = 1.83 \text{ and } D_2 > 2.5 \text{ mm} \quad \text{Verified}$$

$$D_1 + D_2 = 31.5 \text{ mm} > 7.32 \text{ mm} \quad \text{Verified}$$

Section (f) addresses ligaments within the minimum body wall between two adjacent holes that do not exist in Figure 4.5. Therefore, the thickness values provided in the machining drawing are sufficient and fulfil ASME B16.34 requirements. To sum up, options B and C are correct.

Further Reading

American Petroleum Institute 609 (2004). *Butterfly Valves: Double Flanged Lug and Wafer*, 6e. Washington, DC: American Petroleum Institute (API).

American Society of Mechanical Engineers (ASME) (2012). *Design and Fabrication of Pressure Vessels. Boiler and Pressure Vessel Code. ASME Section VIII Div.02.* New York: American Society of Mechanical Engineers (ASME).

American Society of Mechanical Engineers (ASME) B16.34 (2017). *Valves–Flanged, Threaded, and Welding End.* New York: American Society of Mechanical Engineers (ASME).

American Society of Mechanical Engineers B1.1 (2019). *Unified Inch Screw Threads (UN, UNR, and UNJ Thread Forms).* New York: American Society of Mechanical Engineers (ASME).

American Society of Mechanical Engineers B31.3 (2020). *Process Piping*. New York: American Society of Mechanical Engineers (ASME).

Nayyar, M.L. (2000). *Piping Handbook*, 7e. New York: McGraw-Hill Education.

Nesbitt, B. (2007). *Handbook of Valves and Actuators: Valves Manual International*, 1e. Oxford: Elsevier.

NORSOK L-001 (2017). *Piping and Valves*, 4e. Lysaker: NORSOK.

NORSOK M-001 (2004). *Materials Selection*, 4e. Lysaker: NORSOK.

Parisher, R.A. and Rhea, R.A. (2002). *Pipe Drafting and Design*, 2e. Austin, TX: Gulf Professional Publishing.

Skousen, P.L. (2011). *Valve Handbook*, 3e. New York: McGraw-Hill.

Smit, P. and Zappe, R.W. (2004). *Valve Selection Handbook*, 5e. New York: Elsevier.

Sotoodeh, K. (2019). Wafer design valves verification based on ASME B16.34. *Springer Nature Applied Science* 1: 1476. https://doi.org/10.1007/s42452-019-1344-0.

Sotoodeh, K. (2022). *Cryogenic Valves for Liquified Natural Gas Plants*, 1e. Austin, TX: Elsevier (Gulf Professional Publishing).

5

Material and Corrosion

5.1 Introduction

Valves are essential components of piping systems in the oil and gas industry. An oil and gas plant's efficiency, safety, and reliability largely depend on fluid handling and transportation through the piping system, including industrial valves. Material failure because of corrosion is known as one of the significant causes of valve failure. Some of the negative impacts of valve failure in the oil and gas industry, especially the offshore sector, can be summarized as loss of assets and production and Safety and Environmental issues (HSE), including problems like hydrocarbon (oil and gas) spillage and environmental pollution, loss of human life in some cases, jeopardizing safety and reliability, etc. Many valves fail every year due to poor material selection and corrosion. So, proper material selection and corrosion prevention are important aspects of the valve design.

Corrosion occurs when a material deteriorates due to its interaction with its surrounding environment. Corrosive oil and gas in upstream units such as wellhead and separation as well as downstream plants such as refineries contain a high number of undesirable corrosive byproducts such as carbon dioxide (CO_2) and hydrogen sulfide (H_2S). The most critical and complicated material calculation for industrial valves is corrosion allowance. It is important to know that corrosion allowance is selected to mitigate the corrosion produced by carbon dioxide (CO_2). Carbon dioxide is an odorless, nonflammable, and nontoxic substance, unlike hydrogen sulfide, and colorless like hydrogen sulfide. The second material equation covered in this chapter addresses the pitting corrosion resistance of materials. As a result of the presence of chloride in fluids such as seawater, pitting corrosion occurs. The third aspect to discuss is the carbon equivalent, which is relevant since it influences the weldability of carbon steels. The last section of this chapter discusses hydrogen-induced stress cracking (HISC) corrosion and related equations.

Industrial Valves: Calculations for Design, Manufacturing, Operation, and Safety Decisions, First Edition. Karan Sotoodeh.

5.2 Carbon Dioxide Corrosion

5.2.1 Corrosion Mechanism

Sweet corrosion due to carbon dioxide is a serious problem in the oil and gas industry as it causes metal loss and localized corrosion. Therefore, accurate prediction and modeling of carbon dioxide corrosion is an essential task that should be made during the basic design phase of oil and gas projects. Thus, one of the main aims of this chapter is to provide a practical model to predict and calculate the carbon dioxide corrosion rate and corrosion allowance selection. To calculate the corrosion allowance, process parameters such as operating pressure and temperature, fluid pH, the design life of the pipe, and information about glycol, corrosion inhibitor, etc., should be considered. The first important point is to understand the mechanism of carbon dioxide corrosion. The first assumption in the proposed model and other models developed to address carbon dioxide is that CO_2 is considered a cause of uniform corrosion. Uniform corrosion is defined as a type of corrosion that proceeds at the same rate over the whole metal surface. The second important point about CO_2 corrosion is that it is corrosive in the presence of water. Carbon dioxide can form carbonic acid in a chemical reaction with water as follows:

$$CO_2 + H_2O \rightarrow H_2CO_3$$

Therefore, the main conclusion is that there will be no corrosion by carbon dioxide if there is no water. The third key consideration about this type of corrosion is that it causes metal loss from the internal surface of the piping and valves. Figure 5.1 shows carbon dioxide corrosion on a carbon steel pipe. The following paragraph explains the strategies used to prevent this type of corrosion.

Figure 5.1 Carbon dioxide (sweet) corrosion.

5.2.2 Corrosion Mitigation

Sweet corrosion occurrence has been widely experienced in the oil and gas industry. Three main strategies are commonly used in the oil and gas industry to combat and prevent the carbon dioxide or sweet corrosion. The first strategy is to increase the wall thickness of the piping and valves as a form of corrosion allowance. CO_2 corrosion, also known as sweet corrosion, causes metal loss in noncorrosion-resistant materials such as carbon steel. So, adding an extra thickness to the piping and valves – a practice called "corrosion allowance" – is proposed for carbon and low alloy steels. It is typical and logical to add a 1, 1.5, or 3 mm corrosion allowance to the piping and valves made in noncorrosion-resistant alloys like carbon and low alloy steels. However, adding a 6 mm corrosion allowance to the pipe will make it very thick, heavy, and expensive. A standard amount of 3 mm corrosion allowance is proposed to be added to piping and valves in carbon steel material according to Norsok, the Norwegian petroleum standard. Adding more corrosion allowance such as 6 mm is not proposed since it makes the piping and valves thicker and heavier. Many different models and software may be used to calculate the CO_2 corrosion rate, all of which are based on the "Dewaard and Milliams" model. Additionally, the Iranian Petroleum Standard (IPS) as well as Norsok standard M-506 provide a calculative model for CO_2 corrosion prediction.

The alternative solution and the second strategy to combat sweet corrosion is using corrosion-resistant alloys (CRAs) such as stainless steel or nickel alloys like Inconel 625 instead of carbon steel. This practice is common in the Norwegian offshore industry. Nickel alloys such as Inconel 625 can be used as a form of solid pipe or cladding (weld overlay). A cladding carbon steel pipeline with Inconel 625 is proposed for both hydrogen sulfide and carbon dioxide-containing oil and gas services to mitigate both types of corrosion. The third sweet corrosion combat strategy is injecting the corrosion inhibitor or anticorrosive, a chemical compound inside the oil and gas services. Different corrosion inhibitors may be injected into the fluid with different chemistry, physical properties, and solubility. Corrosion inhibitor protection is usually from one or more of the following mechanisms: first, the inhibitor molecule is absorbed by the metal surface; second, the inhibitor makes a protective film; and the last is that the inhibitor reacts with corrosive substances in the piping system. The selection of the best and most effective corrosion inhibitor is performed as per laboratory results, considering specific parameters such as type of protected material, application method, efficiency, cost, HSE considerations, and product availability. Like corrosion inhibitors, glycol or methanol can be injected into the piping system to absorb the water and reduce the corrosion rate. The following section is about carbon dioxide corrosion rate as well as corrosion allowance calculations.

5.2.3 Corrosion Rate Calculation

Researchers and oil companies have developed several corrosion estimation models. Some of the known and commonly used models in the oil and gas industry are listed as follows:

- DeWaard and Milliams Model (DWM);
- Predict Model (Intercorr);
- Norsok (M-506) (Statoil, Saga Petroleum, Hydro);
- LIPUCOR (Total);
- Hydrocorr (Shell);
- Casandra (British Petroleum);
- CorPos (Corr Ocean);
- Cormed (Elf Aquitaine).

There is no totally valid model for carbon dioxide corrosion estimation due to the complexity of this type of corrosion. The suggested model in this book uses various techniques listed here to deliver a practical way of predicting carbon dioxide corrosion in pipelines and piping systems, including industrial valves, where carbon steel or low-alloy steel material is used. The essential consideration is that the proposed model can only be used for hydrocarbon services such as oil and gas. Therefore, this section provides a model for corrosion rate calculation that is not applicable for piping and valves in non-hydrocarbon fluids such as seawater, drinking water, air, hydraulic oil, and caustic services. After calculating the CO_2 corrosion rate, it is possible to select the most suitable material and the corrosion allowance.

5.2.3.1 Basic CO_2 Corrosion Rate

The evaluation of the basic corrosion rate takes into account the effects of operating temperature and pressure, as well as the mole fraction of CO_2. Basic carbon dioxide corrosion rate prediction is calculated through Eq. (5.1).

Basic CO_2 Corrosion Rate Calculation

$$\text{Log}(\text{CR})_{\text{base}} = 5.8 - \frac{1710}{T} + 0.67\log(P_{\text{CO}_2}) \tag{5.1}$$

where:

CR_{base}: Base corrosion rate based on DWM (mm/year);
T: Operating temperature (K);
P_{CO_2}: Partial pressure of CO_2 (bar).

5.2.3.1.1 Effect of Carbon Dioxide Dry carbon dioxide is noncorrosive, whereas it is corrosive when dissolved in water. Thus, water presence is necessary for sweet corrosion to happen, as carbonic acid forms and decreases the pH of the water. The final pH of the fluid service depends on the temperature and carbon dioxide partial pressure.

5.2.3.1.2 Effect of Temperature According to Figure 5.1, some laboratory studies show that the CO_2 corrosion rate increases by increasing the temperature up to almost 70 °C, probably due to an increase in mass transfer rate. Above that temperature, the corrosion rate starts to decrease, associated with a corrosion-protective layer that is explained more in detail later in this chapter (Figure 5.2).

5.2.3.1.3 Effect of Pressure The partial pressure of CO_2 depends on the mole fraction of CO_2 and is calculated according to Eq. (5.2):

CO_2 Partial Pressure Calculation

$$P_{CO_2} = P_{\text{operation}} \times x_{CO_2} \tag{5.2}$$

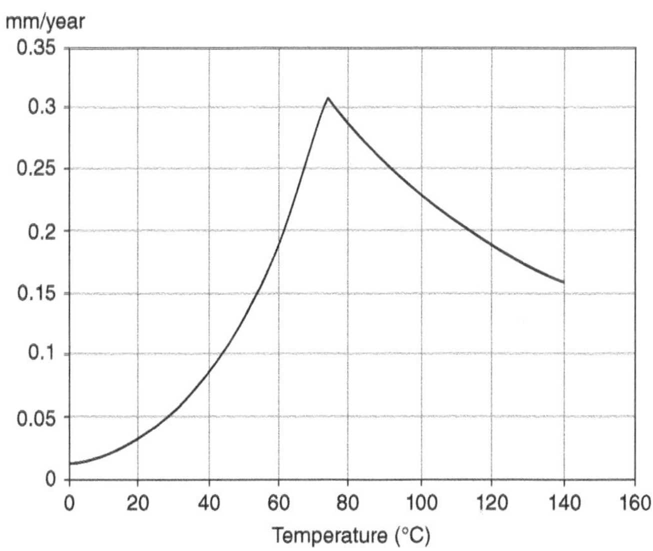

Figure 5.2 Effect of temperature on carbon dioxide corrosion rate based on DWM.

where:

P_{CO_2}: Partial pressure of CO_2 (bar);
$P_{operation}$: Operation pressure (bar);
X_{CO_2}: Mole fraction of CO_2.

The mole fraction represents the ratio of the number of molecules of a particular component to the total number of molecules.

Example 5.1 The operating pressure of a 10″ ball valve used for the stop/start of the gas is 25 bar, and the carbon dioxide mole fraction is 0.008. The valve-operating temperature is 120 °C. Calculate the basic carbon dioxide corrosion rate in millimeters per year.

Answer
The partial pressure of the carbon dioxide is calculated as follows:

$$P_{CO_2} = P_{operation} \times x_{CO_2} = 25 \times 0.008 = 0.2\ \text{bar}$$

120 °C is equal to 393 K.

$$\rightarrow \text{Log(CR)}_{base} = 5.8 - \frac{1710}{T} + 0.67\log(P_{CO_2}) = 5.8 - \frac{1710}{393} + 0.67\log 0.2$$
$$= 5.8 - 4.35 - 0.468 = 0.98 \rightarrow \text{(CR)}_{base} = 9.55\ \text{mm/year}$$

DeWaard and Milliams generated a nomogram (see Figure 5.3) in 1975 (DWM) that can be used to estimate the basic carbon dioxide corrosion rate as an alternative method to Eq. (5.1). Connecting the operating temperature of 120 °C on the left side to the CO_2 partial pressure on the right equal to 0.2 passes through the corrosion rate of 10 mm/year, which is very close to the result provided by the calculations above. The basic corrosion rate of 9.55 mm/year is a high number that should be reduced to a realistic corrosion rate by corrective factors, as explained in the next section.

5.2.3.2 Corrective CO_2 Corrosion Rate

Additional corrective process parameters shall modify the initially calculated corrosion rate to deliver more accurate estimations of the corrosion rate. This section suggests effective process parameters to adjust the basic corrosion rate as follows:

- *F* (System): This parameter is related to the nonideality of hydrocarbon gas on the calculated corrosion rate that is designated as the "fugacity coefficient." Nonideality of hydrocarbon gas is explained more in detail further in this chapter.

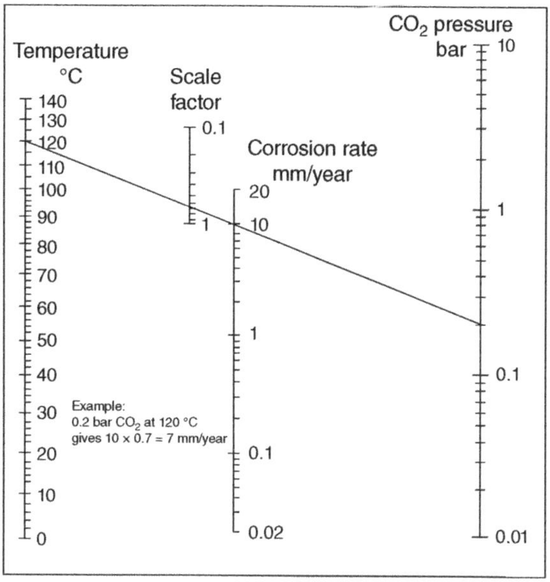

Figure 5.3 Carbon dioxide basic corrosion rate prediction nomogram as per the DWM model.

- *F* (Glycol): This parameter considers the effect of glycol injection (typically monoethylene glycol or MEG) on the corrosion rate calculations. Glycol will reduce the corrosivity of the hydrocarbon by absorbing water. Besides, glycol may be injected into the piping system to prevent hydrate formation. When natural gas is transported with water in low-temperature and high-pressure conditions, ice-like crystals mainly formed from water can cause flow assurance problems called hydrate.
- *F* (Scaling): This parameter evaluates the chance of protective anticorrosion film of $FeCO_3$ production, which is very likely at high temperatures (above 70 °C) when the pH is high.
- *F* (Water cut): The presence of free water is absolutely essential for corrosion to occur. Higher corrosion rates occur when the water amount is high.
- *F* (Inhibitor): This parameter demonstrates the mitigation impact of any corrosion inhibitor injections on the initially calculated corrosion rate. For many years, it has been common practice in the oil and gas industry to inject inhibitors into carbon dioxide-containing fluids carried in the carbon steel piping and valves.
- *F* (Condensate): It demonstrates the effect of water condensation formation for nominally dry gas lines in which the corrosion rate is reduced substantially.
- *F* (pH): It is generally accepted that pH has a critical impact on the corrosion rate in such a way that low pH values of less than five are considered acidic and produce an acidic environment that largely extends the corrosion rate.

Thus, the corrective corrosion rate (mm/year) is calculated according to Eq. (5.3).

CO_2 Corrective Corrosion Calculation

$$(CR)_{correct.} = (CR)_{base} \times F(\text{System}) \times F(\text{Glycol}) \times F(\text{Scale}) \\ \times F(\text{Water cut}) \times F(\text{Inhibitor}) \times F(\text{Condensate}) \times F(\text{PH}) \tag{5.3}$$

5.2.3.2.1 Effect of System F (System) is the outcome of the wet gas nonideality due to both pressure and temperature conditions that can affect the CO_2 activity in piping and valves in a negative manner. The partial pressure of CO_2 is assumed to be 100% efficient (completely ideal) in the base corrosion rate equation of DWM, which is not realistic in practice. Thus, it is a common practice to replace the CO_2 partial pressure by the CO_2 fugacity or activity of $F(CO_2)$ which is calculated in Eq. (5.4):

CO_2 Fugacity or Activity Calculation

$$F_{CO_2} = a \times P_{CO_2} \tag{5.4}$$

where:

F_{CO_2}: CO_2 fugacity or activity;
α: CO_2 activity coefficient, known as F (system); calculated based on Eq. (5.5);
P_{CO_2}: Partial pressure of CO_2 (bar).

CO_2 Activity Coefficient or F (System) Calculation

$$\text{Log}(a) = \left(0.0031 - \frac{1.4}{T}\right) \times P \tag{5.5}$$

where:

a: CO_2 activity coefficient or F (system);
T: Operating temperature (K);
P: Operating temperature (bar).

It should be noted that if the calculated value of F (system) or α is more than one, the CO_2 is assumed to be 100% active and ideal, and, therefore, F (system) = 1.

Example 5.2 Calculate the activity of carbon dioxide in the piping system in the previous example where the basic corrosion rate was calculated to be equal to 9.55 mm/year.

Answer

$$\text{Log}(a) = \left(0.0031 - \frac{1.4}{T}\right) \times P$$

$$\rightarrow \text{Log}(a) = \left(0.0031 - \frac{1.4}{120+273}\right) \times 25 \rightarrow \log(a) = -0.01156 \rightarrow a = F(\text{System})$$
$$= 10^{-0.01156} = 0.97$$

$$F_{CO_2} = a \times P_{CO_2} = 0.97 \times 0.2\ \text{bar} = 0.194\ \text{bar}$$

5.2.3.2.2 ***Effect of Glycol (Methanol)*** Glycol or methanol has a very effective inhibitive effect; it functions by absorbing and reducing the water, which results in a lower corrosion rate and reduced hydrate formation. The reduction of the corrosion rate due to the presence of glycol, F (glycol), is calculated based on Eq. (5.6), as follows:

Effect of Glycol Injection on Corrosion Rate

$$\text{Log}(F(g)) = 1.6 \times \log(W(g)) - 3.2 \tag{5.6}$$

where:

F(g): Glycol reduction effect on the corrosion rate;
W(g): Water concentration percentage in the water/glycol mixture.

Note: F (glycol) = 1 for lines without any glycol injection.

Example 5.3 Glycol is injected into two piping systems carrying corrosive CO_2 containing gas mixed with water. The water concentration in the water and glycol mixture is 40% for the first case and 80% for the second case. Calculate the glycol reduction effect on each case's carbon dioxide corrosion rate.

Answer

$$W(g)_1 = 40\% \rightarrow \log(F(g)_1) = 1.6 \times \log(W(g)_1) - 3.2 = 1.6 \times \log 0.4 - 3.2$$
$$= -3.84 \rightarrow F(g)_1 = 1.445e^{-4}$$

$$W(g)_2 = 80\% \rightarrow \log(F(g)_2) = 1.6 \times \log(W(g)_2) - 3.2 = 1.6 \times \log 0.8 - 3.2$$
$$= -3.36 \rightarrow F(g)_2 = 4.365e^{-4}$$

$$\frac{F(g)_2}{F(g)_1} = \frac{4.365e^{-4}}{1.445e^{-4}} = 3.02$$

Thus, the CO_2 corrosion reduction effect is approximately three times higher in the second case, where the percentage of water in the glycol and water mixture is 80%. It is noticeable that the concentration of water absorbed by the glycol in the second case is double that in the first case.

5.2.3.2.3 Effect of Scaling The scaling effect is connected to producing a protecting layer on the metal surface at temperatures higher than 70 °C. High temperatures will activate the CO_2 and lead to its reaction to the iron, which will produce a protective layer of ferrous carbonate ($FeCO_3$) on the metal surface. Scaling formation reduces the corrosion rate seriously by the corrective factor of F (scaling) based on Eq. (5.7), as follows:

Effect of Scaling Calculation

$$\text{Log}(F(S)) = \frac{2400}{T} - 0.6 \times \log(F_{CO_2}) - 6.7 \tag{5.7}$$

where:

$F(S)$: Scaling effect coefficient;
F_{CO_2}: CO_2 fugacity/activity;
T: Operating temperature (K).

It should be noted that scaling effect exists if ($F(S)$) is less than 1; otherwise, there is no possibility of scale formation and $F(S) = 1$ is utilized for corrosion rate calculations.

Example 5.4 An 8″ valve handles the corrosive crude oil service in 100 bar operating pressure and 120 °C operating temperature. Considering the carbon dioxide mole fraction is 0.002, can scaling be produced in the valve?

Answer

The partial pressure of the carbon dioxide is calculated as follows:

$$P_{CO_2} = P_{\text{operation}} \times x_{CO_2} = 100 \times 0.002 = 0.2\ \text{bar}$$

$$\text{Log}(a) = \left(0.0031 - \frac{1.4}{T}\right) \times P$$

$$\rightarrow \text{Log}(a) = \left(0.0031 - \frac{1.4}{120 + 273}\right) \times 100 \rightarrow \log(a) = -0.046$$

$$\rightarrow a = 10^{-0.046} = 0.8994$$

$$\rightarrow F(\text{System}) = 0.9$$

$$\rightarrow F_{CO_2} = a \times P_{CO_2} = 0.9 \times 0.2\,\text{bar} = 0.18\,\text{bar}$$

$$\text{Log}(F(\text{Scale})) = \frac{2400}{T} - 0.6 \times \log(F_{CO_2}) - 6.7 = \frac{2400}{393} - 0.6 \times \log(0.18) - 6.7$$
$$= 6.1 + 0.447 - 6.7 = -0.153 \rightarrow F(\text{Scale}) = 10^{-0.153} = 0.70$$

Since the effect of scaling is equal to 0.7, which is less than one, the scaling can be produced in the pipes and valves given in Example 5.4.

5.2.3.2.4 Effect of Water Cut Dry hydrocarbons without water are noncorrosive because the presence of water in oil and gas is vital for the occurrence of corrosion. The presence of oil is typically deemed beneficial in corrosion prevention due to the formation of a protective oil film. But the stability of the protective layer depends on the water quantity (water cut). The oil-forming film on a metal surface can be stable at up to a 20–40% water cut, but higher water quantities eliminate its protective effect. However, it is not actually possible to calculate and define the minimum volume of water required for corrosion. Since the calculation of the water cut is not possible, the effect of the water cut should be assumed to be equal to 1 in. the calculations unless, otherwise, the water cut value is known.

5.2.3.2.5 Effect of Inhibitor For many years, it has been commonplace in the oil and gas industry to inject various chemicals as corrosion inhibitors in pipes and valves handling corrosive services to lower the possibility of corrosion by creating a strong film on the metal surface. It is generally accepted to define the inhibitor's capability to reduce the corrosion attack by a parameter known as "inhibitor efficiency," calculated as per Eq. (5.8):

Corrosion Inhibitor Efficiency

$$\text{Inhibitor efficiency} = 1 - \frac{\text{CR with inhibitor}}{\text{CR without inhibitor}} \tag{5.8}$$

Inhibitor efficiency is usually described as a percentage. For example, if the corrosion rate is 3 mm/year without corrosion inhibitor injection and 0.3 mm regarding the inhibition outcome, the calculated corrosion inhibitor efficiency is 90%. Inhibition efficiency above 85% is generally considered ideal; it means an 85% corrosion rate reduction compared to the corrosion rate obtained from Dewaard and Milliams nomogram based on Figure 5.3. F (inhibitor), which is equal to the corrosion reduction factor due to inhibitor injection according to the DWM, can be obtained by Eq. (5.9) as follows:

Effect of Inhibitor Injection

$$(F(i)) = 1 - \text{Inhibitor efficiency} \tag{5.9}$$

It is known to consider 90% efficiency or availability, F (inhibitor) = 0.1 for the corrosion inhibitor during the engineering design phase. However, corrosion availability shall be regarded as 85% for condensate, 90% for gas lines, and 95% for oil streams.

Example 5.5 The valve and material engineers have decided to inject a corrosion inhibitor inside the piping systems with a corrosion rate of 1 mm/year to decrease the corrosion rate. The corrosion rate is reduced to 0.05 mm/year due to inhibitor injection. Calculate the inhibitor efficiency and the effect of inhibitor injection.

Answer

$$\text{Inhibitor efficiency} = 1 - \frac{\text{CR with inhibitor}}{\text{CR without inhibitor}} = 1 - \frac{0.05\ \text{mm/year}}{1\ \text{mm/year}} = 95\%$$

$$\text{Effect of inhibitor injection} = 1 - 0.95 = 0.05$$

5.2.3.2.6 Effect of Condensate Condensate development is taken into account for gas services with humidity, where cooling below the gas dew point results in condensate droplets formation. This situation will likely produce a protective film on the pipe and substantially lower the corrosion rate. If condensate is made, its effect will be 0.1; otherwise, F (condensate) equals 1.

5.2.3.2.7 Effect of pH The pH has an essential effect on the corrosion rate because an acidic environment with a low pH below five essentially boosts the corrosion rate. The pH can be calculated as a function of different parameters like the carbon dioxide partial pressure, operation temperature, bicarbonate, and other dissolved gases and chemicals. To calculate the effect of pH on the corrosion rate, first, pH_{sat} is calculated using Eqs. (5.10) and (5.11).

pH_{sat} Calculation

$$pH_{Sat1} = 1.36 + \frac{1307}{T} - 0.17 \times \log(F(CO_2)) \tag{5.10}$$

$$pH_{Sat2} = 5.4 - 0.66 \times \log(F(CO_2)) \tag{5.11}$$

The lowest value obtained from Eqs. (5.10) and (5.11) is selected for the following step, in which the effect of pH is calculated using Eq. (5.12) or (5.13) depending on the relationship between pH_{sat} and the effect of pH:

pH Effect Calculation

$$\text{If } pH_{Sat} > pH_{actual}$$
$$\rightarrow \text{Log}(F(pH)) = 0.32\,(pH_{Sat} - pH_{actual}) \tag{5.12}$$

$$\text{If } pH_{Sat} \leq pH_{actual}$$
$$\rightarrow \text{Log}(F(pH)) = 0.13\,(pH_{Sat} - pH_{actual})^{1.6} \tag{5.13}$$

The effect of pH shall be considered equal to one in two cases as follows:

- If the pH of the fluid is unknown;
- Scaling is produced in piping and valves when the scaling effect is less than one.

5.2.3.3 Final CO_2 Corrosion Rate

Equation (5.14) is used to calculate the final CO_2 corrosion rate by multiplying the corrected corrosion rate in millimeters per year by the design life of the plant. The design life of a valve or plant could be 10, 15, 20, 25, or even 30 years.

Final Corrosion Rate Calculation

$$CR_{Final} = \text{Design life (year)} \times CR_{corrective}(\text{mm/year}) \tag{5.14}$$

The final corrosion rate should be rounded up to 1, 3, or 6 mm to obtain the corrosion allowance as per Table 5.1:

Table 5.1 Final corrosion rates.

Corrosion severity level	Average of corrosion rate (mm/year)	Final selected corrosion allowance
Mild	<0.04	1.0
Medium	0.04–0.12	3.0
Severe	0.12–0.24	6.0

Note 1
The design life of the plant and valve is considered 25 years in Table 5.1. The average corrosion rates differ for a design life other than 25 years.

Note 2
For offshore piping and valves on the platforms and ships, the NORSOK standard proposes standardizing the corrosion allowance for carbon steel piping and valves to 3 mm. A corrosion allowance of 6 mm on carbon steel should be avoided as it would increase the piping and valve wall thickness and make them heavy. However, it is practical to use a 6 mm corrosion allowance for piping and valves in onshore units such as refineries and petrochemical plants.

Note 3
For onshore piping and valves in the refineries and chemical plants, if the calculated final corrosion rate exceeds 6 mm, various solutions are recommended; upgrading the material to a CRA such as stainless steel, applying Inconel 625 on the internal surface of carbon steel in contact with the fluid service, or injection of corrosion inhibitor.

Note 4
No corrosion allowance is required for CRA materials such as austenitic stainless steels (e.g. 304, 316), duplex stainless steels, and nickel alloys.

Note 5
It is possible for a piping or valve specification to call for minimum corrosion allowances of 1.5 or 1 mm in onshore carbon steel applications such as steam and water that do not contain hydrocarbons.

5.3 Pitting Corrosion

Figure 5.3 illustrates a type of local corrosion known as pitting corrosion that creates holes inside exposed facilities and susceptible materials. In valves and piping systems, pitting corrosion may occur internally as a result of fluid service and

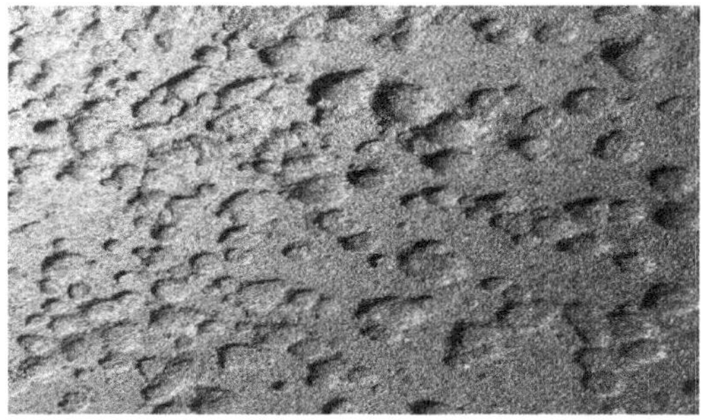

Figure 5.4 Pitting corrosion in carbon steel. *Source:* MDPI.

externally as a result of exposure to the corrosive offshore environment (marine). Internal pitting corrosion is mainly caused by the presence of chloride in corrosive seawater as well as oil and gas services. Chloride can impair the passive layer on sensitive materials, such as austenitic stainless steel, where pitting corrosion initiates (Figure 5.4).

Two essential parameters contribute to the severity of pitting corrosion: the metallurgy of the metallic valve material and the severity of operating conditions. It is common to use various approaches in the industry to prevent pitting corrosion, including selecting materials with high pitting resistance – typically with a pitting resistance equivalent number (PREN) higher than 40 – pitting corrosion test, control of pitting-effective process parameters, such as pH, temperature, and chloride; and lastly, cathodic protection. PREN depends on the three main elements in a material and is calculated according to Eq. (5.15).

Pitting Resistance Equivalent Number (PREN) Calculation

$$\text{PREN} = \text{Cr} + 3.3\%\text{Mo} + 16\text{N} \tag{5.15}$$

where:

Cr: Chromium;
Mo: Molybdenum;
N: Nitrogen.

Example 5.6 Calculate the maximum PREN values for the materials listed in Table 5.2 and identify those with high PREN values.

Table 5.2 Chemical compositions of some of the most commonly used stainless steels and nickel alloys in the oil and gas industry.

	C	Mn	P	S	Cr	Ni	Mo	N
Super austenitic								
6MO (UNS S31254)	Max. 0.02	Max.1	0.03	0.01	20	18	6	0.21
Nickel alloys								
Inconel 600 (UNS N06600)	0.15	1	—	0.015	15.5	72	—	—
Inconel 625 (UNS N06625)	0.1	0.5	0.015	0.01	21.5	58	8.5	—
Incoloy 800 (UNS N08800)	0.10	1.5	—	0.015	21	32.5	—	—
Incoloy 825 (UNS N08825)	0.05	1	—	0.03	21	42	—	—
Duplex								
22Cr duplex (UNS S31803)	Max. 0.03	Max.2	Max. 0.03	Max. 0.020	25	5.5	3	0.08–0.20
25Cr super duplex (UNS S32750)	Max. 0.03	Max. 1.20	Max. 0.035	Max. 0.020	25	7	4	0.24–0.32
25Cr super duplex (UNS S32760)	Max. 0.03	Max. 1	Max. 0.03	Max. 0.010	25	7	3.5	0.24–0.32
Austenitic								
304 (UNS S30400)	0.08	0–2	Max. 0.045	Max. 0.03	18–20	8–11		Max. 0.1
304L (UNS S30403)	Max. 0.03	0–2	Max. 0.045	Max. 0.03	18–20	8–13		Max. 0.1
316 (UNS S31600)	0.08	0–2	Max. 0.045	Max. 0.03	16–18	10–14	2–3	Max. 0.1
316L (UNS S31603)	Max. 0.03	0–2	Max. 0.045	Max. 0.03	16–18	10–14	2–3	Max. 0.1
321 (UNS S32100)	0.08	0–2	Max. 0.045	Max. 0.03	17–19	9–12		Max. 0.1
347 (UNS S34700)	0.08	0–2	Max. 0.045	Max. 0.03	17–20	9–13		Max. 0.1

Answer

$PREN_{6MO} = Cr + 3.3\%Mo + 16N = 20 + 3.3 \times 6 + 16 \times 0.21 = 43.16 > 40$
$\rightarrow$ 6MO is a pitting resistant material

$PREN_{\text{Inconel 600}} = 15.5 < 40 \rightarrow$ Inconel 600 is not a pitting resistant material

$PREN_{\text{Inconel 625}} = 21.5 + 3.3 \times 8.5 = 49.55 > 40$
$\rightarrow$ Inconel 625 is a pitting resistant material

$PREN_{\text{Incoloy 800}} = 21 < 40 \rightarrow$ Incoloy 800 does not resist pitting

$PREN_{\text{Incoloy 825}} = 21 < 40 \rightarrow$ Incoloy 825 is not a pitting resistant material

$PREN_{\text{22Cr duplex}} = 25 + 3.3 \times 3 + 16 \times 0.20 = 38.1 < 40$
$\rightarrow$ 22Cr duplex is not a pitting resistant material

$PREN_{\text{25Cr super duplex UNS S32750}} = 25 + 3.3 \times 4 + 16 \times 0.32 = 43.32 > 40$
$\rightarrow$ 25Cr super duplexUNS S32750 is a pitting resistant material

$PREN_{\text{25Cr super duplex UNS S32760}} = 25 + 3.3 \times 3.5 + 16 \times 0.32 = 41.67 > 40$
$\rightarrow$ 25Cr super duplex UNS S32760 is a pitting resistant material

$PREN_{SS304} = 20 + 16 \times 0.1 = 21.6 < 40$
$\rightarrow$ SS304 is not a pitting resistant material

$PREN_{SS304L} = 20 + 16 \times 0.1 = 21.6 < 40$
$\rightarrow$ SS304 is not a pitting resistant material

$PREN_{SS316} = 18 + 16 \times 0.1 = 19.6 < 40$
$\rightarrow$ SS316 is not a pitting resistant material

$PREN_{SS316} = 18 + 16 \times 0.1 = 19.6 < 40$
$\rightarrow$ SS316 is not a pitting resistant material

$PREN_{SS321} = 19 + 16 \times 0.1 = 20.6 < 40$
$\rightarrow$ SS321 is not a pitting resistant material

$PREN_{SS347} = 20 + 16 \times 0.1 = 21.6 < 40$
$\rightarrow$ SS347 is not a pitting resistant material

5.4 Carbon Equivalent

One of the most significant considerations when using carbon steel piping and welded valves is their ability to be welded. Among other things, carbon and certain other elements in carbon steel increase its hardness and can make welding difficult or impossible. Additionally, a high hardness at the weld joints makes it more likely that corrosion will occur, such as a hydrogen sulfide attack. In fact, reducing the hardness of materials is a key factor in reducing the risk of sour corrosion. For these reasons, it is recommended that the carbon content and equivalent carbon content of carbon steel piping and welded valves be limited. The carbon equivalent equation is a method of converting alloy elements' percentages to an equivalent carbon percentage for carbon steel. Equations (5.16) and (5.17) demonstrate how the carbon equivalent (C_E) can be calculated as described earlier. Pipes and welded valves made from carbon steel should ideally have a carbon equivalent value of less than 0.40 in order to ensure acceptable weldability and corrosion resistance.

Carbon Equivalent Calculation

$$C_E = \mathrm{C} + \frac{\mathrm{Mn}}{6} + \frac{\mathrm{Cr} + \mathrm{Mo} + \mathrm{V}}{5} + \frac{\mathrm{Ni} + \mathrm{Cu}}{15} < 0.40 \tag{5.16}$$

$$C_E = \mathrm{C} + \frac{\mathrm{Mn}}{6} \tag{5.17}$$

where:

C: Carbon;
Mn: Manganese;
Cr: Chromium;
Mo: Molybdenum;
Ni: Nickel;
Cu: Copper.

If values of other alloys (other than carbon and manganese) are not known, Eq. (5.17) can be used to calculate the carbon equivalent. It should be noted that the accepted value of carbon equivalent is usually stated in the project specifications according to the requirements of the end user. In this book, although the proposed and ideal value for carbon steel C_E is less than 0.40, other maximum values of carbon equivalent have been observed by the author in different projects, such as 0.39 (stricter than 0.40) or 0.42 (less strict than 0.4), to meet the clients' or end users' requirements.

Example 5.7 The chemical composition percentages for carbon steel welded valves in three grades, ASTM A106 Gr. A, B, and C are provided in Table 5.3. What grades offer good weldability?

Table 5.3 Chemical properties percentages of carbon steel pipe grades.

	C	Mn	Cr	Mo	V	Ni	Cu
A 106 Gr. A	0.25	0.5	0.3	0.1	0.05	0.4	0.3
A 106 Gr. B	0.30	0.5	0.3	0.1	0.05	0.4	0.3
A 106 Gr. C	0.35	0.5	0.3	0.1	0.05	0.4	0.3

Answer

$$C_E(\text{A106Gr}\cdot\text{A}) = \text{C} + \frac{\text{Mn}}{6} + \frac{\text{Cr}+\text{Mo}+\text{V}}{5} + \frac{\text{Ni}+\text{Cu}}{15}$$

$$= 0.25 + \frac{0.5}{6} + \frac{0.3+0.1+0.05}{5} + \frac{0.4+0.3}{15}$$

$$= 0.25 + 0.0833 + 0.09 + 0.0466 = 0.47$$

$$C_E(\text{A106Gr}\cdot\text{B}) = \text{C} + \frac{\text{Mn}}{6} + \frac{\text{Cr}+\text{Mo}+\text{V}}{5} + \frac{\text{Ni}+\text{Cu}}{15}$$

$$= 0.30 + \frac{0.5}{6} + \frac{0.3+0.1+0.05}{5} + \frac{0.4+0.3}{15}$$

$$= 0.30 + 0.0833 + 0.09 + 0.0466 = 0.52$$

$$C_E(\text{A106Gr}\cdot\text{C}) = \text{C} + \frac{\text{Mn}}{6} + \frac{\text{Cr}+\text{Mo}+\text{V}}{5} + \frac{\text{Ni}+\text{Cu}}{15}$$

$$= 0.35 + \frac{0.5}{6} + \frac{0.3+0.1+0.05}{5} + \frac{0.4+0.3}{15}$$

$$= 0.35 + 0.0833 + 0.09 + 0.0466 = 0.57$$

In accordance with Eq. (5.16), the carbon equivalent of these three grades is 0.47, 0.52, and 0.57 for grades A, B, and C of A106, respectively. C_E values exceed 0.4 for all three grades, which means none of them can be considered acceptable for welding.

5.5 Hydrogen-Induced Stress Cracking (HISC) Corrosion

Over the last 15–20 years, the subject of HISC has received considerable attention in the subsea oil and gas industry because of a number of very costly failures of subsea components. In order to understand the nature of this form of corrosion, it is imperative to understand that HISC is categorized as the result of an environment-aided cracking mechanism caused by either applied or residual stress. For HISC to occur, three factors are required: susceptible material, stress, and hydrogen formation due to the use of cathodic protection.

5.5.1 HISC and Vulnerable Materials

The risk of HISC applies to duplex, super duplex, hard nickel alloys, and high-strength low-alloy steel materials. Ferritic–austenitic stainless steels, such as duplex and super duplex, offer high mechanical resistance, corrosion resistance, and low cost, making them suitable for applications in the subsea oil and gas industry. They have been used in subsea piping and industrial valves for over 20 years. Nonetheless, these steels are also known to be vulnerable to HISC in subsea environments. High-strength carbon and low-alloy steels, which are widely used for the bodies and bolting of subsea valves, are thought to be susceptible to HISC due to their large mechanical strengths. Material microstructure plays a significant role in determining HISC severity. Hydrogen atoms are generally believed to diffuse between the metallic matrix grain boundaries due to their smaller size.

5.5.2 HISC and Stress

Materials subjected to tensile stress exhibit three distinct reactions in relation to the occurrence of HISC. In the first step, the material is deformed, and additional space is provided for hydrogen to enter into the microstructure. Furthermore, the stress increase in the material and crack initiation provide additional space for hydrogen to enter. The material stresses and cracks become more intense as a result. In HISC analysis, it is critical to take into account a variety of loads, such as the external loads applied by the connected piping, such as axial, bending, and compression loads, internal loads primarily caused by the internal pressure of the piping system, accidental or shock loads caused by accidents such as dropped objects or natural disasters, such as hurricanes, and finally residual stresses due to fabrication, manufacturing, and welding processes.

5.5.3 HISC and Cathodic Protection

In this form of corrosion, hydrogen atoms produced by cathodic protection move into the material to cause it to crack. In a subsea environment, all subsea components such as pipes, valves, and structures are subject to external corrosion. Cathodic protection is one method of protecting steel against external corrosion. It is through the transfer and loss of electrons from the anode that metals corrode or rust. By passing a direct current through a pipe, valve, or structure that serves as a cathode, corrosion can be mitigated. By reason of the potential difference between the anode and the cathode, electrons leave the anode and are transferred to the cathode; this process is known as cathodic protection. The cathodic protection method also produces atomic hydrogen on the metallic surface, which may result in HISC corrosion in materials which are vulnerable to corrosion. It is a common practice to use zinc, magnesium, or aluminum as sacrificial anodes that will easily corrode and release electrons for the protection of the cathode, which may consist of pipes, valves, structures, or other components. There is another critical factor, which is the electrolyte, or seawater because it is necessary in order for electrons to be able to flow from the anode to the cathode. Cathodic protection is not the only method for mitigating external corrosion in subsea environments; it can also be combined with another approach, such as coating.

5.5.4 HISC and DNV Standard

A great deal of effort has been put into identifying the failure mechanisms associated with HISC recently. The result is that Det Norske Veritas (DNV) has developed guidelines for the design of duplex components based on the DNV-RP-F112 standard to prevent HISCs. The gap in the DNV standard is to consider only duplex and super duplex as susceptible materials to HISC. To mitigate HISC failures, there are two general criteria for stress and strain limits in the DNV standard. The first step is creating stress limits based on finite element analysis, also known as linear elastic analysis. The strain limit, also referred to as elastic-plastic analysis, is another result obtained from hand calculations. DNV proposes a linear elastic stress-strain model to evaluate HISC. As a result of linear elastic analysis, or load analysis, three main principles are observed: stress is always proportional to strain, the material is assumed to be continually deformed by an increasing load, and the deformation stops and the material returns to its original shape upon releasing or removing the load. For HISC analysis, stress is linearized over the thickness of piping and valves in two ways: membrane stress and bending stress. Peak stress, another type of stress, can be ignored in HISC analysis. Defining membrane stress δ_m as the average, uniform distribution of stress across its thickness, it could be either compressive or tensile in nature. The bending stress δ_b, unlike the membrane stress, varies across the thickness of a component; it is applied along its longitudinal axis and is either tensile or compression-type. As shown in Figure 5.5, the distribution of membrane, bending, and membrane plus bending across the thickness of a component (e.g. pipe or valve) can be seen. According to Eqs. (5.18) and (5.19), DNV guidelines set two limitations for membrane

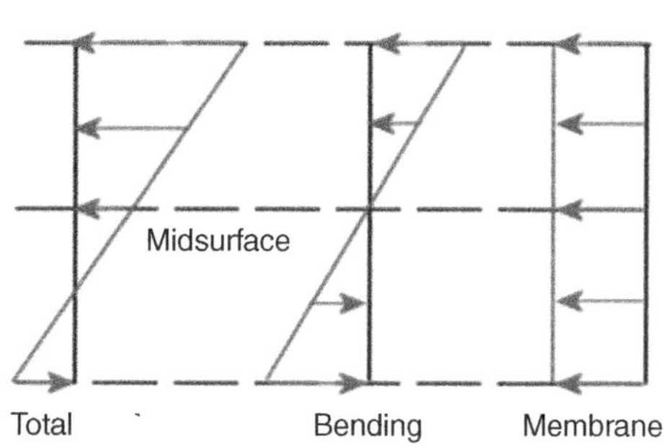

Figure 5.5 The distribution of bending and membrane stresses across the thickness of a component under stress.

δ_m and membrane plus bending stress δ_{m+b} for duplex and super duplex materials. According to these equations, the stress in the component subjected to risk must be limited to a percentage of the specified minimum yield strength (SMYS). In the oil and gas industry, SMYS is a material mechanical term or characteristic commonly used to specify the minimum stress that can be supported without permanent deformation. It is critical to understand that the value of SMYS decreases with increased temperature, and so the value of SMYS must be adjusted based on the temperature.

Membrane Stress δ_m Limitation as Per DNV Guideline for Duplex and Super Duplex Steels

$$\delta_m < \alpha_m \cdot \gamma_{HISC} \cdot SMYS \tag{5.18}$$

Membrane Plus Bending Stress δ_{m+b} Limitation as Per DNV Guideline for Duplex and Super Duplex Steels

$$\delta_{m+b} < \delta_{m+b} \cdot \gamma_{HISC} \cdot SMYS \tag{5.19}$$

where,

δ_m: Membrane stress (psi, Pa);
δ_{m+b}: Membrane plus bending stress (psi, Pa);
γ_{HISC}: In HISC material quality factor formula, coarse grain materials are rated as 85% and fine grain materials as 100%. A microstructure with fine spacing is more resistant to HISC, as discussed previously. As a result, DNV RP allows a higher HISC material factor of 100% for fine spacing compared to 85% for coarse spacing in the stress criteria. With DNV's model for mitigating HISC, the material quality factor is taken into account in such a way that fine-grained materials with fewer spaces are more resistant to HISC than coarse-grained materials. More space between material grains actually facilitates the entry of hydrogen atoms into the grains. There is an important point to remember that materials with excessive grain spacing and grain flow perpendicular to the main loading are nonconservative and cannot be analyzed with respect to the HISC issue by the conservative model and method provided by DNV;
α_m: Allowable SMYS factor for membrane stress (= 0.8; dimensionless);
α_{m+b}: Allowable SMYS factor for membrane plus bending stress (= 0.8 or 0.9 or 1 depending on location). The allowable SMYS factor for membrane plus bending stress with increased stress and weld toes within L_{Res} is 0.8. As shown in Figure 5.6, the weld toe is the point where the weld face joins the base metal. The allowable SMYS factor for membrane and bending stress for high-stress areas outside L_{Res} is 0.9. It is equal to 0.9 and 1 for smooth areas within L_{Res} and outside L_{Res}, respectively. Figure 5.7 display different values of the

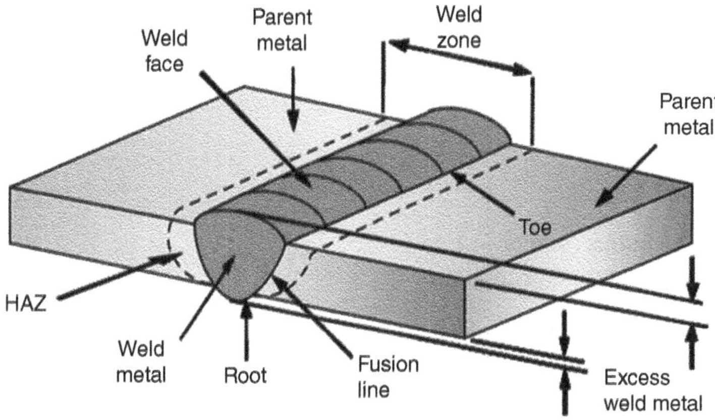

Figure 5.6 A buttweld joint that includes the toe.

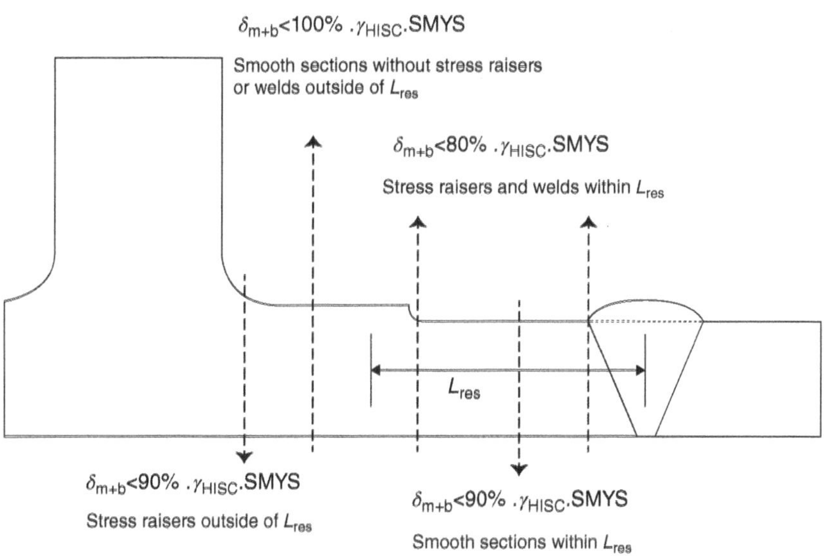

Figure 5.7 Values of SMYS factor for membrane plus bending stress from DNV standard.

allowable SMYS factor for membrane plus bending stress in accordance with DNV standards.

The distance L_{res} from the weld centreline is the distance at which residual strain and stress must be taken into account. In Eq. (5.20), L_{res} is the distance from the weld where HISC poses a higher risk of attack.

Calculation of Distance from Weld Joint Where the Risk of HISC Attack is Higher as Per DNV-RP-F112

$$L_{Res} = 2.5\sqrt{Rt} \tag{5.20}$$

where:

R = Nominal pipe radius (mm or in.);
t = Pipe wall thickness (mm or in.).

Example 5.8 A 10″ side entry subsea ball valve with a pressure rating of 7500 psi will be installed at a depth of 2300 m. Through its pup piece, the valve is attached to the connected subsea piping. The pup piece is, as its name implies, a piece of pipe that is either integrated with the valve body or welded to it. In essence, the pup piece is designed to keep the soft (nonmetallic) materials inside the valve away from the heat generated during the welding process between the valve and the piping, which will reduce the possibility of the soft materials being melted. A pipe connected to a valve has the same internal diameter and thickness as the pup piece at the location where they are welded together. Figure 5.8 shows the dimensional drawing of the pup piece. In accordance with DNV-RP-F112, calculate the distance from the weld joint between the piping and the valve pup piece where the risk of HISC attack is most likely.

Answer
The pipe size is 10″ (254 mm), which is also called the nominal pipe size. The nominal pipe radios are half of the pipe diameter, which is equal to 127 mm in this example. According to the drawing, the pup piece at the connection point to the pipe is 18.3 mm thick. The thickness value is taken from the figure on the right where the weld termination details are described in accordance with ASME B16.25. ASME B16.25 is a standard for the preparation of buttweld joints. Similarly, the thickness of the pup piece and the attached piping is the same. Based on the measurement of the radios of piping equal to 127 mm and its thickness equal to 18.3 mm, the value of L_{Res} indicates the distance from welding where HISC is likely to occur based on Eq. (5.20) as follows:

$$L_{Res} = 2.5\sqrt{Rt} = 2.5\sqrt{127 \times 18.3} = 2.5 \times \sqrt{2324.1} = 2.5 \times 48.21 = 120.52\,\text{mm}$$

120.52 mm is the distance from the weld at which HISC is more risky and less stress is allowed in HISC analysis.

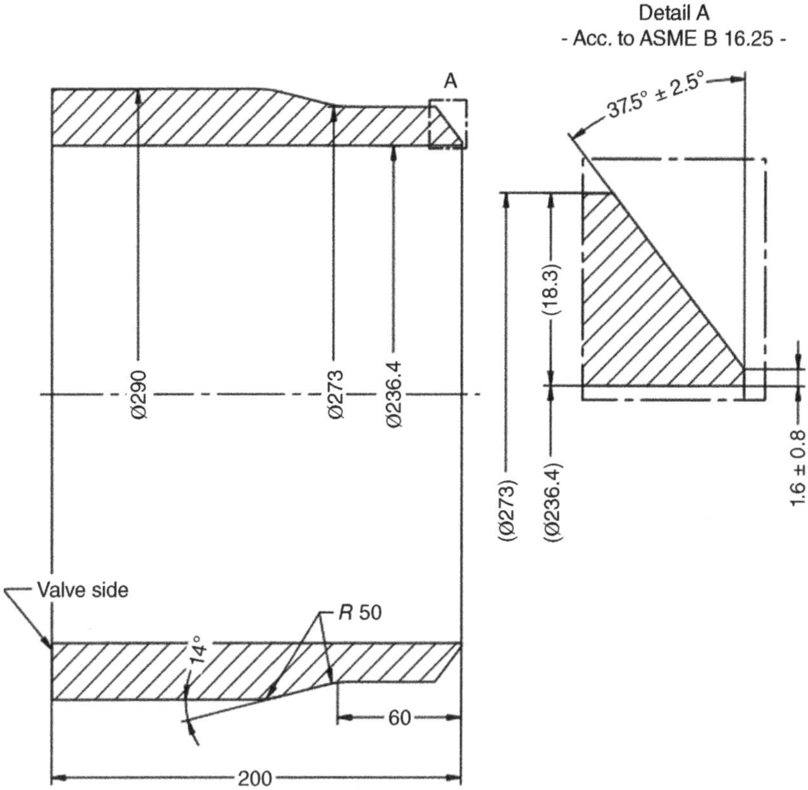

Figure 5.8 Drawing of the 10″ subsea ball valve pup piece.

Example 5.9 Consider the 10″ subsea ball valve in the previous example to be made of forged duplex ASTM A182 F51 and operated at ambient temperature. In terms of metallurgy, the ball valve body and the pup piece are made of coarse grains. During the FEA, the maximum membrane stress is found to be 226.14 MPa on the transition piece within 120.52 mm of the weld joint to the pipe. Similarly, the maximum bending stress is associated with the maximum membrane stress of 46.2 MPa. In this case, what is the maximum allowable stress for the valve material? Will the allowable stress be considered in the HISC analysis? With the ultimate tensile stress and yield strength of duplex equal to 620 MPa and 450 MPa, respectively, can the valve fail due to HISC by considering the applied loads in this case?

Answer

The first step is to determine the allowable stress of the forged duplex material in ASTM A182 F51 based upon its tensile and yield strengths. Allowable stress is defined as the maximum stress (tensile, compression, or bending) that is allowed to apply to a component. As a general rule of thumb, as per different ASME codes and standards, such as ASME B31.3, for process piping, the allowable stress can be selected from 1/3 tensile stress or 2/3 yield stress, whichever is lower.

$$\frac{2}{3}\text{yield stress} = \frac{2}{3} \times 450 = 300\ \text{MPa}$$

$$\frac{1}{3}\text{tensile stress} = \frac{1}{3} \times 620 = 206.7\ \text{MPa}$$

Thus, the maximum allowable stress for duplex valve materials is 206.7 MPa. It should be noted that the allowable stresses in HISC analysis based on the DNV guideline standard is greater than the value of allowable stress equal to 206.7 MPa. For HISC analysis, the allowable stress values (membrane and membrane plus bending) are calculated from Eqs. (5.18) and (5.19) as follows:

$$\text{Allowable membrane stress for HISC according to DNV standard}$$
$$= \alpha_{\text{m}} \cdot \gamma_{\text{HISC}} \cdot \text{SMYS} = 0.8 \times 0.85 \times 450 = 306\ \text{MPa}$$

Due to the coarse grain in the duplex material, the material quality factor γ_{HISC} is 0.8. From the results of FEA, it is possible to determine the maximum membrane stress to be 226.14 MPa < 306 MPa, which means the valve is safe against HISC while the membrane stress is produced within it. Next, the valve must be checked for its resistance to HISC during membrane plus bending stress production.

$$\text{Allowable membrane plus bending stress for HISC according to DNV standard}$$
$$= \alpha_{\text{m + b}} \cdot \gamma_{\text{HISC}} \cdot \text{SMYS} = 0.8 \times 0.85 \times 450 = 306\ \text{MPa}$$

Because the maximum membrane plus bending stress occurs around $L_{\text{Res}} = 120.52$ mm , where HISC attack is more likely, the allowable SMYS factor for membrane plus bending stress $\alpha_{\text{m + b}}$ is 0.8. Based on FEA calculation, the maximum membrane plus bending stress is calculated as 226.14 MPa (membrane stress) + 46.22 MPa (bending stress) = 272.34 Mpa < 306 MPa, which means the valve is safe against HISC while the membrane plus bending stresses are produced within it.

The result of these calculations indicates that there is no risk of HISC occurrence for the subsea valve with respect to the applied stress. However, we cannot conclude that the valve will operate safely under water since the valve must also be checked for strain during HISC analysis according to the DNV guideline.

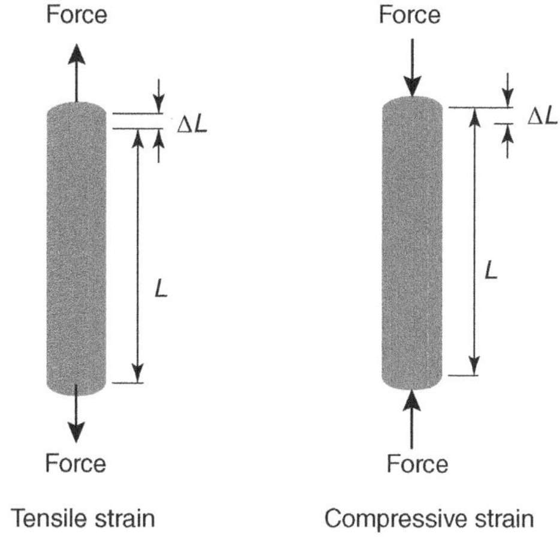

Figure 5.9 A component's length changes due to tensile and compressive stress.

The second parameter that must be checked for HISC after stress is strain, which relates to the mechanics of materials. Strain calculation is, in essence, the method of determining the deformation of an object as a result of applied (tensile or compression) stress, as depicted in Figure 5.9, or residual stress.

A component's strain is defined as the ratio of the change in length to its original form. Strain is calculated from Eq. (5.21). In fact, strain is a term used to describe the amount of deformation per unit of dimension. It is a dimensionless quantity without a unit of measurement. The amount of deformation and strain depends on many factors, including the magnitude and duration of the stress as well as material characteristics such as hardness and mechanical strength.

Strain Calculation

$$\mathcal{E} = \frac{\Delta L}{L} = \frac{\text{Change in length or elongation}}{\text{Original length}} \tag{5.21}$$

FEA can be divided into two types: linear and nonlinear. There is a linear relationship between applied force and component displacement in linear analysis. Structures and components that are subjected to this type of analysis are normally those whose stresses remain within the linear elastic range that is maximum equal to their SMYS. Figure 5.10 shows the stress–strain curve for ductile steel. Ductile steel is able to go under a large amount of plastic deformation before fracturing. A yield point, also known as SMYS, is a point of stress at which a small increase in stress results in a large increase in strain such that the material continues to deform despite no increase in load. At the yield point, the material will have permanent deformation.

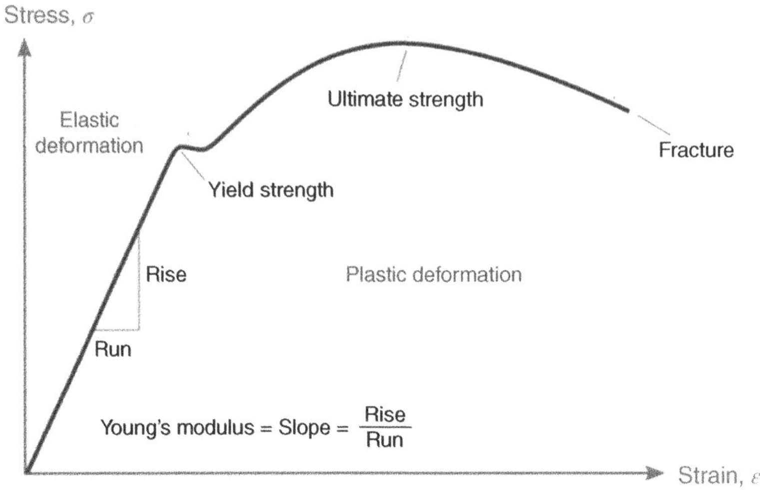

Figure 5.10 The stress–strain curve.

A large displacement is evaluated and considered by DNV through nonlinear FEA using nonlinear strain criteria. Nonlinear strain criteria take into account both elastic and plastic strains. Based on the DNV HISC guideline, residual strain values are 0.15% for weld toes and 0.25% for areas except for weld toes. An illustration of a buttweld joint with different weld areas is shown in Figure 5.7. According to Table 5.4, the allowable maximum strain from all allowable loads is calculated after allocating the residual strain values. Accordingly, the maximum strain obtained through FEA should be lower than the maximum allowable stain limit shown in Table 5.4.

Table 5.4 Maximum allowable strain limit according to DNV guideline.

	Maximum allowable strain limit	
Location in component	**Within L_{res} from weld**	**Outside L_{res} from weld (%)**
Outside 5% WT from any surface	Min. (0.3%; 0.5% $- \mathcal{E}_{res}$])	0.3
Within 5% WT from surface	1.00% $- \mathcal{E}_{res}$	1.00
	0.6% $- \mathcal{E}_{res}$	0.6

Example 5.10 The maximum measured principal strain for the 10″ subsea ball valve in the previous example was 0.0024980 at a distance of 100 mm from the weld joint, and outside of 5% of the wall thickness from the surface. Does high strain concentration cause the valve to fail?

Answer

The maximum strain happens at a distance of 100 mm from the weld that is within the L_{res} = 120.52 calculated in Example 5.8. Moreover, the location of the maximum strain is outside 5% of the wall thickness from any surface and, hence, the maximum allowable strain limit is selected from a minimum of 0.3% and a maximum of 0.5% − $\mathcal{E}_{res}$. As the highest amount of strain is 100 mm away from the weld joint, it is not on the weld toe, and $\mathcal{E}_{res} = 0.25\%$. The maximum allowable strain limit is calculated using Table 5.4:

$$\text{Maximum allowable strain limit} = \text{Min} \cdot [0.3\%; 0.5\% - \mathcal{E}_{res}]$$
$$= \text{Min} \cdot [0.3\%; 0.5\% - 0.25\%] = 0.25\% = 0.0025$$

Due to the maximum strain value of 0.0024980 < 0.0025, the subsea valve will not fail due to HISC and strain concentration.

Questions and Answers

5.1 How is carbon dioxide corrosion prevented in the oil and gas industry?

A By adding thickness to the facilities such as piping and valves as a corrosion allowance

B Using corrosion-resistant alloys such as stainless steel as a form of cladding or solid

C Injection of corrosion inhibitor into the corrosive fluid

D All answers are correct.

Answer

All answers are correct, so option D is the right choice.

5.2 In which case must corrosion allowance be considered for the industrial valve?

A A carbon steel body valve is cladded with three millimeters of Inconel 625.

B Martensitic stainless steel is used for the valve internals.

C Either carbon steel or low-alloy steel is used for the valve body.

D The valve body is constructed of austenitic stainless steel.

Answer

It is incorrect to choose option A because the carbon steel body has an Inconel 625 layer that protects it from corrosion. Option B is also incorrect since corrosion allowance is typically applied to the valve's body rather than its internals. The correct answer is Option C. Option D is incorrect because there is no corrosion allowance needed on CRA materials such as austenitic stainless steel.

5.3 On the outlet piping of a pressure vessel in a refinery that handles carbon dioxide-containing corrosive gas at 57 °C and 119 bar, a 10″ valve is installed. Keeping in mind that carbon dioxide is the only corrosive compound in the service with a mole fraction of 0.018 and pH equal to 4 what is the correct statement?

Assumptions:

- No corrosion inhibitor injection is performed;
- Facilities and components in the refinery, including valves, shall be designed for 20 years of design life;
- The amount of water in the fluid service is low, less than 1% and the effect of water cut and condensate are both 0.1.

A Carbon steel is a suitable material for the valve.
B A three-millimeter corrosion allowance is required.
C There is no chance of scale formation in this case.
D The carbon dioxide is deemed 100% active.

Answer

The first step is to calculate the basic CO_2 corrosion rate in millimeters per year.

The partial pressure of the CO_2 is calculated as follows:

$$P_{CO_2} = P_{\text{operation}} \times x_{CO_2} = 119 \times 0.018 = 2.14\ \text{bar}$$

57°C is equivalent to 330 K·

$$\rightarrow \text{Log}(\text{CR})_{\text{base}} = 5.8 - \frac{1710}{T} + 0.67 \log(P_{CO_2}) = 5.8 - \frac{1710}{330} + 0.67 \log 2.14$$
$$= 5.8 - 5.182 + 0.22 = 0.838 \rightarrow (\text{CR})_{\text{base}} = 6.89\ \text{mm/year}$$

By considering the corrective parameters, it is possible to reduce the basic corrosion rate, which is very high. The first step is to calculate the activity of CO_2 in the industrial valve.

$$\text{Log}(a) = \left(0.0031 - \frac{1.4}{T}\right) \times P = \left(0.0031 - \frac{1.4}{330}\right) \times 119 = -0.136$$
$$\rightarrow a = F(\text{System}) = 0.73$$

Option D is incorrect because the activity of carbon dioxide is 0.73 or 73%, so it is not 100% active.

$$F_{CO_2} = a \times P_{CO_2} = 0.73 \times 2.14\,\text{bar} = 1.56\,\text{bar}$$

The glycol and corrosion inhibitor effects are equal to one because no glycol or corrosion inhibitor is injected into the valve. The effect of water cut and condensate are both assumed as 0.1.

Two other effects need to be calculated; one is the effect of scaling and the other is the effect of pH.

$$\begin{aligned}\text{Log}(F(\text{Scale})) &= \frac{2400}{T} - 0.6 \times \log(F_{CO_2}) - 6.7 = \frac{2400}{330} - 0.6 \times \log(1.56) - 6.7 \\ &= 7.27 + 0.193 - 6.7 = 0.763 \rightarrow F(\text{Scale}) = 10^{0.763} = 5.79\end{aligned}$$

The scaling effect exists if $(F(S))$ is less than one, which is not happening in this case; therefore, there is no possibility of scale formation, and option C is correct, and $F(S) = 1$ is utilized for corrosion rate calculations.

The next step is to calculate the effect of pH. To calculate the effect of pH on the corrosion rate, the first step is to calculate the "pH_{sat}" from Eqs. (5.10) and (5.11).

$$\begin{aligned}pH_{Sat1} &= 1.36 + \frac{1307}{T} - 0.17 \times \log(F(CO_2)) \\ &= 1.36 + \frac{1307}{330} - 0.17 \times \log(1.56) \\ &= 1.36 + 3.96 - 0.033 = 5.29\end{aligned}$$

$$\begin{aligned}pH_{Sat2} &= 5.4 - 0.66 \times \log(F(CO_2)) = 5.4 - 0.66 \times \log(1.56) \\ &= 5.4 - 0.127 = 5.27\end{aligned}$$

The lowest pH_{sat} value is selected for calculating the effect of pH; in this case, it is 5.27. The pH_{sat} value is higher than the actual pH; therefore, the effect of pH is calculated from Eq. (5.12) as follows:

$$\text{If } pH_{Sat} > pH_{actual}$$

$$\begin{aligned}\rightarrow \text{Log}(F(\text{pH})) &= 0.32(pH_{Sat} - pH_{actual}) \rightarrow \log(F(\text{pH})) \\ &= 0.32 \times (5.27 - 4) \\ &= 0.4064 \rightarrow F(\text{PH}) = 10^{0.4064} = 2.55\end{aligned}$$

The next step is to calculate the corrective corrosion rate as follows:

$$\begin{aligned}(\text{CR})_{\text{correct.}} &= (\text{CR})_{\text{base}} \times F(\text{System}) \times F(\text{Glycol}) \times F(\text{Scale}) \\ &\quad \times F(\text{Watercut}) \times F(\text{Inhibitor}) \times F(\text{Condensate}) \times F(\text{PH}) \\ &= 6.89\,\text{mm/year} \times 0.73 \times 1 \times 1 \times 0.1 \times 1 \times 0.1 \times 2.55 \\ &= 0.128\,\text{mm/year}\end{aligned}$$

Considering the design life of the valve equals 20 years, the final corrosion rate is $0.128 \times 20 = 2.56$ mm $\rightarrow$ Carbon steel with a corrosion allowance of 3 mm can be considered for this case. Thus, both options A and B, in addition to option C, are correct. To sum up, option D is the only wrong statement.

5.4 What is one effect that could increase the basic carbon dioxide corrosion rate calculated using Eq. (5.1)?

A Scaling
B Fugacity
C pH
D Corrosion inhibitor

Answer

Option C is correct. All other options either reduce the corrosion rate or do not affect it at all.

5.5 Identify the correct statement about pitting corrosion mechanisms and prevention.

A Pitting corrosion is a form of uniform corrosion.
B Pitting corrosion is only a type of internal corrosion caused by corrosive fluids, such as seawater.
C The only way to prevent pitting corrosion is to select materials with a higher PREN.
D The pitting corrosion process is not confined to austenitic stainless steel.

Answer

Solution A is incorrect since pitting corrosion occurs locally and is not uniform. Option B is wrong since pitting corrosion caused by corrosive marine environments can occur externally. In addition, option C is not correct because other approaches, such as testing for pitting corrosion as well as controlling effective process parameters on pitting corrosion such as pH, temperature, and chloride content, can prevent pitting corrosion. Accordingly, Option D is the most appropriate response because carbon steels, nickel alloys, and duplex stainless steels are all vulnerable to pitting corrosion, as well as austenitic stainless steels.

5.6 Regarding the calculations in Example 5.6, select the pitting-resistant material.

A SS347
B 25Cr super duplex
C Incoloy 800
D 22Cr duplex

Answer

The correct answer is Option B, 25Cr super duplex since its PREN is greater than 40.

5.7 A welded valve is constructed of forged low-temperature carbon steel (LTCS), material grade ASTM A350 LF2. The chemical composition percentage of this material grade can be found in Table 5.5. Which of the following statement is true regarding the material's weldability and carbon equivalent?

A As far as the chemical composition and calculation of carbon equivalent are concerned, ASTM A350 LF2 is weldable.

B It is proposed that ASTM A350 LF2 be welded in this example in a sour service containing hydrogen sulfide.

C Carbon equivalent is equal to 0.45.

D To achieve an acceptable level of weldability, the carbon content must be reduced.

Answer

Carbon equivalent of A350LF2 is calculated as follows:

$$
\begin{aligned}
C_E(\text{A350LF2}) &= \text{C} + \frac{\text{Mn}}{6} + \frac{\text{Cr} + \text{Mo} + \text{V}}{5} + \frac{\text{Ni} + \text{Cu}}{15} \\
&= 0.30 + \frac{0.8}{6} + \frac{0.3 + 0.12 + 0.08}{5} + \frac{0.4 + 0.4}{15} \\
&= 0.30 + 0.1333 + 0.1 + 0.0533 = 0.59
\end{aligned}
$$

ASTM A350LF2 cannot be welded because its carbon equivalent exceeds 0.40, so option A is incorrect. As ASTM A350LF2 in this example should not be welded in a sour service due to the high risk of cracking due to hydrogen sulfide present, option B is incorrect as well. Carbon equivalent equals 0.59, so Option C is incorrect. Choice D is the correct answer.

Table 5.5 Chemical compositions of the welded valve in A350LF2 material.

Carbon	0.30	Molybdenum	0.12
Manganese	0.8	Phosphorous	0.035
Silicon	0.30	Sulfur	0.040
Copper	0.40	Vanadium	0.08
Chromium	0.30	Columbium	0.02
Nickel	0.40		

5.8 According to the DNV guideline for HISC evaluation, which statement is correct?

A HISC analysis is covered in the guideline for components in the subsea sector, as well as refineries and chemical plants.
B The stress analysis for HISC evaluation considers plastic deformation.
C Material types susceptible to risk, such as low-alloy steels, hard nickel alloys, and duplex stainless steels, are all covered by the DNV criteria.
D The strain analysis considers both elastic and plastic conditions.

Answer

Option A is incorrect since HISC only applies to subsea oil and gas components and is not relevant to refineries and petrochemical plants. Additionally, option B is not correct since the DNV guideline for stress analysis in HISC considers only linear and elastic deformation. Option C is incorrect since DNV criteria only apply to duplex stainless steel. The correct answer is D.

5.9 In this application, a subsea check valve in 2″ size will be constructed from duplex stainless steel and operated at a pressure of 10,000 psi at a depth of 2300 m. For this valve, two load scenarios are considered: operation and testing. Operating and testing membrane stress values are respectively 172.92 MPa and 190 MPa. Membrane plus bending values for these two operating and testing conditions are 231.69 MPa and 254.86 MPa, respectively. Based on the stress analysis, which of the following statements is true regarding the occurrence of HISC for the valve? (*Note*: Duplex has an ultimate tensile strength of 620 MPa as well as a yield strength of 450 MPa.)

A The possibility of HISC occurrence in this case cannot be determined based on the stress analysis because the material quality factor and the exact location of the maximum load values are unknown.
B Membrane and bending loads are not large enough to cause HISC in the valve.
C HISC can tolerate maximum stress of 306 MPa in a worst-case scenario.
D In the worst-case scenario, the maximum allowable stress for HISC differs between membrane and membrane plus bending stress conditions.

Answer

As the worst-case scenario for HISC occurrence is when the material contains coarse grains, the value of the material quality factor should be 0.85, the minimum value. The probability of HISC occurring is highest at the worst-case scenario when the SMYS factor for both the membrane and the membrane plus the bending are equal to 0.8. Based upon the worst-case scenario, the allowable membrane stress, as well as membrane plus bending stress, are calculated using Eqs. (5.18) and (5.19) as follows:

$$\text{Allowable membrane stress to prevent HISC at the worst} - \text{case scenario} = \alpha_{\mathrm{m}} \cdot \gamma_{\mathrm{HISC}} \cdot \mathrm{SMYS} = 0.8 \times 0.85 \times 450 = 306\ \mathrm{MPa}$$

$$\text{Allowable membrane plus bending stress to prevent HISC at the worst} - \text{case scenario} = \alpha_{\mathrm{m+b}} \cdot \gamma_{\mathrm{HISC}} \cdot \mathrm{SMYS} = 0.8 \times 0.85 \times 450 = 306\ \mathrm{MPa}$$

$$\text{Membrane stress during operation} = 172.92 < 306\ \mathrm{MPa}\ (\text{No HISC risk})$$

$$\text{Membrane stress plus bending during operation} = 231.69 < 306\ \mathrm{MPa}\ (\text{No HISC risk})$$

$$\text{Membrane stress during testing} = 190 < 306\ \mathrm{MPa}\ (\text{No HISC risk})$$

$$\text{Membrane stress plus bending during operation} = 254.86 < 306\ \mathrm{MPa}\ (\text{No HISC risk})$$

As a result, even without all the data, by assuming the worst-case scenario, HISC cannot happen in this case, so option B is correct, and option A is incorrect. Option C is right since in the worst-case scenario, 306 MPa is the maximum stress that is tolerated before HISC occurs. Option D is incorrect because the maximum allowable stress in a worst-case scenario is equal to 306 MPa in both membrane and membrane plus bending stress conditions. Thus, both options B and C are correct.

5.10 The maximum strain is 0.00178 for the subsea check valve in the previous example. Based on strain value, which of the following statements is correct regarding the risk of HISC?

A Due to the strain's unknown location, the possibility of HISC occurrence cannot be determined.
B There is no risk of HISC in this case.
C In the worst-case scenario, the maximum allowable strain is 0.0025.
D All options are incorrect.

Answer

According to Table 5.4, the maximum allowable strain in the worst-case scenario is 0.0025, so option C is appropriate. This happens when the critical strain occurs in areas other than weld toes, which indicates the residual strain ε_{res} value is 0.25%. Further, the principle strain must be located within the L_{res} from weld and outside 5% of the wall thickness from the surface. Considering that the maximum strain is 0.00178 and less than the maximum allowable strain in the worst-case scenario, HISC cannot occur and option B is also correct. Lastly, both options A and D are incorrect.

Further Reading

American Society of Mechanical Engineers (ASME) (2017). Buttwelding ends. ASME B16.25, New York, USA.

De Waard, C., Lotz, U., and Milliams, D.E. (1991). Predictive model for CO_2 corrosion engineering in wet natural gas pipelines. *Corrosion* 47 (12): 976–985.

Det Norske Veritas (DNV) RP-F112 (2008). Design of duplex stainless-steel subsea equipment exposed to cathodic protection, Hovik, Norway.

Iranian Petroleum Standard (IPS) (2008). Engineering standard for corrosion consideration in material selection. IPS-E-TP-740, Tehran, Iran.

Nesbitt, B. (2007). *Handbook of Valves and Actuators: Valves Manual International*, 1e. Oxford: Elsevier.

Norsok (2005). *CO_2 corrosion rate calculation model.* Lysaker: Norsok M-506, Rev.02.

Nustad, G. (2015). Tackling corrosive challenges in the oil and gas industry. *Valve World* 20 (11): 56–60.

Skousen, P.L. (2011). *Valve Handbook*, 3e. New York: McGraw-Hill.

Smit, P. and Zappe, R.W. (2004). *Valve Selection Handbook*, 5e. New York: Elsevier.

Sotoodeh, K. (2020). Requirement and calculation of corrosion allowance for piping and valves in oil and gas industry. *Journal of Bio and Tribo Corrosion* 6 (21): https://doi.org/10.1007/s40735-019-0319-4.

Sotoodeh, K. (2020). A review and analysis of industrial valve material failures due to corrosion and proposals for prevention measures based on industrial experiences in the offshore sector of oil and gas industry. *Journal of Failure Analysis and Prevention* https://doi.org/10.1007/s11668-020-01064-9.

Sotoodeh, K. (2021). *A Practical Guide to Piping and Valves for the Oil and Gas Industry*, 1e. Austin, TX: Elsevier (Gulf Professional Publishing).

Sotoodeh, K. (2021). *Safety and Reliability Improvement of Valves and Actuators for the Offshore Oil and Gas Industry Through Optimized Design*. University of Stavanger. Doctor Philosophiae (Dr. Philos) UiS no.573. Faculty of science and technology. ISBN:978-82-7644-987-7.

Sotoodeh, K. (2021). HISC analysis for valves in the subsea oil and gas industry. *Safety in Extreme Environments* 3: 1–8. https://doi.org/10.1007/s42797-021-00030-4.

Sotoodeh, K. (2022). *Piping Engineering: Preventing Fugitive Emission in the Oil and Gas Industry*, 1e. New York: Wiley.

Sotoodeh, K. (2022). *Case Studies of Material Corrosion Prevention for Oil and Gas Valves*, 1e. Austin, TX: Elsevier (Gulf Professional Publishing).

6

Noise

6.1 Introduction to Sound

In physics, sound is the result of vibrations and is accompanied by acoustic waves in liquids, gases, and solids. It is true that when an object vibrates, it imparts energy to the particles around it, causing them to move and this movement is known as sound waves. The waves caused by vibrations of molecules follow sine functions characterized by their amplitude and wavelength or frequency. The two most significant characteristics of a sound are its frequency and intensity (amplitude or volume or loudness). A sound can be described by its low or high frequency. The frequency is determined by the speed at which particles vibrate in the medium, which is defined as the number of waves emitted in a given amount of time. Frequency is the number of cycles per second measured in Hertz (Hz) units. For example, a frequency of 1000 Hz means that the waveform is 1000 cycles/s. The human hearing range is between 20 and 20,000 Hz. Each cycle represents one sequence of waves where the intensity changes from zero to the highest value, then to the lowest value, and then back to zero, as shown in Figure 6.1. The length of a wave or wavelength can be measured from one point on the waveform to the exact point at which the next cycle begins (see Figure 6.1). Frequency and wavelength are inversely proportional to each other; therefore, a wave with the greatest frequency will have the shortest wavelength, and twice the frequency will have half the wavelength. Figure 6.2 depicts sound waves of equal amplitude (intensity) and increasing frequency from top to bottom. The amplitude of a sound wave is a measure of its height, which is determined by the distance between its peaks and troughs. In general, the larger the amplitude of the wave, the louder the sound, indicating that it contains more energy and may travel further at a lower speed. Pitch or frequency is shown by the spacing of displayed waves in such a way that the closer the waves are to one another, the higher the frequency.

Industrial Valves: Calculations for Design, Manufacturing, Operation, and Safety Decisions, First Edition. Karan Sotoodeh.

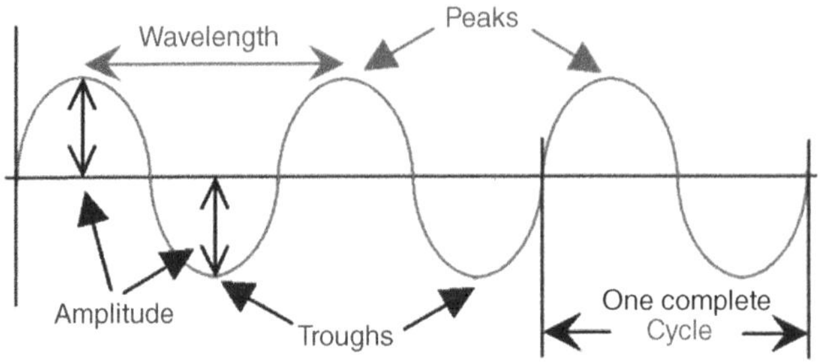

Figure 6.1 Illustration of amplitude, wavelength, and cycle in a sound wave.

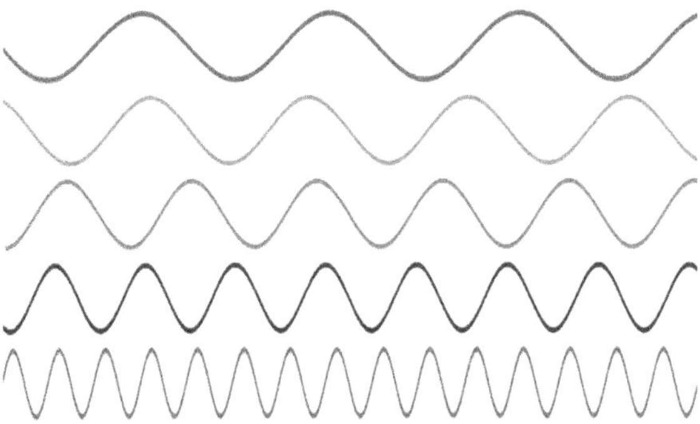

Figure 6.2 From top to bottom, sound waves are equal in amplitude and increasing in frequency.

The temperature, elasticity, and density of the medium determine the speed of sound. Higher temperatures and moisture cause sound to travel faster, while lower temperatures and dry air cause sound to travel slower. In any medium, sound can travel, although the greatest speed occurs in solids and the slowest in gases.

6.2 Introduction to Noise

Generally, the term "noise" is used to refer to a loud, unpleasant, and disruptive sound that produces undesirable psychological effects on an individual such as depression and irritation by interfering with their activities such as work, rest, sleep, etc., as illustrated in Figure 6.3. From a physics point of view, there is no difference between noise and desired sound. This is because both pertain to vibrations that can move through a medium such as air or water. The difference occurs when the brain receives and interprets the sound. Noise pollution is projected to

Figure 6.3 Illustrations of the generated noise and its effects on the human brain.

become the third most significant threat to the human environment after air pollution and water pollution.

Sound power level and *sound pressure level* are two terms which are commonly confused with one another when discussing acoustic energy. The sound power level (SWL, L_W) is the level of energy emitted by the sound source. The value is an absolute value that is unaffected by the environment, and it is independent of distance. The sound pressure level (SPL, L_p) is what ears hear and what sound meters measure. This parameter is influenced not only by the strength of the source but also by the surroundings and the distance from the source to the receiver. As a general rule of thumb, doubling the distance is equivalent to a 6-decibel (dB) reduction in noise levels. While the two parameters differ, they share the same unit, which is the dB.

Example 6.1 When a worker drills the ground, the sound power level produced is 100 dB. There are two individuals standing far away from the worker; one is 100 m away and the other is 50 m away. What is the value of the sound power level at the locations of these two individuals? If the sound pressure level at 50 m from the worker is 58 dB, what is the sound pressure level at 100 m where the other person is standing?

Answer

At the location of these two individuals, the sound power level (SWL) is equal to 100 dB. As doubling the distance reduces the sound pressure level (SPL) by 6 dB, the SPL at 100 m, where the other person stands, is equal to $58 - 6 = 52$ dB.

Figure 6.4 A plot of sound pressure and sound pressure level values for different locations.

A plot of sound pressure and sound pressure values for different locations is shown in Figure 6.4. Furthermore, Table 6.1 provides a summary of the relative noise levels.

Equation 6.1 can be used to calculate the relationship between sound pressure level (SPL) and pressure level. As a general rule of thumb, an increase of 6 dB doubles the level of pressure, whereas an increase of 20 dB tens it.

Table 6.1 Relative noise levels.

130 dB	Jet aircraft on take-off
120 dB	Threshold of feeling
110 dB	Train
100 dB	Loud highway
90 dB	Loud truck
80 dB	Plant site
70 dB	Vacuum cleaner
60 dB	Conversation
50 dB	Offices

Relationship Between SPL and Pressure Level

$$\text{Sound pressure level (SPL)} = 20\log_{10}\left(\frac{\text{Pressure level (Pa)}}{20^{-6}}\right)^2 \tag{6.1}$$

A dB is a logarithmic unit of measurement of sound intensity. This unit is also used for communications, electronics, and signals. Let us consider two sound-generating devices. The first produces a sound with a power of P_1 and the other produces a louder sound with a power of P_2. Assuming that all other parameters are equal, the dB difference between these two is equal to $10\log(P_1/P_2)$. In the case where the second device produces twice as much power as the first, the difference is $10\log 2 = 3$ dB. Decibels may be used to determine the sound level of a single source rather than a ratio. As a result, a reference must be chosen, usually an air sample with a very low sound power level of 20 μPa.

6.3 Noise in Industrial Valves

The safety and instrument engineers analyze the noise generated by various types of valves to minimize its impact in diverse sectors of the oil and gas industry, such as offshore, refineries, and petrochemical plants. The high noise and vibration levels caused by pressure-reducing devices such as pressure safety valves (PSVs) and control valves during operation can increase stress levels in the piping system and cause damage. As a result, it is essential to design piping systems to withstand noise, vibration, and acoustic fatigue from a safety perspective. In addition, as a result of the noise and vibration generated at the same time, wear, pits, and fractures of valve connections occur which significantly impair the performance of the valve. The Occupational Safety and Health Administration (OSHA) has focused its attention on the noise produced by valves in the field of control equipment. Noise in industrial valves can be attributed to two main sources: mechanical work, movement, and vibration within the valves, as well as fluids moving through the valves.

There are two negative effects of noise: acoustic fatigue and hearing damage. The noise, however, may be beneficial as a sign of an abnormal condition within the valve. Due to the negative impacts of noise, measuring the noise level generated by industrial valves is very important and valuable. The experience in gas production, petrochemical and other industries has shown that acoustic energy in high-capacity gas pressure reduction systems, such as pressure safety and control valves, can lead to piping fatigue failure after a few hours of operation. Noise exposure must be limited in accordance with safety standards, such as occupational health standards. It is very important to control the noise level in control valves as well as PSVs. Although control valves and PSVs are known as the main noise-producing valves, check valves such as dual plate and swing check valves are also capable of producing noise due to disk slamming and chattering.

6.3.1 Mechanical Noise and Vibration

Mechanical noises and vibrations are those caused by the movement and vibration of valve components. There are two possible reasons for mechanical noise and vibration. Most valve internals react to turbulent flow flowing through them, resulting in mechanical noise. Valves that undergo rapid changes in flow or high-pressure drops can also generate noise and vibration in their components. The second vibration condition is caused by mechanical loads imposed by the connected piping or by the operation of the valve itself, which is a result of the internal movement of the valve. The noise generated by vibration and mechanical work of valves is typically unpredictable, which is considered the second concern of process engineers. Turbulent flow can cause vibrations in some valve components in contact with nearby surfaces, leading to an intensification of noise and vibration as a result. The mitigation of this type of noise involves improving the design of the valve, such as changing the valve internals, increasing the stem diameter, and balancing the applied forces. The next subsection describes the noise caused by the fluid inside the piping system.

6.3.2 Fluid Noise

Fluid noise is primarily associated with control valves. A possibility of noise generation exists in situations where mass flow rate and pressure drop are high. Control valve noises are typically associated with cavitation and high-pressure drop gas applications. Generally speaking, there are two types of fluid service noise: aerodynamic caused by compressible fluids and hydrodynamic caused by non-compressible fluids. It is undeniable that liquids are less compressible than gases. However, the compressibility effect of liquids cannot be ignored if the liquid pressure is high and its pressure drop is large. Correct trim selection and/or appropriate insulation can reduce the effect of fluid noise in control valves. The calculation of noise in control valves caused by fluids is done in accordance with the International Electrotechnical Commission (IEC) standard.

6.3.2.1 Aerodynamic Noise

Aerodynamic noise is the result of the fluid converting its mechanical energy into acoustic energy as it passes through the restriction in the valve. It represents the most severe form of noise resulting from a high-speed or unsteady (turbulent) flow of fluid. As an example, this type of noise can be generated downstream of vena contracta, which is the point in a fluid stream where the flow diameter is the smallest, and fluid velocity is at its greatest. By utilizing low-noise trim valves and silencers as well as increasing the thickness of the pipe walls, this type of noise can be mitigated.

6.3.2.2 Hydrodynamic Noise

Among hydrodynamic noises are those generated by liquid flow, cavitation, and flashing. In Chapter 3, cavitation and flashing are discussed in more detail. The most common cause of hydrodynamic noise is cavitation, which occurs when part of a liquid undergoes two changes of state. It is important to recognize that when the liquid pressure falls below the vapor pressure in cavitation, the fluid velocity increases considerably. This increases noise and vibration, as discussed later in this chapter.

6.3.3 Noise Control Strategies

In general, there are six kinds of measures used to mitigate or prevent noise in piping systems, including industrial valves, which can be summed up as follows:

- **Selection and installation of noise-reducing valves:** This method is primarily used for noise control. For instance, several types of valves can be used to control fluid flow, including globe, axial, butterfly, V-notch ball, and angle globe valves. Among all these valves, axial and globe valves can offer excellent noise control levels. It is therefore recommended to choose a globe or axial valve over a butterfly, V-notch, or other option when noise is a concern.
- **Process optimization:** Noise levels are strongly influenced by operating conditions. As described below, noise levels can be reduced by reducing the system

pressure, differential pressure, or increasing the size or reducing fluid velocity at the trim exit.

- **Modification of valve design:** Many different types of low-noise valve trims (closure members) are available that create multistep pressure reduction in valves.
- **Piping insulation:** The noise emissions from the piping system, including valves, are kept within the noise limit requirements by using acoustic piping insulation.
- **Using thicker pipe:** When choosing a thicker pipe with a higher schedule number, noise and vibration can be reduced.
- **Improving the piping connection integrity:** The stress on the pipe is reduced by minimizing the number of pipe fittings and attachments that produce high stress concentrations at the connection. It can be accomplished by reducing the number of small bore branch connections (e.g. vents, low point drains), making branch connections with fittings that ensure a smooth transition from one branch to the next (e.g. by using tees rather than olets), ensuring that the header seam does not cross the connection weld line, avoiding pipe-to-pipe connections (also referred to as stub-in connections), etc.
- **Using noise reduction components:** When installed after or downstream of control valves, silencers can be very effective in reducing noise generated by these valves. A silencer is typically installed in a horizontal piping system and has a preferred flow direction. Turbulent and high velocity fluid exiting the control valve expands in the silencer, and the silencer absorbs a significant proportion of the noise generated.

6.4 Noise Calculations for Pipes and Valves

The noise analysis covered in this section comprises four types; the first is acoustic fatigue analysis, which includes the calculation of SWL and Mach number for all types of valves, including control and pressure relief valves. It is the second part that deals with control valves and the types of fluid noises produced by these valves, namely hydrodynamic and aerodynamic noises. The final set of calculations is focused on the noise associated with PSVs according to the method outlined in API 521.

6.4.1 Acoustic Fatigue Analysis

The method outlined in this chapter to measure acoustic fatigue is applicable to all types of valves, both control valves and pressure relief valves.

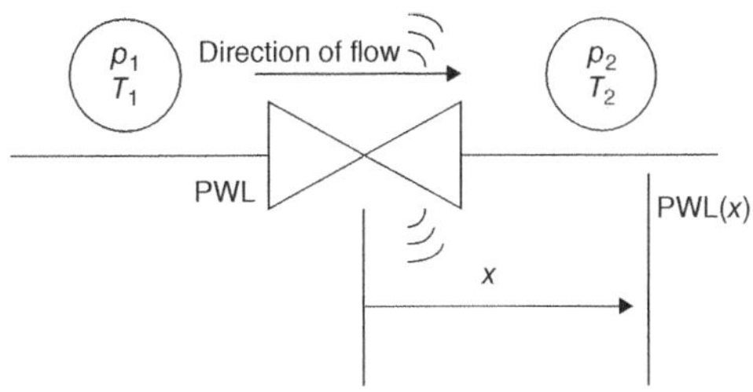

Figure 6.5 Schematic of valve and piping system for acoustic fatigue calculations.

6.4.1.1 Sound Power Level Calculations

The American Society of Mechanical Engineers (ASME) publication PVP-Vol. 328, flow-induced vibration, ASME 1996, by Eisinger may be used to evaluate acoustic fatigue. The sound power level (SWL or PWL) from the source of noise and vibration, which is typically a valve or orifice, is calculated according to Eq. (6.2) (see Figure 6.5).

Calculation of Acoustic Fatigue at the Source of Noise

$$\mathrm{PWL} = 10\log\left[\left(\frac{P_1 - P_2}{P_1}\right)^{3.6} \times \left(\frac{W}{3600}\right)^2 \times \left(\frac{T_1 + 273}{m}\right)1.2\right] + 126.1 + \mathrm{SFF} \tag{6.2}$$

where:

PWL: the sound power level in dB;
P_1: the upstream pressure of the valve (pressure let down source) in bar;
P_2: the downstream pressure of the valve (pressure let down source) in bar;
T_1: the upstream temperature of the valve in °C;
W: Gas or liquid flow rate in kg/h;
M: Molecular weight;
SFF: Sonic flow correction factor. In the presence of sonic conditions, Mach number is greater than one, so SFF = 6. In the absence of sonic conditions, SFF = 0. Mach number is discussed further in this chapter.

For valves equipped with low-noise trim, as explained further in this chapter, Eq. (6.2) is not applicable. As an alternative, the manufacturer of low-noise trim valves can provide the value of PWL.

The next step is to calculate the attenuation of acoustic fatigue or vibration at a distance "x" from the valve that is producing the noise using Eq. (6.3). The design calculations suggest that a distance of 50 times the inside diameter of the valve or the pipe connected to the valve can reduce the PWL by 3 dB in comparison to the noise produced within the valve.

Calculation of Acoustic Fatigue at a Distance of "*x*" from the Source of the Noise

$$\mathrm{PWL}_{(X)} = \mathrm{PWL} - \frac{0.06x}{D} \tag{6.3}$$

where:

$\mathrm{PWL}_{(X)}$: Level of sound power in dB at a distance of "*x*" from the source of the noise (valve);
x: Distance from the valve;
D: Internal diameter of the valve or pipe connected to the valve.

Whenever more than one source generates noise, the noise should be added at the pipe junctions where the piping from the sources meets as per Eq. (6.4).

Calculation of Acoustic Fatigue at a Distance of "*x*" from the Source of the Noise by Considering Multiple Sources of Noise

$$\sum \mathrm{PWL}_X = 10\log\left[10^{\mathrm{PWL}_1/10} + 10^{\mathrm{PWL}_2/10} + \cdots\right] \tag{6.4}$$

where:

$\mathrm{PWL}_{1,2,\mathrm{etc.}}$: The power level for each source at a distance of "*x*" in the piping.

Equation (6.5) is used to calculate PWL's acceptance level.

Allowable Level of Acoustic Fatigue

$$\mathrm{PWL}_\mathrm{A} = 173.6 - 0.125 \times \left(\frac{D_\mathrm{i}}{t}\right) \tag{6.5}$$

where:

PWL_A: Accepted Sound level (dBA);
D_i: Inside diameter of the pipe (mm);
t: Thickness of the pipe (mm).

Note 1

Some major oil and gas end users, such as British Petroleum (BP), may set stricter limits on the allowable noise levels than those determined by Eq. (6.5), which is equal to 155 dB. So even if the calculated value of PWL is less than x but greater than 155 dB, the design of the valve and connected pipe may not be reliable regarding noise generation. In this case, it is critical to know LOF, or the likelihood of failure. The design of piping and valves is safe in such a case as long as the LOF does not exceed 0.5, according to the BP noise guidelines.

Note 2

The fact that valves are not the only components that can cause noise is important to keep in mind. The noise intensity can also be increased by piping components such as bends, tees, and reducers. A discussion of methods for calculating noise levels produced by components other than valves does not fit the scope of this chapter.

Note 3

The noise level associated with a PSV can be increased by 5 dB under conditions where an improvement in design is applied to the downstream of the PSV as follows:

- In piping systems, olets larger than 2″ should be replaced with tees made according to ASME B16.9. Both olets and tees are branching fittings.
- Pipe-to-pipe connections should be avoided when taking branches from the header. In the event that it is not possible to avoid pipe-to-pipe connections, an attachment plate or reinforcement plate should be positioned around the circumference of the pipe.

The design improvement is commonly required for a distance corresponding to the calculated sound power in excess of the accepted limit, as calculated according to Eq. (6.5). In the event that the accepted sound level still cannot be attained despite the above-mentioned design improvements, additional options such as selecting a valve with low-noise trim (applicable to control valves), evaluating the use of multiple valves in parallel, or reducing the pressure steps can be considered. Other approaches to mitigate noise impacts include increasing the thickness of the piping walls, reducing the flow capacity, and applying insulation around pipes.

Example 6.2 The outlet of a pressure relief valve is connected to an 8″ pipe with a standard (STD) wall thickness. Upstream and downstream pressures at the safety valve are 30 bara and 1 bara, respectively, while the temperature at the valve inlet and outlet is 39.85 °C. The mass flow rate of the gas through the valve is 20, 000 kg/h, and the molecular weight of the gas is 25. Assuming there is no risk of sonic flow, is there a risk of acoustic fatigue in this instance?

Answer

The provided process data for the valve and outlet piping are summarized as follows:

P_1: the upstream pressure of the relief valve = 30 bara
P_2: the downstream pressure of the relief valve = 1 bara
T_1: the upstream temperature of the valve = 39.85 °C
W: Gas or liquid flow rate = 20,000 kg/h
M: Molecular weight = 25
SFF = 0

Using Eq. (6.2):

$$\text{PWL} = 10\log\left[\left(\frac{P_1 - P_2}{P_1}\right)^{3.6} \times \left(\frac{W}{3600}\right)^2 \times \left(\frac{T_1 + 273}{m}\right)1.2\right] + 126.1 + \text{SFF}$$

$$= 10\log\left[\left(\frac{30 - 1}{30}\right)^{3.6} \times \left(\frac{20{,}000}{3600}\right)^2 \times \left(\frac{39.85 + 273}{25}\right)1.2\right] + 126.1$$

$$= 10\log(0.8851 \times 30.86 \times 20.74) + 126.1 = 153.63\ \text{dB}$$

Equation (6.5) is used to calculate PWL's acceptance level.

$$\text{PWL}_\text{A} = 173.6 - 0.125 \times \left(\frac{D_\text{i}}{t}\right)$$

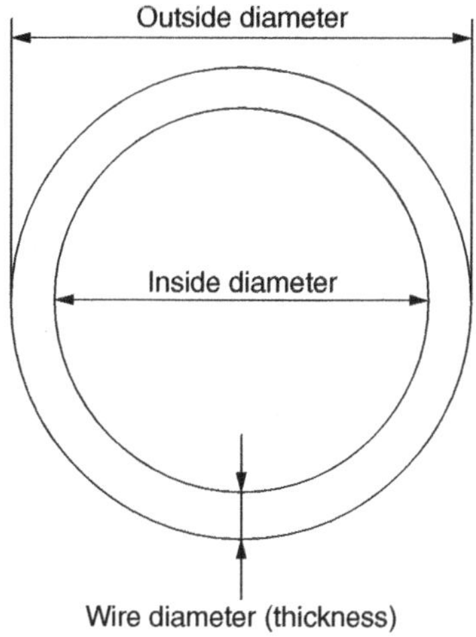

Figure 6.6 A schematic of a pipe showing the internal and external diameters, as well as the wall thickness.

In order to determine the PWL's acceptance level, we should know the internal diameter and thickness of the outlet piping. As per ASME B36.10/19, standards for carbon, alloy, and stainless steel pipes, an 8″ pipe with a standard (STD) thickness has an outside diameter of 8.625″ and a thickness of 0.322″. Equation (6.6) gives the relationship between the pipe's internal and external diameters as well as its wall thickness, as shown in Figure 6.6.

Relationship Between the Internal and External Diameters of a Pipe, As Well As Its Wall Thickness

$$\mathrm{ID} = \mathrm{OD} - 2 \times t \tag{6.6}$$

where:

ID: pipe internal diameter (in., mm);
OD: pipe outside diameter (in., mm);
t: Pipe wall thickness (in., mm);
$t = 0.322''$ and $\mathrm{OD} = 8.625'' \rightarrow \mathrm{ID} = 8.625'' - 2 \times 0.322'' = 7.981''$.

$$\mathrm{PWL_A} = 173.6 - 0.125 \times \left(\frac{7.981}{0.322}\right) = 170.50\ \mathrm{dB}$$

The noise level produced by the safety relief valve is 153.63 dB, which is lower than both 155 and 170.50 dB. Thus, there is no risk of acoustic fatigue in the relief valve.

6.4.1.2 Mach Number

In order to avoid vibration and noise in control and PSVs, fluid velocity must be considered an extremely crucial parameter. By reducing the valve trim exit velocity, it is logical to achieve low noise and vibration. It has been a practice in the valve industry to use a Mach number at the valve outlet to measure the velocity limitation of the fluid. In fluid mechanics, the Mach number is defined as the ratio of fluid velocity to sound velocity at a given temperature. The Mach number was named after Austrian physicist Ernst Mach (1838–1916). The Mach Number is an extremely crucial number in valves for reducing not only noise and vibration but also erosion damage. Most valve manufacturers recommend a maximum Mach number of 0.3 to avoid noise and vibration issues. Reducing the Mach number can be achieved by increasing the valve size or using a silencer. In the case of PSVs which have a high velocity at their outlet, an expander should be installed close to the PSVs in order to increase the diameter of the line and limit the velocity and Mach number. Generally, the Mach number is taken into account for the outlet of valves according to Eq. (6.7).

Calculation of the Mach Number at the Valve's Outlet

$$M_{\text{outlet}} = \frac{V_{\text{outlet}}}{C} \tag{6.7}$$

where:

M_{outlet}: Outlet Mach number;
V_{outlet}: Fluid velocity at the trim exit or valve outlet velocity (m/s);
C: Sonic or sound velocity (m/s) calculated using Eq. (6.8) for gas services.

Calculation of Sonic Velocity in a Gas

$$C = \sqrt{\frac{\gamma.R.T}{M}} \tag{6.8}$$

where:

C: Sonic velocity (m/s);
γ: Gas-specific heat ratio (J/(kg · K));
R: Universal gas constant = 8315 J/(kg · K);
M: Gas molecular weight (g/mol);
T: Gas temperature (K).

Example 6.3 Calculate the speed of sound in air at 0 °C and 1 bar absolute pressure. The specific heat ratio of air is 1.4 J/(kg · K) and the air molecular weight is 28.97 g/mol. Calculate the sound velocity if the air temperature exceeds to 20 °C.

Answer

Speed of sound in air at

$$0°\text{C}(273.15°\text{K}) = C_1 = \sqrt{(\gamma.R.T_1)/M} = \sqrt{(1.4 \times 8315 \times 273.15)/28.97} = 331.3\,\text{m/s}$$

Speed of sound in air at

$$20°\text{C}(293.15°\text{K}) = C_2 = \sqrt{(\gamma.R.T_2)/M} = \sqrt{(1.4 \times 8315 \times 293.15)/28.97} = 343.23\,\text{m/s}$$

It is possible to calculate the speed of sound in gases, liquids, and solids using Eq. (6.9).

Calculation of Sonic Velocity in Gases, Fluids, and Solids

$$C = \sqrt{\frac{K}{\rho}} \tag{6.9}$$

where:

C: Sonic velocity (m/s);
K: Bulk modulus of elasticity (Pa);
ρ: Density of gas, fluid or solid (kg/m^3).

The bulk modulus of elasticity is a measure of an object's resistance to being deformed elastically within its volume. Table 6.2 provides density and bulk modulus of elasticity values for four substances: water, oil, ethylene alcohol and mercury at 1 bar and 0 °C.

Table 6.2 The density and bulk modulus of elasticity of four substances: water, oil, ethylene alcohol, and mercury at 0 °C and 1 bar.

Substance	Bulk modulus of elasticity (K) 10^9 Pa	Density ρ (kg/m^3)
Water	2.15	999.8
Oil	1.35	920
Ethylene alcohol	1.06	810
Mercury	28.5	13,595

Example 6.4 Calculate the speed of sound in water at 0 °C.

Answer

$$C = \sqrt{\frac{K}{\rho}} = \sqrt{\frac{2.15 \times 10^9}{999.8}} = 1466 \text{ m/s}$$

The fluid velocity at the trim exit or valve outlet velocity is calculated according to Eq. (6.10):

Fluid Velocity at the Trim Exist or Valve Outlet Velocity

$$V_{\text{outlet}} = \frac{w}{A.\rho} = \frac{w}{\pi r^2 \rho} \quad (6.10)$$

where:

V_{outlet}: Fluid velocity at the trim exit or valve outlet velocity (m/s);
w: Mass flow rate (kg/m^3);
ρ: Density of gas or liquid (kg/m^3);
A: A valve outlet area equal to the diameter of the outlet pipe connected to the valve (m^2);
r: A valve outlet radius equal to the internal radius of the outlet pipe connected to the valve (m).

The flow velocity head at the outlet of the valve is also calculated by using Eq. (6.11):

Flow Velocity Head at the Valve's Outlet

$$P = 1/2\,\rho V_{\text{outlet}}^2 \quad (6.11)$$

where:

P: Valve exit velocity head or pressure (Pa);
ρ: Density of gas or liquid (kg/m^3);
V_{outlet}: Fluid velocity at the trim exit or valve outlet velocity (m/s).

The density of a gas is defined as the mass of the gas divided by its volume, as per Eq. (6.12).

Basic Gas Density Calculation

$$\rho_g = \frac{m_g}{V_g} \tag{6.12}$$

where:

ρ_g : Density of gas (kg/m^3);
m_g : Mass of the gas (kg);
V_g : Volume of the gas (m^3).

A1-pound mole gas has a weight of m_g, which is equal to the molecular weight of the gas, m_g. For a 3-lb mole gas, the gas weight, m_g, would be triple the molecular weight, m_g, of the gas. As a general rule, the gas weight is equal to the number of moles multiplied by the molecular weight of the gas, as per Eq. (6.13). The next step is to convert the volume of the gas to other properties, such as pressure and temperature according to the gas law, using Eq. (6.14).

Calculation of Number of Moles in a Gas

$$n = \frac{m}{M} \tag{6.13}$$

where:

n: Number of lb-mol;
m: Mass of the gas (lb or g);
M: Molecular weight of the gas (lb · mol or g · mol).

Gas Law Calculation

$$\begin{aligned} &PV = ZnRT = Z\frac{m}{M}RT \\ &\rightarrow \frac{V}{m} = \frac{ZRT}{MP} \rightarrow \frac{m}{V} = \frac{MP}{ZRT} \rightarrow \rho = \frac{MP}{ZRT} \end{aligned} \tag{6.14}$$

ρ: Density of gas (kg/m^3)
V: Volume of the gas (m^3);
Z: Gas compressibility factor (dimensionless);
R: Gas universal constant;
P: Gas pressure (Pa);
T: Gas temperature (K)

Based on Eq. (6.7), since the Mach number is measured at the outlet of the valve, this equation for calculating gas density can be modified in the following manner:

Calculation of Gas Density Based on Pressure and Temperature

$$\rho = \frac{MP2}{ZRT2} \quad (6.15)$$

Using Eqs. (6.7), (6.8), (6.10), and (6.15) together, the calculation of the Mach number at the valve outlet can be modified as follows:

$$M_{\text{outlet}} = \frac{V_{\text{outlet}}}{C} \quad (6.7)$$

$$C = \sqrt{\frac{\gamma.R.T}{M}} \quad (6.8)$$

$$V_{\text{outlet}} = \frac{w}{A.\rho} = \frac{w}{\pi r^2 \rho} \quad (6.10)$$

$$\rho = \frac{MP2}{ZRT2} \quad (6.15)$$

$$\rightarrow M = \frac{w}{\pi r^2} \times \frac{ZRT2}{MP2} \times \sqrt{\frac{M}{\gamma.R.T}} = \left(\frac{4Z\sqrt{R}}{\pi} \times \frac{w}{P_2 D_i^2} \times \sqrt{\frac{T_2}{\gamma \cdot m}}\right) \quad (6.16)$$

In the case of an ideal gas with $Z = 1$, and $R = 8315$, the calculation of the Mach number is simplified as shown by Eq. (6.17).

Calculation of Mach Number for Ideal Gases Based on the Process Parameters

$$M = 116 \times \frac{W}{P_2 D_i^2} \times \sqrt{\frac{T_2}{\gamma \cdot m}} \quad (6.17)$$

where:

W: Flow rate of gas and liquid (kg/s);
P_2: Downstream pressure (Pa);
D_i: Outlet pipe inside diameter (m);
T_2: Downstream temperature (K);
γ: Ratio of specific heat (dimensionless);
m: Gas molecular weight (g/mol).

Example 6.5 Determine the Mach number for the gas service passing through the safety valve in Example 6.2. The ratio of specific heat is 1.21 and the gas is ideal.

Answer

P_2: the downstream pressure of the relief value = 1 bara = 10^5 Pa
T_1: the upstream temperature of the valve = 39.85 °C = 39.85 + 273.15
= 313 °K
W: Gas or liquid flow rate = 20,000 kg/h = 5.56 kg/s
M: Molecular weight = 25
γ: Ratio of specific heat = 1.21

$$\rightarrow \text{Diameter} = D_\text{i} = \text{ID} = 8.625'' - 2 \times 0.322'' = 7.981'' = 0.2027\ \text{m}$$

z factor = 1

By using Eq. (6.17) $\rightarrow M = 116 \times \left(W/\left(P_2 D_\text{i}^2\right)\right) \times \sqrt{T_2/(\gamma \cdot m)} = 116 \times (5.56/(10^5 \times 0.2027^2)) \times \sqrt{313/(1.2 \times 25)} = 0.1570 \times 3.22 = 0.50$

6.4.2 Noise in Control Valves

As discussed in Chapter 3, noise may be generated as a result of cavitation in control valves. Reducing pressure with valves, in particular control valves, can cause unwanted and bothersome noises. As a matter of fact, control valves often generate a lot of noise and contribute significantly to the overall noise level at a plant. Most control system and valve designers specify a maximum noise level of 90 dB in order to comply with noise standards set by OSHA, the United States Department of Labour's Occupational Safety and Health Administration. When calculating control valve noise, fluid noise is the most critical factor taken into account, including aerodynamic noise and hydrodynamic noise.

6.4.2.1 Aerodynamic Noise in Control Valves

According to the earlier discussion, aerodynamic noise is caused by the conversion of mechanical energy associated with the turbulent flow into acoustic energy through valve restrictions. Many oil and gas companies use IEC 60534-8:3: *Industrial Process Control Valves – Part 8 – Noise Considerations – Section 3 – Control Valve Aerodynamic Noise Prediction Method.* It is worth mentioning that the parameters that are listed in Table 6.3 are those that are used in the aerodynamic noise prediction methodology based on the aforementioned standard. A sequential method is described in this section for determining the aerodynamic noise from control valves using a sequential model. It should be noted that, although this method is mainly associated with regime I, it also gives an overview of how to calculate this type of noise in general. In the next paragraph, a variety of flow regimes based on IEC 60534-8:3 will be briefly described.

Table 6.3 Nomenclature in IEC 60534-8:3.

A: Flow area	P: Pressure
C_v: Valve flow coefficient	p_a: Actual pressure outside pipe
C: Gas sound speed	p_s: Standard pressure outside pipe (1 Atmosphere)
C_p: Sound speed in piping	R: Universal gas constant
d_p: Outlet pipe inner diameter	T: Gas temperature
d_v: Outlet valve inner diameter	TL: Pipe transition loss
d_{vc}: Trim jet vena contracta diameter	TL_{fr}: Pipe transition loss at ring frequency
F_d: Valve style modifier	t_p: Pipe wall thickness
F_L: Pressure recovery coefficient	U: Velocity
f: Sound frequency	W_a: Sound power
f_o: Acoustic-structural coincident frequency	W_m: Stream power of mass flow
f_p: Flow peak frequency	γ: Ratio of specific heat
f_r: Pipe ring frequency	η: Acoustical efficiency factor
L_e: Expander and pipe flow noise sound pressure level, A-weighted and 1 m from the pipe wall	ρ: Gas density
L_g: Pipe Mach number correction factor	1: Upstream of valve or vena contracta
L_{pAe}: A weighted sound pressure level	2: Downstream of valve or vena contracta
$\dot{m}$: Mass flow rate	e: Expander
L_{pAe}: A-weighted sound-pressure level, 1 m from the pipe wall on the outlet side	v: Valve outlet
L_{pi}: Pipe internal sound-pressure level	vc: Vena contracta
M: Mach number M_{vc}: Mach number at vena contracta M_w: Molecular weight of gas	I: Regime I

In the IEC standard, there are five flow regimes that are defined based on process parameters such as inlet pressure, downstream pressure, fluid physical data, and valve pressure recovery factor. These flow regimes are as follows:

- **Regime I**: Subsonic
- **Regime II**: Sonic with turbulent flow mixing
- **Regime III**: No compression but with flow shear mechanism
- **Regime IV**: Sock cell turbulent flow interaction
- **Regime V**: Constant acoustical efficiency that produces maximum noise

According to the standard, the methodology consists of seven basic steps as follows:

1) Calculation of total mechanical stream power W_m at the vena contracta or fluid restriction point. In fact, the aerodynamic noise is caused by valves inside and downstream of the valve.
2) If the total fluid power at the vena contracta is calculated, then the fraction that accounts for noise power may be determined according to the acoustic efficiency factor η.
3) In this step, the acoustic power ratio r_w, also known as the ratio between the sound power in the downstream pipe and the sound power inside the valve, is determined.
4) Calculation of the flow peak frequency f_p for the internal noise inside the pipe.
5) Calculation of the internal sound pressure level L_{pi}.
6) Calculation of the transmission loss, TL.
7) Calculation of the sound pressure level $L_{pAe.1\,m}$ 1 m away from the pipe on the outlet side.

For the purpose of implementing these steps, the first step is to calculate the noise source magnitude W_m according to Eq. (6.18) as follows:

Noise Source Magnitude

$$W_m = \frac{U_{vc}^2 \times \dot{m}}{2} \tag{6.18}$$

The acoustic efficiency factor η can be calculated as per Eq. (6.19) in the following manner:

Calculation of Acoustic Efficiency Factor

$$\eta = 0.0001 \times M_{vc}^{3.6} \tag{6.19}$$

It is important to note that the peak noise frequency depends on the velocity at vena contracta and the jet vena contracta diameter. Jet vena contracta diameter is a function of jet pressure recovery and valve style modifier. The next step is to calculate noise peak frequency f_p according to Eq. (6.20) as follows:

Calculation of Noise Peak Frequency

$$f_p = \frac{0.2U_{vc}}{d_{vc}} \tag{6.20}$$

Table 6.4 Valve noise propagation values.

Valve type and style	r_w
Globe valve	0.25
Rotary globe (varimax) valve	0.25
Eccentric rotary plug valve	0.25
Ball valve	0.5
Butterfly valve	0.5

In the next step, we need to calculate the noise propagation through the valve downstream of the valve that is designated with parameter r_w. The ratio of noise propagation through the valve varies depending on the type and style of valve as indicated in Table 6.4.

Using Eq. (6.21), it is possible to determine the sound pressure level in the downstream piping.

Calculation of Sound Pressure Level in the Downstream Piping

$$L_{pi} = 10 \log \frac{3.2 \times 10^9 W_a \rho_2 C_2}{d_p^2} \tag{6.21}$$

Earlier, it was explained that increased noise levels can result from increasing Mach numbers M_2. In fact, a higher Mach number in the downstream piping can increase the noise level by the value of L_g as shown in Eq. (6.22) as follows:

Calculation of the Noise Reduction Obtained by Reducing the Mach Number

$$L_g = 16 \log \left(\frac{1}{1 - M_2} \right)_{10} \tag{6.22}$$

By using Eq. (6.23), we are able to determine the loss of sound transmission due to the downstream piping. In addition to this, it is important to pay attention to the fact that increasing a pipe's wall thickness leads to an increase in noise loss.

Calculation of the Basic Sound Transmission Loss Through Piping Wall

$$\mathrm{TL} = 10 \log 7.6 \times 10^{-7} \times \left[\left(\frac{C_2}{t_p f_p} \right)^2 \times \frac{G_X}{(\rho_2 C_2 / 415 G_Y) + 1} \times \left(\frac{p_a}{p_s} \right) \right] \tag{6.23}$$

Transmission loss is determined by frequencies that are calculated according to Eqs. (6.24) and (6.25). There are two types of frequencies that are calculated in order to determine the value for transmission loss; one is pipe ring frequency f_r and the other is pipe coincidence frequency (Minimum Transmission Losses) f_0.

Calculation of the Pipe Ring Frequency

$$f_r = \frac{\text{Pipe material sound speed}}{\text{Pipe circumference}} = \frac{C_p}{\pi d_p} \tag{6.24}$$

Calculation of the Pipe Coincidence Frequency

$$f_0 = \frac{\text{Pipe circumferencial bending wave speed}}{\text{Pipe acoustic wave speed}} = \frac{f_r}{4} \times \frac{C_2}{C_{air}} \tag{6.25}$$

It should be noted that f_0 is less than f_r ($f_0 < f_r$) as illustrated in Figure 6.7.

There is a possibility that three conditions could arise regarding the relationship between frequency and transmission loss as follows:

1) If $f_p < f_0$, there will be a greater amount of transmission loss. This condition can happen in the standard valves.
2) If $f_p = f_0$, there will be the smallest amount of transmission loss. This condition can happen in the standard valves.
3) If $f_p > f_0$, there is a significant increase in transmission loss. This condition can happen in low-noise trims.

In order to calculate the slope of the transmission loss in each of the three regimes, we can use the following relationships, which are given in Eq. (6.26).

Figure 6.7 Relationship between frequency and transmission loss.

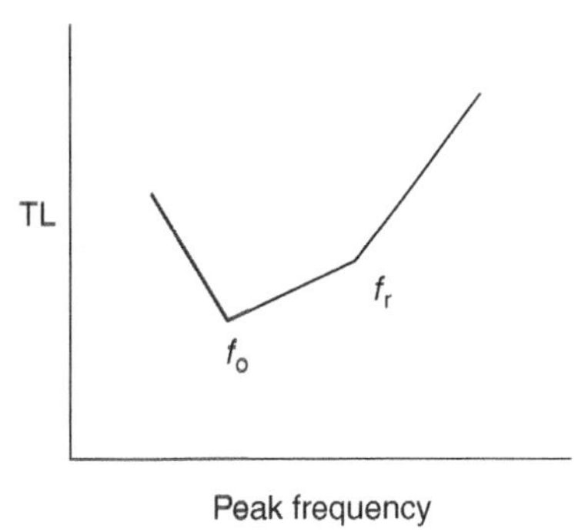

Calculation of Transmission Losses in Three Regimes

$$\begin{aligned}
&\text{If } f_p < f_0 \rightarrow \Delta\text{TL} = 20\log\left(\frac{f_0}{f_p}\right) + 13\log\left(\frac{f_0}{f_r}\right)\\
&\text{If } f_0 < f_p < f_r \rightarrow \Delta\text{TL} = 13\log\left(\frac{f_0}{f_p}\right)\\
&\text{If } f_r < f_p \rightarrow \Delta\text{TL} = 20\log\left(\frac{f_p}{f_r}\right)
\end{aligned} \tag{6.26}$$

There is an important reason for knowing that a higher f_p (smaller d_{vc}) can increase the damping of the piping and reduce the noise associated with the control valve throttling.

The net aerodynamic sound level at the pipe wall is converted to dB as per Eq. (6.27) as follows:

Calculation of the Net Aerodynamic Sound Level at the Pipe Wall

$$L_{pAe} = 5 + L_{pi} + \text{TL} + L_g \tag{6.27}$$

The aerodynamic noise level at 1 m from the pipe wall is calculated with Eq. (6.28).

Calculation of the Net Aerodynamic Sound Level at the 1 m from the Pipe Wall

$$L_{\text{pAe,1 m}} = L_{\text{pAe}-10\log\left(\frac{d_p 2\times t_p + 2}{d_p + 2\times t_p}\right)} \tag{6.28}$$

6.4.2.2 Hydrodynamic Noise in Control Valves

It is common to use the IEC 60534-8:4: *Industrial Process Control Valves – Part 8 – Noise Considerations – Section 4 – Prediction of Noise Generated by Hydrodynamic Flow, for Control Valves*. Restriction of the valve converts fluid pressure energy into turbulence, heat, and mechanical waves that produce noise and vibration inside the valve and the pipes that connect to it. Even though cavitation and flashing are the main causes of hydrodynamic noise generation, as discussed earlier in this chapter, a turbulent flow can also result in hydrodynamic noise. Note that the hydrodynamic noise generated by flashing, unstable, and unpredictable flow conditions, as well as mechanical vibrations, is not included in the aforementioned standard, and as such, this standard only takes into account noise caused by cavitation and hydraulic turbulence. The standard provides a methodology for noise prediction based on fundamental principles of acoustics, fluid mechanics, and mechanics. Moreover, the method described in this section for noise prediction can be applied to all types of valves used for flow control or throttling such as globe, butterfly, and V-notch ball valves. In this standard, the prediction noise method has been validated by using a water test for nearly all kinds of control valves, with an inlet pressure of up to 15 bar. Test results indicate that this method is accurate in the majority of cases within the range of plus or minus dB. The hydrodynamic noise calculations are divided into two categories: preliminary calculations and noise calculations. There are nine preliminary calculations:

6.4.2.2.1 Preliminary Calculations

Pressure and Pressure Ratios In order to predict and measure noise, different pressures and pressure ratios are needed. Differential pressure ratios x_F for liquids depend on the differential pressure across the valve $P_1 - P_2$ and the difference between the inlet and vapor pressure $P_1 - P_v$, as calculated in Eq. (6.29). As discussed in Chapter 3, the pressure ratio is calculated in much the same way as the cavity index is calculated in order to evaluate cavitation.

Differential Pressure Ratio Calculation

$$x_F = \frac{P_1 - P_2}{P_1 - P_v} \tag{6.29}$$

The differential pressure for starting the choked flow is approximately equal to $F_L^2(P_1 - P_v)$ where F_L is the liquid pressure recovery factor of the valve without attached fitting. Choked flow is a fluid dynamic phenomenon associated with the venturi effect. A liquid pressure recovery factor predicts how much pressure recovery will happen after the vena contracta at the valve outlet. This experimentally determined coefficient is a dimensionless number that accounts for the influence of the internal geometry of the valve on the maximum capacity of the valve. From Eq. (6.30), it can be calculated what the differential pressure is at the vena contracta ΔP_c.

Differential Pressure Calculation at the Vena Contracta

$$\Delta P_c = \text{Lower than } x_F(P_1 - P_v) \text{ or } F_L^2(P_1 - P_v) \tag{6.30}$$

In low-pressure differentials, noise is typically generated by turbulent flow. Whenever x_F exceeds the characteristic pressure ratio x_{Fz}, cavitation noise overwhelms the turbulent noise. The following subsection discusses the characteristic pressure ratio.

Characteristic Pressure Ratio The valve characteristic pressure ratio x_{Fz} identifies the pressure ratio at which cavitation is detected. x_{Fz} is dependent on a number of factors, including the type of valve, its trim shape and type, and the valve flow capacity. The value of x_{Fz} can be estimated using either Eq. (6.31) or Eq. (6.32), depending on the type of valve trim (internal).

Characteristic Pressure Ratio Calculation

$$x_{FZ} = \frac{0.90}{\sqrt{1 + 3F_d\sqrt{C/(N_{34} \times F_L)}}} \text{(For valve types except multi-hole trims)} \tag{6.31}$$

$$x_{FZ} = \frac{1}{\sqrt{4.5 + 1650 \times \left((N_0 d_H^2)/(F_L)\right)}} \text{(For multi-hole trims)} \tag{6.32}$$

where:

F_d: Based on the type of valve and closure member as well as the flow coefficient, a valve style modifier is determined. This dimensionless value must be provided by the valve manufacturer or calculated according to Eq. (6.33);

C: Valve flow coefficient (C_v or K_v);

N_0: Number of independent and identical flow passages in the valve trim or number of stages of throttling in the valve trim;

d_H: Multihole trim hole diameter (m);

N_{34}: It is a numeric constant whose value depends on the type of flow coefficient (C_v or K_v) used in the calculation of the characteristic pressure ratio. $N_{34} = 1$ for K_v, and $N_{34} = 1.17$ for C_v.

Calculation of Valve Style Modifier

$$F_d = \sqrt{\frac{1}{N_0}} \tag{6.33}$$

If x_{Fz} is obtained from testing at an inlet pressure of 6×10^5 Pa, then Eq. (6.34) must be applied to correct for the actual inlet pressure.

Calculation of Differential Pressure Ratio Corrected for Inlet Pressure

$$x_{\mathrm{Fzp1}} = x_{\mathrm{Fz}}\left(\frac{6\times 10^{5}}{P_1}\right)^{0.125} \tag{6.34}$$

Fluid Jet Diameter and Velocity Jets are streams of fluid that are projected into nozzles or orifices in piping systems and can travel a considerable distance due to their high momentum. In reality, the velocity of the fluid is increased within the orifice or nozzle while its pressure is reduced. A great deal of energy is contained within the fluid which is converted into kinetic energy and high-speed fluid jets. The diameter of the jet fluid can be predicted from Eq. (6.35).

Calculation of Fluid Jet Diameter

$$D_{\mathrm{J}} = N_{14}F_{\mathrm{d}}\sqrt{CF_{\mathrm{L}}} \tag{6.35}$$

where:

$N_{14} = 4.9 \times 10^{-3}$ When K_{v} is used in the equation. Otherwise, $N_{14} = 4.6 \times 10^{-3}$ if C_{v} is used.

In Eq. (6.36), the vena contracta velocity U_{vc}, which is an important parameter for determining the mechanical power, is calculated.

Calculation of Fluid Jet Velocity

$$U_{\mathrm{vc}} = \frac{1}{F_{\mathrm{L}}}\sqrt{\frac{2\Delta P_{\mathrm{c}}}{\rho_{\mathrm{L}}}} \tag{6.36}$$

where:

U_{vc}: Vena contracta velocity (m/s);
F_{L}: Liquid pressure recovery factor of the valve without attached fitting (dimensionless);
ΔP_{c}: Differential pressure at the vena contracta calculated from Eq. (6.30) (Pa);
ρ_{L}: Density of liquid or fluid (kg/m^3).

By using Eq. (6.37), the mechanical energy or power W_{m} dissipated in the orifice or valve vena contracta can be determined.

Calculation of Mechanical Fluid Stream Power

$$W_{\mathrm{m}} = \frac{m \times U_{\mathrm{vc}}^{2}F_{\mathrm{L}}^{2}}{2} \tag{6.37}$$

where:

W_m: Mechanical power (W);
m: Mass flow rate (kg/s).

Example 6.6 When a single-seat 4″ globe valve without multihole trim is installed on a 4″ water pipe, it has a flow coefficient C_v of 90. In addition, the valve style modifier is 0.42 while the pressure recovery factor is 0.92. The valve has inlet and outlet pressures of 10 bar and 8 bar, respectively. However, the liquid passing through the valve has a vapor pressure of 2.32×10^3. Knowing that the density of the water is 997 kg/m^3 and its mass flow rate is 30 kg/s, calculate the fluid jet diameter, the velocity of the water at the vena contracta as well as the mechanical power produced by the fluid stream. What is the cause of the noise produced in the valve? Is it mainly due to turbulent flow or cavitation?

Answer

$$P_1 = 10 \text{ bar} = 10^6 \text{ Pa}, \quad P_2 = 8 \text{ bar} = 8 \times 10^5, \quad P_v = 2.32 \times 10^3$$

Differential pressure ratio: $x_F = (P_1 - P_2)/(P_1 - P_v) = (0.2 \times 10^6)/9{,}997{,}680 = 0.0200$

Pressure differential or choked pressure across the vena contracta: $\Delta P_c =$ Lower than $x_F(P_1 - P_v)$ or $F_L^2(P_1 - P_v)$

$$x_F(P_1 - P_v) = 0.020 \times (9{,}997{,}680) = 2 \times 10^5 \text{ Pa}$$

$$F_L^2(P_1 - P_v) = (0.92)^2 \times (9{,}997{,}680) = 8.46 \times 10^5 \text{ Pa}$$

$$\rightarrow \Delta P_c = 2 \times 10^5 \text{ Pa}$$

Characteristic pressure ratio: $x_{FZ} = 0.90/\left(\sqrt{1 + 3F_d\sqrt{C/(N_{34} \times F_L)}}\right)$ (For valve types except for multihole trims) $\rightarrow x_{FZ} = 0.90/\left(\sqrt{1 + 3 \times 0.42\sqrt{90/(1.17 \times 0.92)}}\right) = 0.9/3.54 = 0.254 > x_F = 0.0200$

Since the characteristic pressure ratio is more than the differential pressure ratio, turbulence flow is the primary cause of noise in the valve.

Characteristic pressure ratio corrected for the inlet pressure: $x_{Fzp1} = x_{Fz}((6 \times 10^5)/P_1)^{0.125} = 0.254((6 \times 10^5)/10^6)^{0.125} = 0.238$

Water jet diameter: $D_J = N_{14}F_d\sqrt{CF_L} = 4.6 \times 10^{-3} \times 0.42\sqrt{90 \times 0.92} = 0.01758$ m

Vena contracta velocity: $U_{vc} = 1/F_L\sqrt{(2\Delta P_c)/\rho_L} = 1/0.92 \times \sqrt{(4 \times 10^5)/997} = 21.77$ m/s

Mechanical water stream power: $W_m = (m \times U_{vc}^2 F_L^2)/2 = (30 \times (21.77)^2 (0.92)^2)/2 = 6017\text{W}$

6.4.2.2.2 Secondary Calculations (Noise Prediction)

Based on the function of acoustic efficiency, an amount of the mechanical power W_m computed from Eq. (6.37) is converted to the valve internal noise W_a. Acoustic conversion efficiency, also called acoustic efficiency factor η, describes the percentage of fluid power that is converted to the noise inside the valve. The acoustic power ratio r_w represents the fracture of sound power radiated into the pipe. There are two different sets of equations and conditions that are used to calculate the sound noise; the first is during flow turbulent conditions (Eq. 6.38), while the second is during cavitation (Eq. 6.39).

Noise Calculation for Turbulent Flow

$$\text{If } \Delta P < x_{Fzp1}(P_1 - P_v) \rightarrow W_a = \eta_{turbulant} \times W_m \times r_w \tag{6.38}$$

Noise Calculation for Cavitation Flow Condition

$$\begin{aligned} &\text{If } \Delta P > x_{Fzp1}(P_1 - P_v) \text{ and } x_{Fzp1} > x_F > 1 \rightarrow W_a \\ &= (\eta_{turbulant} + \eta_{cavitation}) \times W_m \times r_w \end{aligned} \tag{6.39}$$

where:

W_m: Mechanical power (W);
W_a: Sound power (W);
$\eta_{turbulant}$: Acoustic efficiency factor at fluid turbulent condition (dimensionless) calculated from Eq. (6.40);
$\eta_{cavitation}$: Acoustic efficiency factor at fluid cavitation condition (dimensionless) calculated from Eq. (6.41);
r_w: Acoustic power ratio (dimensionless).

r_w is determined by the type of valve, and it can be found in Table 6.5.

Table 6.5 Acoustic power ratio r_w.

Valve or fitting	r_w
Globe	0.25
Butterfly	0.5
Eccentric rotary plug	0.25
Ball	0.25
Expanders	1

Acoustic Efficiency Factor for Turbulent Flow

$$\eta_{\text{turbulant}} = 10^{-4}\left(\frac{U_{\text{vc}}}{C_{\text{L}}}\right) \tag{6.40}$$

C_{L}: Speed of sound in liquid (m/s)

Cavitation is the second part of the process after the liquid is converted to vapor bubbles. This occurs at the vena contracta or at a point with a high-pressure drop and increased fluid velocity. The second part of this process is the collapse of the vapor bubbles due to the fluid pressure exceeding the vapor pressure. As the bubbles collapse, a large amount of jet energy is created which is a source of noise and serious damage to the valve. Water is known to be one of the most destructive fluids during cavitation. In the cavitation region, $x_{\text{Fzp1}} > x_{\text{F}} > 1 \eta_{\text{cavitation}}$ is calculated as follows:

Acoustic Efficiency Factor for Cavitation Flow

$$\eta_{\text{cavitation}} = 0.32\eta_{\text{turbulant}}\sqrt{\frac{P_1 - P_2}{\Delta P_{\text{c}}}\cdot\frac{1}{x_{\text{Fzp1}}}} \times e^{5x_{\text{Fzp1}}} \times \left(\frac{1 - x_{\text{Fzp1}}}{1 - x_{\text{F}}}\right)^{0.5} \times \left(\frac{x_{\text{F}}}{x_{\text{Fzp1}}}\right)^{0.5} \times \left(x_{\text{F}} - x_{\text{Fzp1}}\right)^{1.5} \tag{6.41}$$

Example 6.7 In the previous example, it was necessary to determine whether the fluid was turbulent or cavitating. In addition, what are the values of the acoustic efficiency factor and the sound power?

Answer

$$P_1 = 10\text{ bar} = 10^6\text{ Pa}, \quad P_2 = 8\text{ bar} = 8 \times 10^5$$

$$\Delta P = 0.2 \times 10^6\text{ Pa} = 2 \times 10^5\text{ Pa}$$

$$x_{\text{Fzp1}}(P_1 - P_{\text{v}}) = 0.238\left(10^6 - 2.32 \times 10^3\right) = 0.238 \times 9{,}997{,}680$$
$$= 2.38 \times 10^5\text{ Pa}$$

$$\rightarrow \Delta P < x_{\text{Fzp1}}(P_1 - P_{\text{v}}) \rightarrow \text{The flow is turbulent}$$

Using Eq. (6.40), we can calculate the acoustic efficiency factor for the turbulent flow as follows:

$$\eta_{\text{turbulant}} = 10^{-4}\left(\frac{U_{\text{vc}}}{C_{\text{L}}}\right) = 10^{-4}\left(\frac{21.77}{1482}\right) = 1.47 \times 10^{-6}$$

It is important to know that parameter C_L is the speed of sound in the water that is equal to 1482 m/s in this case.

In terms of acoustic power ratio, r_w.The globe valve has a value of 0.25. It is now possible to calculate the noise power for a turbulent flow as per Eq. (6.38) according to what follows:

$$W_a = \eta_{turbulant} \times W_m \times r_w = 1.47 \times 10^{-6} \times 6017 \times 0.25 = 0.00234 \text{ W}$$

6.4.3 Noise in Pressure Safety or Relief Valves

However, safety valves are not the primary issue when considering noise emission, safety valves are evaluated more and more with regard to noise emission, especially if they discharge to open air, which can cause significant noise pollution within a short space of time. The high noise levels produced by pressure-reducing devices, such as PSVs, contribute to high-stress levels in the piping system and lead to damage as illustrated in Figure 6.8.

The noise calculations are generally based on the expansion of the steam or gas at the end of the pipe. Specific characteristics of safety valves, such as the geometry of the valve outlet, are not taken into account in the calculations. Noise emission tests are not typically conducted on safety valves. Additionally, the frequency of

Figure 6.8 Failure of the relief piping system connected to a PSV due to noise and acoustic fatigue.

the noise is not determined by the noise calculations for the safety valve. Also, worth mentioning is the fact that, unlike control valves, there is no low-noise trim available for safety valves. For some specifications, such as NORSOK L-002, piping system layout, design, and structure analysis, there is a maximum permissible noise limit, as shown in Eq. (6.5). In cases where the calculated noise level exceeds the maximum permissible level, different approaches, such as the use of a silencer, are recommended. Another method of reducing noise is to decrease the mass flow of air passing through the valve by limiting the opening of the valve. There are three ways to calculate the noise emission level. They are as follows:

- ISO 4126-9
- API 521
- VDI 2713

It should be noted that noise calculations based on all three standards are independent of the safety valve designs provided by manufacturers. Due to this, the design of the various safety valves does not affect the noise level as long as they have the same capacity. In general, two physical values that were discussed earlier in this chapter are relevant to the noise evaluation:

- Sound power level (PWL) is a measure of the amount of energy being generated and emitted by the noise source which is the safety valve. The sound power level is independent of the distance from the source of the noise, as previously explained.
- The sound pressure level describes the pressure variation caused by the noise source depending on the distance from it. This is the type of noise that affects human hearing.

6.4.3.1 Calculation of Noise Emission According to ISO 4126-9

According to ISO 4126-9, safety valve noise calculations are based on symbols, designations, and units defined in Table 6.6.

The sound power level of the safety valve, PWL, expressed in dB, can be estimated by Eq. (6.42) as follows:

Calculation of the Sound Power Level of a Safety Valve

$$\mathrm{PWL} = 20\log\left(10^{-3}d_{\mathrm{A}}\right) - 10\log v + 80\log u - 53 \tag{6.42}$$

It is possible to calculate the sound pressure level SPL_r, expressed in dB at a distance r from the point of discharge to the atmosphere by using Eq. (6.43) as follows:

Table 6.6 Symbols, designations, and units used in calculating noise for safety valves as defined in ISO 4126-9.

Used symbols	Designations	Units
d_A	Internal diameter of outlet pipe	mm
v	Specific volume of the stream at relieving pressure and temperature	m^3/kg
u	Velocity of fluid in outlet pipe	m/s
r	Distance from noise source	M

Calculation of the Sound Pressure Level of a Safety Valve

$$\text{SPL}_r = \text{PWL} - 10 \log 2\pi r^2 \tag{6.43}$$

Example 6.8 It has a 6″-diameter pipe with a standard thickness or schedule and it is located on the outlet of a PSV. As the safety valve is dealing with natural gas whose specific volume is 1.2 m^3/kg in this instance, the gas volume will be handled safely. How much sound power level will there be at 2 and 10 m away from the safety valve if the velocity of the gas in the outlet pipe is 15 m/s? What is the sound pressure level near the safety valve at a distance of 2 and 0 m?

Answer

The outside diameter of the 6″ pipe, according to ASME B36.10, is 6.625″ (168.3 mm), which is equal to 168.3 mm. Furthermore, the wall thickness for a 6″ pipe with a standard wall thickness is 7.11 mm. So the internal diameter of the pipe is calculated as follows:

$$\begin{aligned} d_A &= \text{Pipe inside diameter} = \text{Pipe outside diameter} \\ &\quad - 2 \times \text{Thickness} = 168.3 - 2 \times 7.11 = 154.08 \text{ mm} \end{aligned}$$

The sound power level (PWL) that is calculated according to Eq. (6.42) is independent of the distance, meaning it is the same in both a 2-m and a 10-m distance.

$$\begin{aligned} \text{PWL} &= 20 \log(10^{-3} d_A) - 10 \log V + 80 \log u - 53 \\ &= 20 \log(10^{-3} \times 154.08) - 10 \log 1.2 + 80 \times \log 15 - 53 \\ &= -16.245 - 0.79 + 94.09 - 53 = 24 \text{ dB} \end{aligned}$$

Therefore, the sound power level (PWL) is 24 dB both at a distance of 2 and 10 m from the safety valve. As a next step, the noise pressure level (SPL) needs to be calculated at a distance of two and ten meters from the safety valve.

$$\mathrm{SPL}_2 = \mathrm{PWL} - 10\log 2\pi r^2 = 24 - 10\log 2\times\pi\times(2)^2$$
$$= 24 - 10\times 1.4 = 10\,\mathrm{dB}$$

$$\mathrm{SPL}_{10} = \mathrm{PWL} - 10\log 2\pi r^2 = 24 - 10\log 2\times\pi\times(10)^2$$
$$= 24 - 10\times 2.8 = -4\,\mathrm{dB}$$

Accordingly, the PWL at 2 m and 10 m is 10 dB and −4 dB, respectively.

6.4.3.2 Calculation of Noise Emission According to API 521

Noise levels are normally calculated in accordance with API 521 at a distance of 30 m or 100 ft from the point of discharge. However, it is possible to calculate the noise level at other distances from the discharge of the safety valve using an equation which will be discussed further in this chapter. In Table 6.7, you are given a list of the symbols, designations, and units used in API 521 for the sizing of safety valves.

Table 6.7 Symbols, designations, and units used in calculating noise for safety valves as defined in API 521.

Used symbols	Designations	Units	
$L_{30\,(100)}$	Noise or sound pressure level at 30 m (100 ft) from the point of discharge	dB	
L	Noise or sound pressure level at the valve discharge as extracted from Figure 6.9	dB	
L_p	Sound pressure level at distance r	dB	
r	Distance from the source of the sound	m	ft
q_m	Mass flow through the valve	kg/s	lb/s
C	Speed of sound in the gas inside the valve that can be calculated from either Eq. (6.45) or Eq. (6.46)	m/s	ft/s
k	Ratio of the specific heats in the gas	—	
M	Relative molecular mass of the gas	—	
T	Gas temperature	k	°R
PR, X	Pressure ratio across a valve that is calculated as the ratio of relieving pressure to back pressure	—	
Y	Sound pressure level, $L_{30\,(100)}$ that can be determined from Figure 6.9 by considering the value of the pressure ratio or by using Eq. (6.44)	dB	

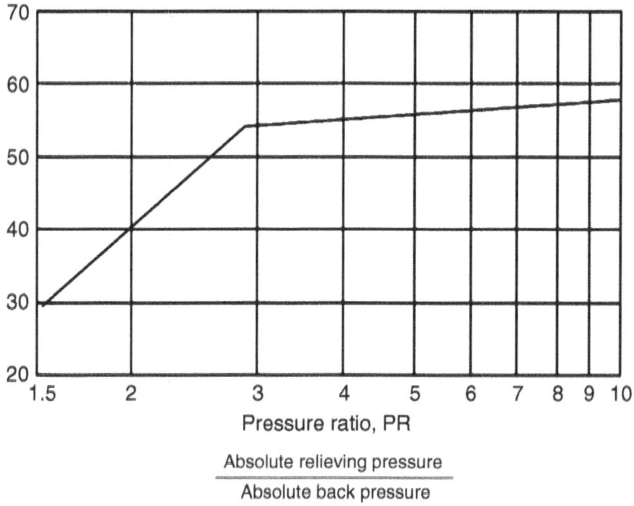

Figure 6.9 Noise intensity at the valve outlet $L = (L_{30} - 10\log_{10}(0.5q_m \times C^2))$ based on the pressure ratio according to API 521 standard.

In order to calculate the noise level at 30 m from the discharge point of the PSV to the atmosphere, Eq. (6.44) is applied:

Calculation of the Noise Level at 30 m from the PSV's Discharge Point to the Atmosphere

$$L_{30} = L + 10\log_{10}(0.5q_m \times C^2) \tag{6.44}$$

Calculation of the Speed of Sound in the Gas Inside the Valve

$$C = 91.2(kT/M^{0.5}) \text{ m/s (SI units)} \tag{6.45}$$

$$C = 223(kT/M^{0.5}) \text{ ft/s (USC)} \tag{6.46}$$

By applying Eqs. (6.47) and (6.48), the noise level L_p can be adjusted for distances that differ from 30 m (100 ft).

$$L_p = L_{30} - \left[20\log\left(\frac{r}{30}\right)\right] \text{ (SI units)} \tag{6.47}$$

$$L_p = L_{30} - \left[20\log\left(\frac{r}{30}\right)\right] \text{ (USC units)} \tag{6.48}$$

Example 6.9 Figure 6.10 illustrates a pressure vessel protection system in 12″ × 10″ for the protection of pressure vessels. The set pressure of the valve is 42 barg. So the inlet of the valve is Class 300, which is equal to 50 barg design pressure. In addition, the outlet line of the valve is Class 150. There is a mass flow rate of 59,507.55 kg/h passing through the PSV and the temperature of the fluid is 110 °C. In addition, the specific heat of the gas is 1.18 and the molecular weight of the gas is 21.88. There is a 7.50 bar back pressure on the valve. What are the sound pressure levels at both one meter and thirty meters away from the safety valve? In this case, it is important to note that there is a 10% overpressure on the safety valve.

Answer

The first step is to get parameter C by using Eq. (6.45). To do so, it is important to convert the temperature from degrees Celsius to Kelvin using the following formula.

$$\text{Gas temperature is } 110^{\circ}\text{C},$$

$$\text{Temperature}_{\text{Kelvin}} = \text{Temperature}_{\text{Centigrade}} + 273.15 \rightarrow \text{Temperature}_{\text{Kelvin}} = 110 + 273.15 = 383.15$$

$$C = 91.2\left(\frac{1.18 \times 383.15}{21.88}\right)^{0.5} = 414.57\ \text{m/s}$$

$$\text{Mass flow rate } q_{\text{m}} = 59{,}507.55\ \text{kg/h} = 16.53\ \text{kg/s}$$

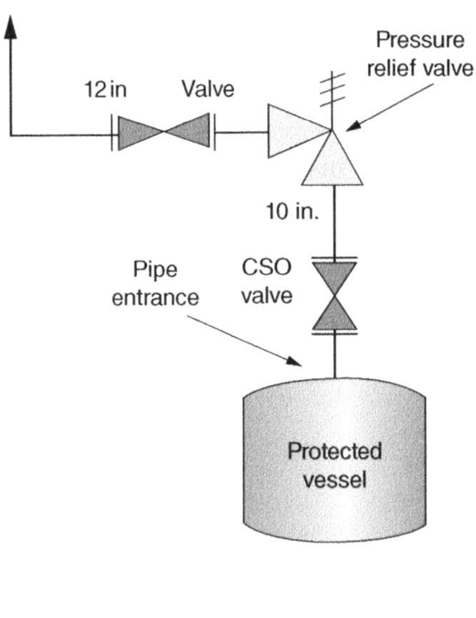

Figure 6.10 12″ × 10″ PSV for protection of the vessel.

$$\text{PR} = \frac{\begin{array}{c}\text{Valve relieving pressure (bara)} = \text{Set pressure} + \\ \text{Overpressure} + \text{Atmospheric pressure}\end{array}}{\begin{array}{c}\text{Valve back pressure (bara)} = \text{Valve backpressure (barg)} \\ + \text{Atmospheric pressure}\end{array}}$$

$$= \frac{42 + 0.1 \times 42 + 1}{7.50 + 1} = \frac{47.2}{8.5} = 5.55$$

Thus, based on Figure 6.9, the noise intensity level or sound pressure level at the discharge of the valve corresponds to the pressure ratio of 5.55. This level is approximately 55.7 dB. Now it is possible to calculate and obtain the value of the noise intensity level at a distance of 30 m from the safety valve at a distance of 30 m by using Eq. (6.44).

$$\begin{aligned} L_{30} &= L + 10\log_{10}\left(0.5q_{\text{m}} \times C^2\right) \\ &= 55.7 + 10\log_{10}\left(0.5 \times 16.53 \times 414.57^2\right) \\ &= 55.7 + 10\log 1{,}420{,}491 \\ &= 55.7 + 61.52 = 117.22\ \text{dB} \end{aligned}$$

Using Eq. (6.47), it is possible to measure the noise level at a distance of 1 m from the valve outlet.

$$\begin{aligned} L_{\text{p}} &= L_{30} - \left[20\log\left(\frac{r}{30}\right)\right] \rightarrow L_1 = 117.22 - \left[20\log\left(\frac{1}{30}\right)\right] \\ &= 117.22 - [20\log(0.03333)] \\ &= 117.22 + 29.54 = 146.76\ \text{dB} \end{aligned}$$

6.4.3.3 Calculation of Noise Emission According to VDI 2713

As a rule, noise levels for steam services are usually calculated according to VDI 2713. In Table 6.8, you will find a list of the symbols, designations, and units used in VDI 2713 to determine the size of safety valves.

The following equation is used in order to calculate the noise level of steam.

$$L_{\text{W}} = 17\log\left(\frac{q_{\text{m}}}{20}\right) + 51\log T - 15 \tag{6.49}$$

The distance-dependent noise level can be calculated in the following way:

$$L_{\text{A}} = L_{\text{W}} \times 10 \times \log A \tag{6.50}$$

Table 6.8 Symbols, designations, and units used in calculating noise for safety valves as defined in VDI 2713.

Used symbols	Designation	Units
L_W	Noise level	dB
L_A	Noise level at the distance of r meter(s)	dB
q'_m	Mass flow rate	kg/h
P	Set pressure	Bar
α_d	Coefficient of discharge	—
T	Temperature	Kelvin
R	Radios of the imaginary hemisphere are used to measure the distance from the source of the noise	m
A	Surface of the "imaginary hemisphere" having a radius of r $A = 2\pi r^2$	m^2

Questions and Answers

6.1 The control valve in question is 4″ in size and connected to 8″ inlet and outlet piping with a schedule 40 wall thickness. The flow passing through the valve is air with a capacity of 30,000 kg/h at a temperature of 100 °C. Valve inlet and outlet pressures are 30 bar and 12 bar, respectively. Which of the following statements is true regarding the noise generated by the valve? The air density and molecular weight are 1.225 kg/m^3 and 28.97 g/mol, respectively.

A The calculated Mach number of the valve is less than 0.3.
B The sonic speed in this case is 290 m/s.
C The valve has an outlet velocity of less than 150 m/s.
D The noise produced by the valve is approximately 150 dB, which does not exceed the allowable level.

Answer
Sonic velocity is calculated by using Eq. (6.8) as follows:

$$C = \sqrt{\frac{\gamma.R.T}{M}} = \sqrt{\frac{1.4 \times 8315 \times (100 + 273.15)}{28.97}}$$
$$= 387.22 \text{ meter per second (m/s)}$$

Thus, option B is incorrect. The next step is to calculate the valve outlet velocity as per Eq. (6.10). Parameter r is the internal radius of an 8″ pipe STD thickness connected to the valve. Using Example 6.2, a pipe with an outside diameter of 8″ and a thickness of 0.322″ has an outside diameter of 8.625″ and a thickness of 0.322″. The internal diameter and radius of the pipe are calculated as follows:

$$t = 0.322'' \text{ and OD} = 8.625'' \rightarrow \text{ID} = 8.625'' - 20.322''$$
$$= 7.981'' \rightarrow \text{Internal radius}$$
$$= 3.995 \text{ in.} = 0.10135$$

$$V_{\text{outlet}} = \frac{w}{A.\rho} = \frac{w}{\pi r^2 \rho} = \frac{30,000/3600}{\pi \times (0.10135)^2 \times 1.225} = 210.91 \text{ m/s}$$

The outlet velocity of the valve is greater than 150 m/s. Therefore, option C is incorrect.

Equation (6.7) can be used to determine the Mach number by knowing the sonic velocity and the valve outlet velocity:

$$M_{\text{outlet}} = \frac{V_{\text{outlet}}}{C} = \frac{210.91}{387.22} = 0.544$$

Mach number is greater than 0.3. So option A is incorrect. The last step is to determine the noise generated in the valve by Using Eq. (6.2):

$$\text{PWL} = 10 \log \left[\left(\frac{P_1 - P_2}{P_1} \right)^{3.6} \times \left(\frac{W}{3600} \right)^2 \times \left(\frac{T_1 + 273}{m} \right)^{1.2} \right] + 126.1 + \text{SFF}$$
$$= 10 \log \left[\left(\frac{30 - 12}{30} \right)^{3.6} \times \left(\frac{30,000}{3600} \right)^2 \times \left(\frac{100 + 273}{28.97} \right)^{1.2} \right] + 126.1$$
$$= 10 \log(0.1590 \times 69.44 \times 21.464) + 126.1$$
$$= 23.747 + 126.1 = 149.85 \text{ dB}$$

Equation (6.5) provides the following formula for calculating the allowable noise level:

$$\text{PWL}_{\text{A}} = 173.6 - 0.125 \times \left(\frac{D_{\text{i}}}{t} \right) = 173.6 - 0.125 \times \left(\frac{7.981''}{0.322''} \right)$$
$$= 173.6 - 3.0982 = 170.50 \text{ dB}$$

There is a stricter noise limit of 155 dB. Since the produced noise level is less than both limits, it is acceptable and option D is the correct response.

6.2 A relief valve, which is connected to a pipe of 8 in. in diameter, has a maximum flow capacity of 1,40,000 kg/h. At the downstream of the valve, there is a developed back pressure of 8 bar and the discharge temperature of the valve is 40 °C. Inlet pressure and temperature of the valve are 30 bar and 40 °C, respectively. The ideal gas passing through the valve has a molecular weight of 25 and a heat ratio of 1.21. Assuming that the outlet pipe has a standard thickness and no sonic condition exists (SFF = 0), which of the following statements is correct regarding noise and acoustic fatigue analysis for the valve?

A Mach number at the valve outlet is calculated to be 0.40.
B The power of the produced sound is 160 dB.
C The produced sound exceeds the acceptable sound power limit.
D All answers are incorrect.

Answer

Consider Example 6.2, which shows that an 8″ pipe with a standard (STD) wall thickness has an outside diameter of 8.625″ (0.2190 m), an inside diameter of 7.981″ (0.2027 m), and a thickness of 0.322″ (0.0082 m). Given here is a summary of all provided piping and process data:

P_1: The upstream pressure of the relief valve = 30 bar;
P_2: The downstream pressure of the relief valve = 8 bar = 8×10^5 Pa
T_1: The upstream temperature of the valve = 39.85 °C = 39.85 + 273.15 = 313 °K
W: Gas or liquid flow rate = 140,000 kg/h = 38.89 kg/s
M: Molecular weight = 25
γ: Ratio of specific heat = 1.21
Outlet pipe inside diameter = D_i = ID = 8.625″ − 20.322″ = 7.981″ = 0.2027 m
Outlet pipe wall thickness = t = 0.0082 m
z factor = 1

In order to calculate the Mach number, we need to use Eq. (6.16), where Z is equal to 1 and R is equal to 8315.

$$M = 116 \times \frac{W}{P_2 D_i^2} \times \sqrt{\frac{T_2}{\gamma . m}} = 116 \times \frac{38.89}{8 \times 10^5 \times 0.2027^2}$$
$$\times \sqrt{\frac{313}{1.21 \times 1.25}} = 0.1372 \times 3.22 = 0.44$$

According to Eq. (6.2), the sound power is calculated as follows:

$$\mathrm{PWL} = 10\log\left[\left(\frac{P_1 - P_2}{P_1}\right)^{3.6} \times \left(\frac{W}{3600}\right)^2 \times \left(\frac{T_1 + 273}{m}\right)^{1.2}\right] + 126.1 + \mathrm{SFF}$$
$$= 10\log\left[\left(\frac{30-1}{30}\right)^{3.6} \times \left(\frac{1{,}40{,}000}{3600}\right)^2 \times \left(\frac{39.85 + 273}{25}\right)^{1.2}\right] + 126.1$$
$$= 10\log(0.885 \times 1512 \times 20.76) + 126.1$$
$$= 44.43 + 126.1 = 170.54\ \mathrm{dB}$$

Equation (6.5) is used to calculate PWL's acceptance level.

$$\text{PWL}_\text{A} = 173.6 - 0.125 \times \left(\frac{D_\text{i}}{t}\right) = 173.6 - 0.125 \times \left(\frac{0.2027}{0.0082}\right) = 170.51$$

Option A is incorrect as the Mach number is equal to 0.44. Option B is incorrect as well since the generated sound power exceeds 160 dB. Since the produced sound is 170.54 dB, it exceeds the acceptance level of 170.51, as well as 155 dB. Therefore, option C is the correct answer. Option D is incorrect since option C is the correct answer.

6.3 In Figure 6.11, three noise sources are illustrated, namely PSVs C, A, and D. Each of these PSVs has an 8″ inlet pipe with a standard (STD) wall thickness. The sound power levels at point "B" located 10 m from each PSV source are as follows:

- Sound power level at point B due to pipe section A – B = 160 dB
- Sound power level at point B due to pipe section C – B = 163 dB
- Sound power level at point B due to pipe section D – B = 165 dB

Which of the following statements is true?

A Due to multiple sources of noise, the sound level produced at point B is greater than 170 dB.

B Since the calculated noise level at point B does not exceed the established limit, the piping at point B can handle the noise safely.

C Depending on the likelihood of piping failure resulting from the noise, the amount of noise produced at point B may be acceptable.

D By reducing the wall thickness and increasing the flow rate, the pipe system's noise problem can be mitigated.

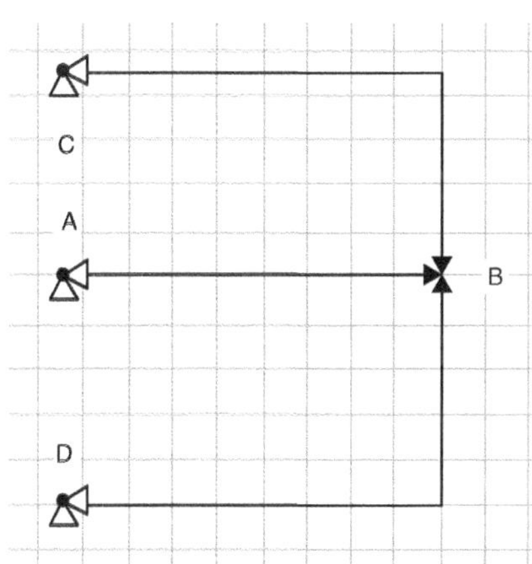

Figure 6.11 The three PSVs are located 10 m from point B.

Answer

From Eq. (6.4), the sound power level generated at point B due to the discharge of multiple sources of noise that are PSVs C, A, and D at the same time is calculated as follows:

$$\begin{aligned}\sum \mathrm{PWL}_X &= 10\log\left[10^{(\mathrm{PWL1})/10} + 10^{(\mathrm{PWL2})/10} + 10^{(\mathrm{PWL3})/10}\right] \\ &= 10\log\left[10^{160/10} + 10^{163/10} + 10^{165/10}\right] \\ &= 10\log\left(10^{16} + 10^{16.3} + 10^{16.5}\right) = 10 \times 16.79 = 167.9\,\mathrm{dB}\end{aligned}$$

Using Eq. (6.5), the sound power at point B, which is located at a distance of 10 m from each PSV source, can be calculated as follows. Referring to Example 6.2, an 8″ pipe with standard (STD) wall thickness has an outside diameter of 8.625″ (0.2190 m), an inside diameter of 7.981″ (0.2027 m), and a thickness of 0.322″ (0.0082 m).

$$\mathrm{PWL_A} = 173.6 - 0.125 \times \left(\frac{D_\mathrm{i}}{t}\right) = 173.6 - 0.125 \times \left(\frac{0.2027}{0.0082}\right) = 170.51$$

It is incorrect to choose option A since the sound level produced at point B as a result of the simultaneous operation of three PSVs is 167.9 dB, which is below 170 dB. Furthermore, option B is incorrect as well, as even if the calculated noise level is below the allowable noise level, the calculated noise levels are still higher than 155 dB. So we cannot conclude that the pipes at point B are safe from noise. The correct answer is C. In this case, even if the predicted noise is above 155 dB, the piping design at point B can be considered safe if the likelihood of failure is less than 0.5. Option D is incorrect because increasing the pipe wall thickness and reducing the flow rate are two methods for mitigating the noise risk in the piping system. Increasing the number of PSVs in the system can reduce the flow rate at the PSVs' outlet piping.

6.4 Water is carried by a 4″ globe control valve with a flow rate of 40 kg/s. The valve has an inlet pressure and an outlet pressure of 10 bars and 6.5 bars, respectively. In this case, water density is assumed to be 997 kg/m^3 and its vapor pressure is 2320 Pa. Considering the speed of sound in water to be 1400 m/s, which of the following statements is wrong regarding the hydrodynamic noise in the control valve? (*Note*: Liquid pressure recovery factor = 0.92, valve style modifier = 0.42, and valve flow coefficient C_v is 90.)

A Acoustic noise is caused primarily by cavitation in the valve.
B The water jet velocity in vena contracta is 28.8 m/s.
C The power of the sound produced is more than 1 W.
D Currently, the Acoustic power ratio stands at 0.25.

Answer

In order to determine whether the hydrodynamic noise in the control valve is primarily caused by cavitation or turbulent flow, the following two conditions should be examined:

1) Whenever the differential pressure ratio x_F exceeds the characteristic pressure ratio x_{Fz}, cavitation noise domains the turbulent noise. According to the following calculations, cavitation is a dominant cause of noise in the control valve.

$$x_F = \frac{P_1 - P_2}{P_1 - P_v} = \frac{10^6 - 6.5 \times 10^5}{10^6 - 2320} = \frac{3{,}50{,}000}{9{,}97{,}680} = 0.3508$$

$$x_{FZ} = \frac{0.90}{\sqrt{1 + 3F_d\sqrt{C/(N_{34} \times F_L)}}} \text{(For valve types exceptmulti-hole trims)}$$

$$x_{FZ} = \frac{0.90}{\sqrt{1 + 3F_d\sqrt{C/(N_{34} \times F_L)}}} = \frac{0.90}{\sqrt{1 + 3 \times 0.42\sqrt{90/(1.17 \times 0.92)}}}$$

$$= \frac{0.90}{3.5385579} = 0.2543$$

Another condition that indicates cavitation is the predominant cause of the noise is the following: $\Delta P > x_{Fzp1}(P_1 - P_v)$. By examining the calculations later, we can see that the second condition for noise generation in the valve due to cavitation is met. Thus, option A is correct.

$$\Delta P = 10^6 - 6.5 \times 10^5 = 350{,}000$$

$$x_{Fzp1} = x_{Fz}\left(\frac{6 \times 10^5}{P_1}\right)^{0.125} = 0.2543\left(\frac{6 \times 10^5}{10^6}\right)^{0.125} = 0.2386$$

$$x_{Fzp1}(P_1 - P_v) = 0.2386\left(10^6 - 2320\right) = 238{,}046$$

Next, it is necessary to calculate the fluid velocity at the vena contracta. In order to accomplish this, the pressure differential ΔP_c at the vena contracta is first calculated as follows:

$$\Delta P_c = \text{Lower than } x_F(P_1 - P_v) \text{ or } F_L^2(P_1 - P_v)$$

$$\rightarrow x_F(P_1 - P_v) = 0.3508\left(10^6 - 2320\right) = 349{,}986.144 \text{ Pa}$$

$$\rightarrow F_L^2(P_1 - P_v) = (0.92)^2\left(10^6 - 2320\right) = 844{,}436.352 \text{ Pa}$$

$$\rightarrow \Delta P_c = 349{,}986.144 \text{ Pa}$$

We can now calculate the velocity at vena contracta as follows:

$$U_{vc} = \frac{1}{F_L}\sqrt{\frac{2\Delta P_c}{\rho_L}} = \frac{1}{0.92}\sqrt{\frac{2 \times 349{,}986.144}{997}} = 28.80 \text{ m/s}$$

Therefore, option B is also valid. Next, it is necessary to calculate the mechanical power generated by noise.

$$W_{\mathrm{m}} = \frac{m \times U_{\mathrm{vc}}^2 F_{\mathrm{L}}^2}{2} = \frac{40 \times 28.80^2 \times 0.92^2}{2} = 14{,}040.76 \text{ W}$$

The following part discusses how to calculate the portion of mechanical power that is converted to sound power under conditions of cavitation. At this stage, three parameters must be determined: the acoustic efficiency factor at turbulent condition $\eta_{\text{turbulant}}$ and acoustic efficiency factor at cavitation conditions $\eta_{\text{cavitation}}$, as well as the acoustic power ratio r_{w}. Referring to Table 6.5, the acoustic power ratio should be 0.25. Thus, option D should be selected.

$$\eta_{\text{turbulant}} = 10^{-4}\left(\frac{U_{\mathrm{vc}}}{C_{\mathrm{L}}}\right) = 10^{-4}\left(\frac{28.80}{1400}\right) = 2.057 \times 10^{-6}$$

$$\eta_{\text{cavitation}} = 0.32\eta_{\text{turbulant}}\sqrt{\frac{P_1 - P_2}{\Delta P_{\mathrm{c}}} \cdot \frac{1}{x_{\mathrm{Fzp1}}}} \times e^{5x_{\mathrm{Fzp1}}} \times \left(\frac{1 - x_{\mathrm{Fzp1}}}{1 - x_{\mathrm{F}}}\right)^{0.5} \times \left(\frac{x_{\mathrm{F}}}{x_{\mathrm{Fzp1}}}\right)^{0.5} \times \left(x_{\mathrm{F}} - x_{\mathrm{Fzp1}}\right)^{1.5} = 1.243 \times 10^{-6}$$

Sound power W_{a} for cavitation dominant flow is calculated as follows:

$$W_{\mathrm{a}} = \left(\eta_{\text{cavitation}} + \eta_{\text{turbulant}}\right) W_{\mathrm{m}} \times r_{\mathrm{w}} = \left(3.3 \times 10^{-6}\right) \times 14{,}040.76 \times 0.25 = 0.01158 \text{ W}$$

Since the sound power is less than one watt, option C is wrong.

6.5 A globe valve is used to control the flow of water in a piping system. The flow through the valve corresponds to a pressure drop across the valve of 1 psi, which is equivalent to 100 gpm. Based on a liquid pressure recovery factor of 0.89 and a valve style modifier of 0.45, what is the diameter of the liquid jet inside the valve?

A 19.53 mm
B 20.20 mm
C 18.19 mm
D 17.50 mm

Answer

The first step is to calculate the flow coefficient C_{v} value of the valve as follows:

$$C_{\mathrm{v}} = Q\sqrt{\frac{\mathrm{SG}}{\Delta P}} = 100\sqrt{\frac{\mathrm{SG} = 1 \text{ for water}}{1}} = 100$$

Now it is possible to calculate the water jet diameter inside the valve by using Eq. (6.35).

$$D_J = N_{14}F_d\sqrt{CF_L} = 0.0046 \times 0.45\sqrt{100 \times 0.89}$$
$$= 0.01953 \text{ m} = 19.53 \text{ mm}$$

Thus, option A is the correct answer.

6.6 In a globe valve, the acoustic efficiency factors are 2×10^{-6} and 1.9×10^{-6} for turbulent and cavity conditions, respectively. What would be the hydrodynamic sound power as a function of the cavitation flow conditions and the mechanical stream power of 14,000 W?

A 0.01365 W
B 0.01546 W
C 0.01687 W
D 0.01837 W

Answer

Cavitation is the type of fluid, and the sound power is calculated using Eq. (6.39) as follows:

$$W_a = (\eta_{\text{turbulant}} + \eta_{\text{cavitation}}) \times W_m \times r_w = (2 \times 10^{-6} + 1.9 \times 10^{-6})$$
$$\times 14{,}000 \times 0.25 = 0.01365 \text{ W}$$

Thus, option A is the correct answer.

6.7 In a liquid service, throttling is achieved via a globe valve. Due to turbulent flow, the valve produces a considerable amount of noise in the piping system. The liquid pressure recovery factor is 0.9, and the pressure loss at vena contracta is 2,00,000 Pa. The mass flow rate of water through the valve is assumed to be 35 kg/s, and the density of water is assumed to be 1000 kg/m³. What percentage of the fluid's mechanical energy is converted into noise? (*Note*: Assume that the sound velocity in water is 1400 m/s.)

A 70%
B 50%
C Less than 1%
D 5%

Answer

In order to determine the fluid mechanical power, the first step is to determine the fluid velocity at the vena contracta area according to Eq. (6.36).

$$U_{vc} = \frac{1}{F_L}\sqrt{\frac{2\Delta P_c}{\rho_L}} = \frac{1}{0.9}\sqrt{\frac{2 \times 200{,}000}{1000}} = 22.22 \text{ m/s}$$

$$W_m = \frac{m \times U_{vc}^2 F_L^2}{2} = \frac{35 \times 22.22^2 \times (0.9)^2}{2} = 6998.60\ \mathrm{W}$$

Considering that noise is generated by turbulent flow, the acoustic efficiency factor for turbulent flow should be calculated according to Eq. (6.40) as follows:

$$\eta_{turbulant} = 10^{-4}\left(\frac{U_{vc}}{C_L}\right) = 10^{-4}\left(\frac{22.22}{1400}\right) = 1.587 \times 10^{-6}$$

The value of r_w is 0.25 for a globe valve. As a result, the sound power generated by turbulent flow can now be calculated using Eq. (6.38) as follows:

$$W_a = \eta_{turbulant} \times W_m \times r_w = 1.587 \times 10^{-6} \times 6998.60 \times 0.25$$
$$= 0.002776\ \mathrm{W}$$

$$\frac{W_a}{W_m} = \frac{0.002776}{6998.60} = 3.9675 \times 10^{-7}$$

Considering that less than 1% of fluid mechanical power is converted to noise, option C is the correct answer.

6.8 A safety valve is used for handling natural gas with a molecular weight of 19 g and a relieving capacity of 7502 ft/h. Inlet relieving temperature is 120 °F, set pressure is 210 psi, overpressure is 10%, and backpressure is 50 psi. What is the noise level at a distance of 500 ft away from the valve discharge when the valve is opened based on API 521 standard? The specific ratio of heat for natural gas is 1.27.

A 89.79 dB
B 103.8 dB
C 54.5 dB
D 28 dB

Answer

The first step is to calculate the absolute relieving pressure as follows:

$$P_1 = \text{Absolute relieving temperature} = \text{Set pressure} + \text{Overpressure}$$
$$+ \text{Atmospheric pressure} = 210 + (210 \times 10\%) + 14.7 = 245.7\ \text{psia}$$

$$\text{Backpressure absolute pressure} = 50\ \text{psig} + 14.7 = 64.7\ \text{psia}$$

$$\text{PR} = \frac{\textit{Valve relieving pressure}}{\textit{Valve back pressure (bara)}} = \frac{245.7}{64.7} = 3.8$$

It has been determined that the noise intensity at the discharge point is 54.5 dB ($L = 54.5$ dB) (refer to Figure 6.9). It's important to understand that the temperature of the gas in SI unit is measured in Kelvin, which is calculated as follows:

Gas temperature is 120 °F, $\text{Temperature}_{\text{Kelvin}} = 322.039$ K

By using Eq. (6.45), the next step is to calculate the speed of sound in the gas inside the valve using the formula as follows:

$$C = 91.2\,(kT/M)^{0.5} = 91.2 \times (1.27 \times 322.039/19)^{0.5} = 423.13\,m/s$$

The mass flow rate is 7502 lb/h = 2.084 lb/s = 0.945 kg/s

Now it is possible to calculate and obtain the value of the noise intensity level at a distance of 30 m (100 ft) from the safety valve at a distance of 30 m by using Eq. (6.44).

$$\begin{aligned} L_{30(100\,ft)} &= \text{L} + 10\ log_{10}\left(0.5q_m \times C^2\right) \\ &= 54.5 + 10\ log_{10}(0.5 \times 0.945 \times 423.13^2) \\ &= 54.5 + 10 \log 84{,}596 = 54.5 + 49.27 = 103.77\,\text{dB} \end{aligned}$$

Using Eq. (6.47), it is possible to measure the noise level at a distance of 500 ft equal to 150 m from the valve outlet.

$$\begin{aligned} L_{150(500\,ft)} &= L_{30(100\,ft)} - \left[20log\left(\frac{r}{30}\right)\right] \rightarrow L_{150(500\,ft)} \\ &= 103.77 - \left[20log\left(\frac{150}{30}\right)\right] = 103.77 - [20\ log(5)] \\ &= 103.77 - 13.98 = 89.79\,\text{dB} \end{aligned}$$

Thus, the correct choice is A.

Further Reading

Ahmed, D. (2011). Acoustic fatigue assessment of piping components by finite element analysis. *Proceeding of ASME Pressure Vessel and Piping Conference*, Baltimore, MA, USA (1–21 July 2011).

American Petroleum Institute (API) (2014). *API Standard 521, Pressure-Relieving and Depressuring Systems*, 6e. Washington, DC: American Petroleum Institute (API).

American Society of Mechanical Engineers (ASME) B16.9 (2018). *Factory-Made Wrought Buttwelding Fittings*. New York: American Society of Mechanical Engineers (ASME).

American Society of Mechanical Engineers (ASME) B36.10/19 (2004). *Carbon, Alloy and Stainless-Steel Pipes*. New York: American Society of Mechanical Engineers (ASME).

Baumann, H.D. (1997). Solve valve noise and cavitation problems. *Hydrocarbon Processing* 76 (3): 45–50.

Dresser Flow Control (2020). Noise control manual. Bulletin OZ3000 01/02.

Fagerlund, A.C. (1986). Recommended maximum valve noise level. *Proceedings of the ISA/86 International Conference and Exhibition*, Houston, TX, USA (November).

Fisher Controls International (2001). *Control Valve Handbook*, 3e. Fisher.

Glaunach (2010). *The Silencer Handbook. A General Introduction Into Noise and Its Prevention*. Glaunach.

International Electrotechnical Commission (IEC) 60534-8-4 (2015). *Industrial Process Control Valves – Part 8-4: Noise Considerations-Prediction of Noise Generated by Hydrodynamic Flow*. Geneva: International Electrotechnical Commission (IEC).

International Organization for Standardization (ISO) 4126-9 (2008). *Safety Devices for Protection Against Excessive Pressure. Application and Installation of Safety Devices Excluding Stand-Alone Bursting Disc Safety Devices*. Geneva: International Organization for Standardization (ISO).

Momber, A.W. (2005). *Hydrodemolition of Concrete Surfaces and Reinforced Concrete*, 67–103. Elsevier https://doi.org/10.1016/B978-185617460-2/50003-2.

Nesbitt, B. (2007). *Handbook of Valves and Actuators: Valves Manual International*, 1e. Oxford: Elsevier.

NORSOK L-002 (2009). *Piping System Layout, Design and Structure Analysis*, 3e. Lysaker: NORSOK.

Norwegian Oil Industry Association (2013). *Valve Technology*. Norsk Olje & Gass.

Pettigrew, M.J., Paidoussis, M.P., Weaver, D.S., and Au-Yang, M.K. (1996). Flow-induced vibration. *Proceeding Paper. Conference: American Society of Mechanical Engineers (ASME) Pressure Vessels and Piping Conference*, Montreal, Canada (21–26 July 1996). Other Information: PBD: 1996.

Randall, E.I.M. (2001). PSV noise – criteria, limits and prediction. *Valve World*. https://docplayer.net/39961976-Psv-noise-criteria-limits-and-prediction.html (accessed 12 March 2022).

Samson (2008). Improvement of IEC 60534-8-3 for noise prediction in control valves. https://www.samsongroup.com/document/w01950en.pdf (accessed 2 June 2022).

Shahda, J. (2010). Predicting control valve noise in gas and steam applications: valve trim exit velocity head vs. valve outlet Mach number. *Dresser Masoneilan*. https://www.plantservices.com/assets/Media/1003/WP_Valve.pdf (accessed 13 March 2022).

Skousen, P.L. (2011). *Valve Handbook*, 3e. New York: McGraw-Hill.

Smit, P. and Zappe, R.W. (2004). *Valve Selection Handbook*, 5e. New York: Elsevier.

Sotoodeh, K. (2019). Noise and acoustic fatigue analysis in valves (case study of noise analysis and reduction for a 12″ × 10″ pressure safety valve). *Journal of Failure Analysis and Prevention* 19: 838–843. 10.1007/s11668-019-00665-3.

Statoil (2014). *Noise Control in Projects. Guideline, GL0563*. 1st revision. Oslo: Statoil.

The Engineering Toolbox (2022). Speed of sound equations. https://www.engineeringtoolbox.com/speed-sound-d_82.html (accessed 13 March 2022).

VDI 2713 (1974). *Noise Reduction in Thermal Power Stations*. Berlin: Deutsches Institut fur Normung E.V. (DIN).

7

Water Hammering

7.1 Introduction

There is a risk of water hammering when fluid parameters, including flow, velocity, and pressure, change suddenly in a piping system due to various circumstances such as pump stopping or starting as well as valve opening or closing. High pressure is generated in liquid services when the velocity of a mass of liquid or large velocity liquid is suddenly reduced. Thus, the mobile energy of the liquid is converted into pressure energy. Water hammering is not a kinetic energy problem but rather an acoustic problem. Waves produced in the pipe as a result of water hammering are much faster than liquid velocity. It has undesirable and devastating consequences such as noise, wear, tear, and eventually the collapse and rupture of the piping system, as well as the associated valves. Consequently, it is extremely important to understand this problem in order to prevent it as well as to calculate and analyze the pressure change in the piping caused by water hammering. More specifically, the primary cause of water hammering is a rapid change in flow rate in the piping system caused by the shutoff or startup of a pump or the opening or closing of a valve. The effects of a water hammer can range from small changes in pressure and velocity to relatively high pressure that can result in the bursting and failure of pipes, pipe fittings, valves, and, in some cases, damage to pumps. For instance, a pump stoppage can cause a severe form of water hammering, and a sudden power failure that causes all the pumps to fail at once can cause an extremely severe case of water hammering. The following section describes in detail what causes water hammering in check valves.

7.2 Water Hammering and Pressure Loss in Check Valves

Nonreturn valves, or check valves, open with the forward flow and close with the reverse flow. Essentially, the check valve's main function is to prevent equipment and facilities, such as pumps, from reverse flow or backflow. In the oil and gas

Industrial Valves: Calculations for Design, Manufacturing, Operation, and Safety Decisions, First Edition. Karan Sotoodeh.

industry, there are several types of check valves, including swing, dual plate, axial, piston, and ball lift. Several factors influence the selection of check valves, including initial and maintenance costs, heat loss and energy costs, nonslam characteristics, location of the valve (e.g. after pumps or compressors), fluid compatibility, sealing ability, and flow characteristics. It is common to install nonslam check valves, such as axial types, following pumps, and compressors. In various industries such as oil and gas, pumps are used to pressurize and move fluid (liquid) through the piping system. In order to pressurize and move gas services, compressors are used. It is important to know that all check valves are operated automatically by the fluid flow inside the piping, without any need for an operator.

The most essential requirement for check valves, particularly those that are installed after equipment such as pumps and compressors, is that they provide less resistance to the flow in the normal direction that opens the valve. In contrast, a check valve should exhibit unlimited resistance to backflow, also referred to as reverse flow. Figure 7.1 illustrates a swing check valve, which has a disk that moves upward as fluid moves through the piping system and the valve. Swing check valves are available, low-cost, and offer relatively low-pressure loss when fully open. The swing check valve is a popular choice of valve in water piping systems and, compared to other check valves, swing check valves are relatively inexpensive. Moreover, the valves provide very low-pressure drops, or head losses, when fully opened. Referring to Eq. (7.1), head loss refers to the loss of fluid pressure inside the valve compared to the pressure inside the piping. Valve pressure drop is typically greater than pipe pressure drop. It means that the pressure drop in the valve is 4 bars if the valve has an inlet pressure of 20 bars and an outlet pressure of 16 bars.

Figure 7.1 A swing check valve in the open position. *Source:* American Journal of Industrial Engineering/Science and Education Publishing Co. Ltd.

Valve Head Loss

$$\Delta P_{\mathrm{v}} = P_{\mathrm{p}} - P_{\mathrm{v}} \tag{7.1}$$

where:

ΔP_{v}: Head loss or fluid pressure drop inside a valve;
P_{p}: Fluid pressure in the pipe;
P_{v}: Fluid pressure in the valve.

The operator company's experience shows that check valves are not entirely open when fluid passes through the valve. For example, the disk of a valve swings only about 60° from the closed position and is approximately 30° from 100% or a fully open position. Swing check valves are closed when the disk closes due to the weight of the disk. So, if the fluid flow is interrupted, the disk quickly closes and slams against the seat during the complete closure of the valve. The long stroke of the disk coupled with the sudden closing of the valve caused by the weight of the disk exacerbates the slamming effect of swing check valves. The sudden slamming of the valve disks causes water hammering, which is a form of hydraulic shock, as shown in Figure 7.2. The valve closes when the fluid cannot move forward. So, its kinetic energy turns into waves that apply pressure and load to the pipe wall. These shocks are actually pressure surges or waves that cause noise and damage to piping systems. The constant opening and closing of the disk make slamming, which causes water hammering and a higher rate of wear on the disk and valve seat.

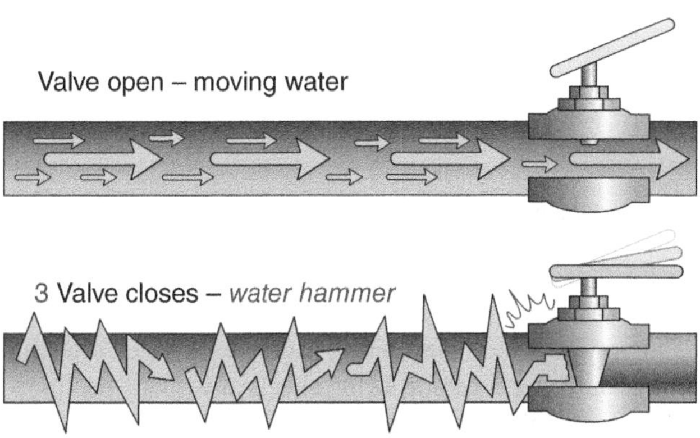

Figure 7.2 Water hammering in the piping system due to rapid valve closure.

Example 7.1 The pressure of the water in the pipes is 12 bars. If 25% of the piping fluid pressure loss occurs in the valve, what is the valve outlet pressure?

Answer

P_p: Fluid pressure in the pipe = 12 bar
ΔP_v: Fluid pressure drop or head loss in the valve = $25\,\% \times 12 = 3$ bar
$\Delta P_v = P_p - P_v \rightarrow P_v = P_p - \Delta P_v = 12 - 3 = 9$ bar

The valve has an outlet pressure of 9 bar.

Alternatively, the head loss can be calculated by using Eq. (7.2) as follows:

An Alternative Method of Calculating Valve Head Loss

$$\Delta H = \frac{K_v V^2}{2g} \tag{7.2}$$

where:

ΔH: The head loss or pressure loss as measured by the water column (m, ft);
K_v: Valve flow resistance coefficient (dimensionless);
V: Fluid velocity (m/s, ft/s);
g: Acceleration due to gravity (9.81 m/s^2, 32.2 ft/s^2).

Equation (7.3) can be used to calculate the valve flow resistance coefficient using the valve internal diameter and valve flow coefficient.

Calculation of the Valve Flow Resistance Coefficient

$$K_v = 980\frac{d^4}{C_v^2} \tag{7.3}$$

where:

K_v: Valve flow resistance coefficient (dimensionless);
d: Valve diameter (in.);
C_v: Valve flow coefficient.

Alternative solutions to prevent water hammering can be used when selecting check valves. Many valve manufacturers equip swing check valves with dashpots, shock absorbers, or cushioning/damping systems, which are practical accessories for dampening swing check valves. In Figure 7.3, the check valve is equipped with a dashpot which counterbalances the effect of slamming so that the disk does not close too quickly. Although it should be noted that adding a dashpot as a slamming

Figure 7.3 A swing check valve fitted with a dashpot. *Source:* DeZurik.

and water hammering prevention solution creates four new problems, the first is that the valve closes slower, which causes a higher rate of backflow and reduces the resistance of the valve. As a result, the pump installed upstream or before the swing check valve must be able to handle some backflow. In addition, the swing check valve with dashpot has a higher pressure drop than a standard check valve. The fluid pressure drop is more critical and problematic when the piping system is vertically installed with upward fluid movement. Thirdly, the swing check valve with the dashpot is more expensive. The dashpot contains high-pressure oil, often exceeding 2000 psi, and is very expensive. Furthermore, since dashpots apply high loads to the hinge pin that connects the disk to the valve body, a check valve with a large hinge pin diameter is required. The last challenge is that swing check valves with dashpots require additional maintenance because they contain more moving parts. Using a dashpot for swing check valves to prevent water hammering is prohibited in many projects.

Studies have demonstrated that in order to prevent a check valve from slamming and creating water hammering problems, the check valve must be closed rapidly before reverse flow begins or the valve must be closed very slowly once reverse flow has developed and reached the valve. As shown in Figure 7.4, a swing check valve disk slams firmly onto the seat by the force of reverse flow, resulting in a strong shock wave and water hammering inside the valve and connected piping. In this case, the valve disk was unable to be closed quickly before reverse flow occurred. Furthermore, it could not be shut slowly as the reverse flow hit the disk with a solid force on the seat. According to the same studies, at least two conditions must be met to achieve a nonslamming feature for check valves. First, the disk or closure

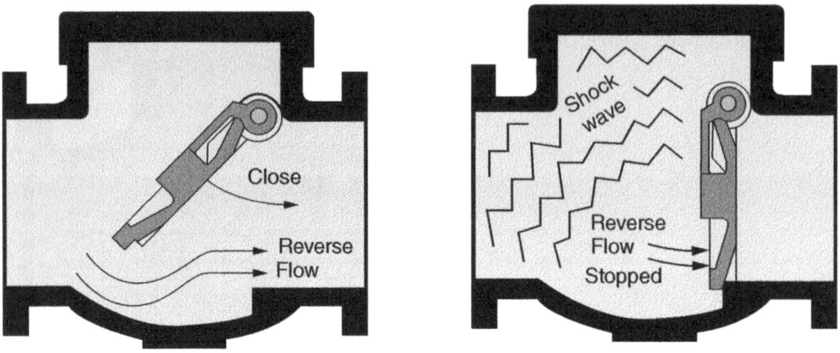

Figure 7.4 Swing check valve slamming.

Figure 7.5 A dual plate check valve.

member of the valve should have low inertia and low friction. Second, the travel distance of the disk must be short, or the spring must assist the motion of the disk.

As shown in Figure 7.5, one alternative solution to minimizing the slamming effect is to switch from swing check valves to dual plate check valves. The dual plate check valve has two disks instead of one. So, each half-disk is less weighty and less likely to slam when closing. The closing of the valve disk is accomplished by the spring force rather than the weight of the disks, in order to reduce the slamming effect and water hammering. Another advantage of dual plate check valves is that their initial cost, total cost, maintenance, and energy costs are lower than swing check valves. The dual check valve has two disadvantages. The first is the fact that there is still a low-to-medium slamming effect present for this type of valve, and the second is that it has a reduced bore, which results in a relatively high-pressure drop and energy loss. Dual plate check valves are manufactured according to the American Petroleum Institute (API) 594 standard, which has the ability to have a bore or port that is approximately 80% of the area connected to the piping.

Example 7.2 After a pump, a 12″ dual plate check valve is installed in pressure class 150 (pressure nominal of 20 bars) with a flow coefficient of 4000. According to the American Society of Mechanical Engineers (ASME B16.34), the internal diameter of the valve is standard for valves to be 304.8 mm in diameter. When the fluid velocity inside the valve is 22 m/s, what is the pressure drop inside the valve based on that information? What would be the pressure drop of a 100-ft pipe run if the pressure drop was equal to 1.5? Then the pressure drop in the valve would be equal to what length of the connected pipe?

Answer

The internal diameter of the valve is 304.8 mm or 12″. In the first step, we must calculate the valve flow coefficient resistance by using Eq. (7.3).

$$K_V = 890\frac{d^4}{C_V^2} = 890 \times \frac{12^4}{(4000)^2} = 1.15344$$

According to Eq. (7.2), now it is possible to calculate the head loss in terms of the length of the water column.

$$\Delta H = \frac{K_V V^2}{2g} = \frac{1.15 \times 22^2}{2 \times 9.81} = 28.4539\text{ m} = 93.35\text{ ft}$$

The water at a 10-m depth has a pressure of 1 bar. So, the loss of pressure in the valve is 2.84 bar.

$$\text{Equivalent pressure loss in the piping} = \frac{1.15344}{1.5} \times 100\text{ ft} = 76.90\text{ ft}$$

The pressure drop produced by a 12″ dual plate check valve in pressure class 150 is equal to 76.90 ft or 23.44 m.

In addition to the issues mentioned earlier, there are several additional issues associated with dual plate check valves, the most significant ones being pressure drop and medium slamming behavior. Consequently, it has become common to use axial flow check valves or nozzle check valves after pumps and compressors in Norwegian offshore projects. There are several characteristics of axial flow check valves that make them nonslamming, with a very low-pressure drop compared to dual plate check valves and a fast closing characteristic. Even though axial check valves are more expensive than swing and dual plate check valves initially, they can reduce pressure drops, protect rotating equipment from damage, and save a lot of energy in the long run. These valves are widely used in offshore platforms, subsea, refineries, pipelines, LNG, and petrochemical plants in the oil and gas industry. Its short axial disk travel to the seat, spring-assisted design, and low-mass disk contribute to the quick closing of these valves. Fast closing reduces the risk of equipment damage caused by backflow and protects expensive mechanical equipment. Figure 7.6 illustrates a nozzle check valve with a low-mass disk that is energized with a spring that can be depressed easily by the hand of the operator to keep the valve closed. Another advantage of axial check valves is their robust design, which can withstand perfectly the loads and vibration generated by upstream equipment and facilities such as pumps and compressors. Typically, axial check valves are fabricated from a single piece without any flange connections, maximizing the valve's resistance to loads and minimizing leakage risk.

While the initial or purchasing cost of an axial check valve is higher than both dual plate and swing check valves, it is the cheapest of the three when considering the total cost of valves. The total cost of valves is not only determined by the initial cost but also by the energy and maintenance costs. Compared to the dual plate and swing check valves, axial check valves require very little maintenance and provide less pressure loss. The cost estimation for 12″ check valves over 40 years is presented in Table 7.1.

Figure 7.6 A nozzle check valve.

Table 7.1 Comparison of the costs of 12″ check valves over 40 years of operation in a plant.

Type of check valve	Purchase (initial) cost	Energy cost	Maintenance cost	Total cost
Nozzle or axial	8000 USD	16,000 USD	1,000 USD	25,000 USD
Dual plate	3000 USD	22,000 USD	6,000 USD	31,000 USD
Swing	5000 USD	31,000 USD	20,000 USD	56,000 USD

Example 7.3 The dual plate check valve is replaced by the nozzle check valve in the previous example. The flow coefficient resistance of the nozzle check valve is 0.83. Based on its flow coefficient resistance, how many meters of the pipe length would be lost due to the nozzle check valve?

Answer

$$\Delta H = \frac{K_v V^2}{2g} = \frac{0.83 \times 22^2}{2 \times 9.81} = 20.475\ \text{m} = 67.175\ \text{ft}$$

Each 10 m of water has a pressure of 1 bar. So, the valve loses 2.04 bars of pressure.

$$\text{Equivalent pressure loss in the piping} = \frac{0.83}{1.5} \times 100\ \text{ft} = 55.33\ \text{ft}$$

A 12″ nozzle check valve in pressure class 150 produces a pressure drop equal to 55.33 ft or 16.86 m of pipe.

Based on Eq. (7.4), the head loss from valves, including check valves, is converted to the electrical power and cost of a pump required to overcome the pressure or head loss from the valve. Two main components typically determine the cost of electrical energy; a demand charge and an energy charge. The energy charge represents the consumption of kilowatt-hours of electricity with a fee of approximately 0.05 USD/kWh. In contrast, the demand charge can be charged at a higher rate and represents the cost of electricity generation capacity, with a cost of approximately 10 USD/kW. Demand charges can be affected by the time of day.

Energy Cost Due to the Valve Head Loss

$$A = \frac{1.65 \times Q \times \Delta H \times S_g \times C \times U}{E} \tag{7.4}$$

where:

A: Annual energy cost, dollars per year;
Q: Flow rate, gallon per minute (gpm);
ΔH: Head loss, feet of water;
S_g: The specific gravity (SG) of fluids in the piping system, including the valves (dimensionless), and the SG for water is equal to one;
C: Cost of electricity in USD per kilowatt per hour;
U: An electricity usage percentage that is equal to 1 if the pump is running 24 hours a day;
E: Pump efficiency which is typically assumed equal to 0.8.

Example 7.4 Calculate the head loss for swing check valves, dual plate check valves, and nozzle check valves in 12″ sizes. Flow resistance coefficients k are respectively 1.58, 1.15, and 0.83 for swing, dual plate, and nozzle check valves. The fluid in the piping is the oil with a density of 800 kg/m^3, a velocity of 12.76 ft/s, and a flow rate of 4500 gpm. Calculate the annual energy costs in three cases of using each type of valve, assuming that electricity is used 50% of the time at the cost of 0.05 USD/kWh?

Answer

According to Eq. (7.2), the first step is to calculate the head loss in terms of the water column length.

$$\text{Loss of pressure in the swing check valve} = \Delta H = \frac{K_v V^2}{2g}$$

$$= \frac{1.58 \times 12.76^2}{2 \times 32.2} = 3.9946 \text{ ft water column}$$

As each foot of water can generate 0.43352 lb of pressure, the pressure generated by a column of 3.9946 ft is 1.73 psi.

$$\text{Loss of pressure in the dual plate check valve} = \Delta H = \frac{K_v V^2}{2g} = \frac{1.15 \times 12.76^2}{2 \times 32.2}$$

$$= 2.9074 \text{ ft water column} = 1.26 \text{ psi}$$

$$\text{Loss of pressure in the nozzle check valve} = \Delta H = \frac{K_v V^2}{2g}$$

$$= \frac{0.83 \times 12.76^2}{2 \times 32.2} = 2.0984 \text{ ft water column} = 0.9097 \text{ psi}$$

The annual energy cost of using a swing check valve (A)

$$= \frac{1.65 \times Q \times \Delta H \times S_g \times C \times U}{E} = \frac{1.65 \times 4500 \times 3.9946 \times 0.8 \times 0.05 \times 0.5}{0.8}$$

$$= 741.50 \text{ USD}$$

The annual energy cost of using a dual plate check valve (A)

$$= \frac{1.65 \times Q \times \Delta H \times S_g \times C \times U}{E} = \frac{1.65 \times 4500 \times 2.9074 \times 0.8 \times 0.05 \times 0.5}{0.8}$$
$$= 539.69\,\text{USD}$$

The annual energy cost of using a nozzle check valve (A)

$$= \frac{1.65 \times Q \times \Delta H \times S_g \times C \times U}{E} = \frac{1.65 \times 4500 \times 2.0984 \times 0.8 \times 0.05 \times 0.5}{0.8}$$
$$= 389.52\,\text{USD}$$

In this example, the density of oil is 800 kg/m^2. So, the SG of oil is equal to 0.8 when calculating energy costs.

7.3 Water Hammering Calculations

Joukowsky's formula is used to measure water hammering to calculate the pressure change that results from a rapid velocity change, as shown in Eq. (7.5). Based on this equation, it is evident that the greater the magnitude of the velocity change, the greater the magnitude of the pressure change, as well as the wave load and speed.

Water Hammering Evaluation as a Column of Fluid Length Pressure According to Joukowsky's Formula

$$\Delta H = \frac{\Delta Q \times C}{g.A} \tag{7.5}$$

where:

ΔH: Fluid (water) pressure change expressed as water column (m, ft);
ΔQ: Flow rate change (m^3/s, ft^3/s);
C: Velocity of the pressure wave (m/s, ft/s) calculated from Eq. (7.8);
g: Acceleration due to gravity (9.81 m/s^2, 32.2 ft/s^2);
A: Pipe area (m^2, ft^2).

Example 7.5 Due to the closing of a swing check valve, the velocity of the water fluid in the piping system is suddenly changed from 3 to 0, causing a wave speed of 1100 m/s. What is the increase in pressure due to the change in velocity?

Answer
In piping, the change in flow rate is dependent on the change in fluid velocity and the area of the pipe. Here is a summary of the calculation of water hammer pressure changes as a function of water column length using Eq. (7.6):

$$\Delta Q = \Delta V \times A \rightarrow \Delta H = \frac{\Delta Q \times C}{g.A} = \frac{\Delta V \times A \times C}{g.A} = \frac{\Delta V \times A \times C}{g.A}$$
$$= \frac{\Delta V \times C}{g} = \frac{3 \times 1100}{9.81} = 336.39\ \text{m}$$

One bar of pressure is produced by every 10 m of water. So, 33.64 bars of pressure are produced by 336.39 m of water. Due to an increase in fluid velocity of 3 m/s, the water pressure is increased by 33.64 bar.

Calculation of Water Hammering Through the Use of a Column of Fluid Pressure by Considering Velocity Changes or Reverse Velocity

$$\Delta H = \frac{V \times C}{g} \tag{7.6}$$

where:

ΔH: Fluid (water) pressure change in the form of a water column (m, ft);
V: Reverse velocity or change in velocity (m/s, ft/s);
C: Pressure wave velocity (m/s, ft/s) calculated from Eq. (7.8);
g: Acceleration due to gravity (9.81 m/s^2, 32.2 ft/s^2).

Note: Based on field experience, a water hammer that occurs in the range of 15–30 m of water column equals approximately 50–100 ft representing a mild slam that is tolerable. In contrast, water hammers over 30 m of the water column, or over 100 ft, are significantly loaded and dangerous and therefore require the selection of a different check valve or the modification of the existing valve design to incorporate a spring or dashpot.

Equation (7.7) measures the water hammering effect for various fluids in the form of maximum pressure variation. A pressure increase in a pipe due to water hammering is determined by a number of factors, including the fluid properties of density and velocity, as well as the generated wave velocity in the pipe.

Calculation of Water Hammering Based on Maximum Pressure Variation by Incorporating Velocity Change or Reverse Velocity

$$\Delta P_{\text{max}} = \rho \times C \times V \tag{7.7}$$

where:

ΔP_{max}: The maximum pressure increase in the pipe as a result of water hammering (Pa);
ρ: Maximum density of the fluid (kg/m^3);

C: Wave velocity (m/s);
V: A reverse velocity or a change in velocity in a pipe (m/s).

A wave speed or acoustic velocity produced in the pipe due to water hammering can be calculated using Eq. (7.8). As was mentioned earlier, the wave speed is much faster than that of the liquid.

Wave Speed or Acoustic Velocity in the Pipe

$$C = \sqrt{\frac{E_b}{\rho}} \text{ if the pipe is rigid} \tag{7.8}$$

$$C = \sqrt{\frac{E_c}{\rho}} \text{ if the pipe is elastic}$$

where:

$$\frac{1}{E_c} = \frac{1}{E_b} + \frac{Dk}{E_p t}$$

where:

C: Wave velocity due to water hammering (m/s or ft/s);
ρ: Water density (1000 kg/m^3 or 0.01347 $slug/in.^3$);
E_b: Bulk modulus of water (2.1×10^9 N/m^2 or 3.0×10^5 psi);
E_c: Effective bulk modulus of water in the elastic pipe (N/m^2, psi);
E_p: Modulus of elasticity of pipe material (N/m^2, psi);
t: pipe wall thickness (m/ft);
D: pipe diameter or NPS (m/ft);
k: This factor is influenced by the pipe anchorage method and Poisson's ratio ε of the pipe material. Depending on the piping material, Poisson's ratio can vary from 0.25 to 0.35. However, it is common to standardize Poisson's ratio at 0.25.

$k = \left(\frac{5}{4} - \mathcal{E}\right)$ for pipes free to move longitudinally;
$k = 1 - \mathcal{E}^2$ for pipes anchored at both sides;
$k = 1 - 0.5\mathcal{E}$ for pipes with expansion joints

From Table 7.2, the modulus of elasticity E_p can be determined based on the piping material.

In general, when a piece of material is stretched in one direction, it tends to become thinner in the lateral direction – and if a piece is compressed in one direction, it tends to become thicker in the lateral direction. In Physics, Poisson's ratio is defined as the ratio between the relative contraction strain (transverse, lateral, or

Table 7.2 Modulus elasticity E_p of piping material.

Pipe material	E_p (N/m^2)	E_p (psi)
Gray cast iron	1.1×10^{11}	16×10^6
Malleable cast iron	1.6×10^{11}	23×10^6
Reinforced concrete	1.6×10^{11}	25×10^6
Copper	9.7×10^{10}	14×10^6
Steel	1.9×10^{11}	28×10^6

radial strain) to the applied load and the relative extension strain (or axial strain) in the direction of the applied load. The term strain refers to the deformation of a material as a result of stress. It is simply a ratio of the changed length to the original length.

Typically, the sound waves generated in the piping exacerbate the slamming effect in the valve. Before the pressure wave is reflected from the end of the pipe and returned to the valve, it cannot affect the pressure process within the valve and the elevated slamming action. In order to measure the time it takes the sound wave to return to the valve, Eq. (7.9) is used. To prevent water hammering, it is imperative to select and design the valve in a way that it can be closed once the sound wave returns to the valve.

The Maximum Amount of Time It Takes for a Sound Wave to Cause Valves to Slam and Water Hammering

$$t = 2 \times \frac{L}{c} \tag{7.9}$$

where:

t: The time required for the sound wave to return (s), and the valve must be closed after that period;
L: Pipe length (m);
C: Wave speed or velocity (m/s).

Example 7.6 Figure 7.7 shows the dynamic characteristic of a dual plate check valve installed after a pump in the water containing piping. In the figure, the horizontal axis represents the deceleration of fluid flow after the pump is stopped, expressed in feet per second squared. A vertical axis represents the maximum reverse velocity or velocity change, expressed in feet per second. Following the pump stoppage, fluid deceleration is approximately 40 ft/s^2. What is the pressure rise in the steel pipe caused by the pump stoppage as well as the slamming of the

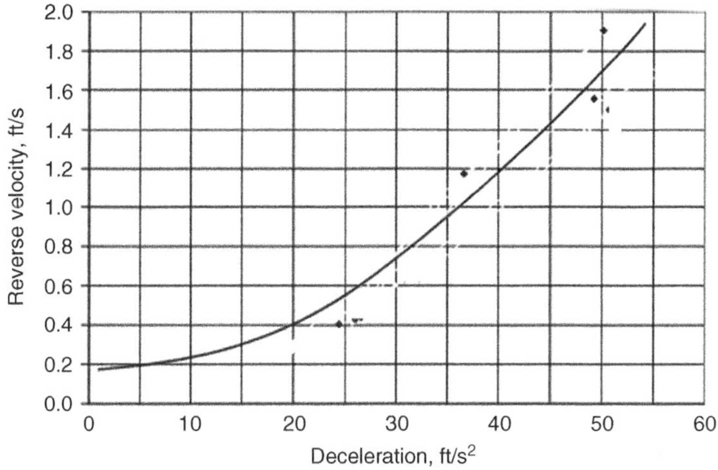

Figure 7.7 Test-based dynamic characteristic of a dual plate check valve.

dual plate check valve if the wave velocity in the steel pipe is 3200 ft/s? Is the amount of calculated slamming acceptable? In this instance, what is the water hammering strategy?

Answer

The reverse flow velocity associated with the fluid deceleration of 40 ft/s^2 is 1.2 ft/s according to Figure 7.7. The water hammering can now be calculated in terms of water column length according to Eq. (7.6) as follows:

$$\Delta H = \frac{V \times C}{g} = \frac{1.2 \times 3200}{32.2} = 119.25\,\text{ft}$$

There is more than 100 ft of water hammer, which indicates dangerous slamming and hammering that should be avoided by changing the valve type or modifying its design (e.g. adding a dashpot). A foot of water has a pressure of 0.43352 psi. So, the amount of water pressure caused by 119.25 ft of water is 51.70 psi. Equation (7.7) may be used as an alternative method to calculate the increase in pressure in the pipe caused by water hammering. It is important to note that the water density is 1000 kg/m^3.

$$\begin{aligned}\Delta P_{\text{max}} &= \rho \times C \times V \\ &= 1000\,\text{kg/m}^3 \times 3200\,\text{ft/s} \times 0.3048\,\text{m/ft} \times 1.2\,\text{ft/s} \times 0.3048\,\text{m/ft} \\ &= 356{,}747.6736\,\text{Pa} = 51.74\,\text{psi}\end{aligned}$$

Example 7.7 The swing check valve is installed in a piping system handling oil with an SG of 0.8. Upon closure of the valve, a reverse flow velocity of 1 m/s and a wave velocity of 1100 m/s are produced. Calculate the increase in oil pressure. If the wave sound must travel through an 1110-m pipe and return to the valve, what is the minimum time for the valve to be closed in order to prevent water hammering?

Answer

According to Eq. (7.7), oil pressure increases due to reverse flow velocity as follows:

$$\begin{aligned}\Delta P_{\max} &= \rho_{\text{oil}} \times C \times V = \text{SG} \times \rho_{\text{water}} \times C \times V\\ &= 0.8 \times 1000 \times 1100 \times 1 = 880,000\ \text{Pa}\\ &= 8.8\ \text{bar} = 127.63\ \text{psi}\end{aligned}$$

The next step is to calculate the length of time necessary for the pressure wave to traverse the pipe and return to the valve based on Eq. (7.9).

$$t = 2 \times \frac{L}{c} \rightarrow 2 \times \frac{1100}{1100} = 2\ \text{seconds}$$

In order to prevent water hammering, the valve must be closed after the arrival of the sound wave. Therefore, the valve must be closed after two seconds.

Using Eq. (7.10), it is possible to calculate the water hammer pressure in a piping system due to sudden valve closure:

Produced Water Hammer Pressure by Knowing the Valve Closing Time

$$\Delta P = 0.070 \frac{\Delta V \times L}{\Delta t} \tag{7.10}$$

where:

ΔP: An increase in pressure or a spike in pressure (psi);
ΔV: Flow velocity change (ft/s);
Δt: The closing time of the valve (second);
L: Length of the upstream pipe (ft).

$1\ \text{ft} = 0.3048\ \text{m}$;

$1\ \text{ft/s} = 0.3048\ \text{m/s}$;

$1\ \text{psi or } (\text{lb/in}^2) = 6894.8\ \text{Pa or } (\text{N/m}^2)$

Example 7.8 Calculate the pressure increase caused by the closure of a valve in a pipe in a time of two seconds, resulting in a decrease in velocity from 20 to 15 m/s. The pipe upstream of the valve is 20 m in length.

Answer

Except for valve closure time, all the given parameters must be converted from SI units to imperial units first.

$$\text{Initial velocity} = 20\ \text{m/s} = 6.096\ \text{ft/s}$$

$$\text{Secondary velocity} = 15\ \text{m/s} = 4.572\ \text{ft/s}$$

$$\text{Velocity change} = 6.096\ \text{ft/s} - 4.572\ \text{ft/s} = 1.524\ \text{ft/s}$$

The length of the pipe upstream of the valve = 20 m = 6.096 ft

$$\Delta P = 0.070\frac{\Delta V \times L}{\Delta t} = 0.070\frac{1.524 \times 6.096}{2} = 0.32\ \text{psi}$$

Questions and Answers

7.1 Which of the following statements about water hammering is true?

A A very common problem with ball and butterfly valves is water hammering.

B The devastating effects of water hammering are restricted to piping systems, including valves.

C Water hammering is directly related to the pipe area.

D Water hammering is directly proportional to changes in pressure and wave velocity.

Answer

Option A is incorrect since water hammering occurs most frequently in check valves and not in ball and butterfly valves. Option B is not acceptable either, as the undesirable effects of water hammering are not limited to piping systems, as it can also harm pumps. According to Joukowsky's formula, option C does not hold true since increasing the pipe area through different methods, such as increasing the wall thickness, reduces the severity of water hammering. The correct option is D.

7.2 A swing check valve is installed after a pump on a water-handling piping system. The speed of water inside the valve is 20 m/s. The port size of the valve is 80% of the connected pipe, creating a flow resistance coefficient of 1.58. To reduce this flow resistance coefficient and energy loss in the swing check valve, a nozzle check valve is installed with a flow resistance coefficient of 0.83. If the fluid velocity is the same in both types of valves, how much pressure drop is reduced by switching from swing to nozzle check valves?

A 1.5 bar
B 2 bar
C 0.5 bar
D 3 bar

Answer

Equation (7.2) can be used to calculate the head loss of the valves as follows:

Drop in pressure at the swing check valve:

$$\Delta H = \frac{K_v V^2}{2g} = \frac{1.58 \times 20^2}{2 \times 9.81} = 32.21\ \text{m}$$

Every 10 m of water produces 1 bar of pressure. So, 32.21 m of water causes a 3.21 bar loss in the swing check valve.

Drop in nozzle check valve pressure:

$$\Delta H = \frac{K_v V^2}{2g} = \frac{0.83 \times 20^2}{2 \times 9.81} = 16.92\ \text{m}$$

A 10-m length of water generates 1 bar of pressure, and thus 16.92 m of water causes 1.692 bar of pressure loss in the nozzle check valve.

The difference in pressure loss generated by the swing and axial nozzle check valves = 3.21 − 1.692 = 1.518 bar. Therefore, the most appropriate option is A.

7.3 According to the following assumptions, how much USD is saved annually by using a nozzle check valve instead of a swing check valve after a pump:

1) The fluid velocity is 20 m/s and the flow rate is 8000 gpm.
2) The fluid service in the pipes and valves is water.
3) Electricity is used at a rate of 70% and the cost of electricity is 0.06 USD/kWh.
4) The pump located upstream of the check valve has an efficiency of 0.8.
5) For swing and nozzle check valves, respectively, the flow coefficient resistances are 1.58 and 0.83.

A 400 USD
B 3200 USD
C 30,000 USD
D 28,000 USD

Answer

As a first step, calculate the head loss difference between the swing and axial nozzle check valves by applying Eq. (7.2) as follows:

$$\Delta H = \frac{K_v V^2}{2g} = \frac{(1.58 - 0.83) \times 20^2}{2 \times 9.81} = 15.29\ \text{m} = 50.16\ \text{ft water}$$

Using a nozzle check valve rather than a swing check valve will result in areduction of energy costs per year (A)

$$= \frac{1.65 \times Q \times \Delta H \times S_g \times C \times U}{E}$$

$$= \frac{1.65 \times 8000 \times 50.16 \times 0.8 \times 0.06 \times 0.7}{0.8} = 27,808.70\ \text{USD}$$

The best answer is choice D.

7.4 A series of experiments were conducted on an 8″ CL1500 dual plate and nozzle check valve installed after a pump in the water system in order to measure water hammering generated as a result of the valve closure. The reverse flow velocity generated by stopping the pump is 0.75 m/s for the dual plate valves and 0.3 m/s for the nozzle check valves. What is the correct answer? (*Note*: assume a wave velocity of 3200 ft/s.)

- **A** In this case, installing a nozzle check valve after the pump is recommended.
- **B** The pressure increase caused by water hammering in the dual plate check valve is 90 psi.
- **C** In comparison with the axial check valve, the severity and value of water hammering in dual plate check valves are more than three times higher.
- **D** Water hammering causes more than 50 psi pressure increase in the axial check valve.

Answer

Equation (7.6) can be used to calculate the water hammering in terms of the water column length as follows:

Case 1

Water hammering in the dual plate check valve: $\Delta H = \dfrac{V \times C}{g}$

$$= \frac{0.75\ \text{m/s} \times 3.28\ \text{ft/m} \times 3200\ \text{ft/s}}{32.2\ \text{ft/s}^2}$$

$$= 244.47\ \text{ft} \rightarrow \text{Pressure increase} = 244.47 \times 0.43352 = 106\ \text{psi}$$

Case 2

$$\text{Water hammering in the axial check valve: } \Delta H = \frac{V \times C}{g}$$

$$= \frac{0.30\ \text{m/s} \times 3.28\ \text{ft/m} \times 3200\ \text{ft/s}}{32.2\ \text{ft/s}^2}$$

$$= 97.79\ \text{ft} \rightarrow \text{Pressure increase} = 97.79 \times 0.43352 = 42.39\ \text{psi}$$

Based on field experience, a water hammer within a range of 50–100 ft is acceptable. Water hammering in the dual plate check valve is equivalent to a water column of 97.79 ft; therefore, it is acceptable. However, water hammers that exceed 100 ft are not tolerated and are considered dangerous. Therefore, the water hammer created by the dual plate check valve is not acceptable, and the dual plate check valve is not a suitable valve choice in this application. There is an error in option B since the water hammering in the dual plate check valve causes a pressure increase of 106 psi, not 90 psi. The intensity of water hammering in a pipe as a result of a dual plate check valve closing compared to an axial check valve closing = 106/42.39 = 2.5 that is less than three. So, option C is incorrect. In addition, option D is incorrect since water hammering increases the pressure in the axial check valve by 42.39 psi, which is less than 50 psi. Therefore, option A is the correct answer.

7.5 Figure 7.8 illustrates the dynamic characteristics of various types of check valves as determined by the results of experimental testing. Based on the given chart, which is the correct statement regarding the risk of valve slamming and water hammering?

A In the case of a reverse flow velocity of 0.1 m/s, the risk of water hammering is exceptionally high for swing and dual plate check valves.

B Tilted check valves are more susceptible to slamming and water hammering than swing check valves.

C The deceleration of 5 m/s squared gives mild slamming for swing check valves but no slamming for nozzle check valves.

D A dual plate check valve will slam mildly with a reverse fluid velocity of 0.45 m/s.

Answer

Based on the chart, option A is incorrect since 0.1 m/s reverse flow velocity does not cause slamming or water hammering for both dual and nozzle check valves. Option B is also not correct since the slamming and hammering associated with a tilted check valve is less severe than that of a swing check valve. The correct choice is option C. Option D is incorrect as the reverse fluid velocity of 0.45 m/s would result in severe slamming of a dual plate check valve.

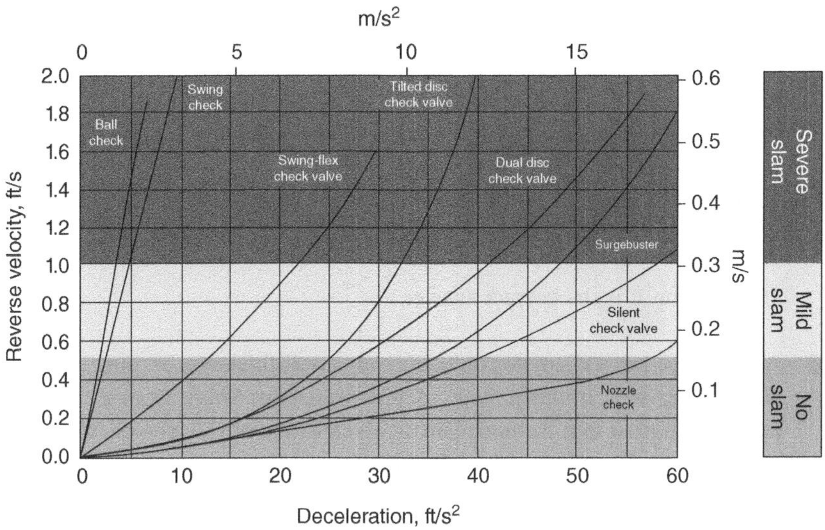

Figure 7.8 A comparison of the dynamic characteristics of various types of check valves.

7.6 A check valve closure produces a wave velocity of 1100 m/s in the piping system. What is the minimum duration of time for a valve to be closed in order to prevent a slamming effect in the piping system if the length of pipe the produced wave must travel and return is 2200 m?

A One second
B Two seconds
C Three seconds
D Four seconds

Answer

$$t = 2 \times \frac{L}{c} \rightarrow 2 \times \frac{2200}{1100} = 4 \text{ seconds}$$

To prevent water hammering, the valve must be closed after the arrival of the sound wave, which means that it shall be closed after four seconds. Therefore, option D is the correct answer.

7.7 A check valve closes in one second, reducing the velocity of water in a piping system from 10 ft/s to zero. Based on the length of the downstream pipe being 50 ft, how much pressure is generated by the water hammering effect in the piping system?

A 30 psi
B 35 psi

C 40 psi
D 45 psi

Answer
For this particular case, Eq. (7.10) is applied to calculate the water hammering as follows:

$$\Delta P = 0.070\frac{\Delta V \times L}{\Delta t} = 0.070 \times \frac{10\text{ ft/s} \times 50\text{ ft}}{1\text{ second}} = 35\text{ psi}$$

Thus, option B is the correct answer.

7.8 A steel pipe has a diameter of 18 in. and a wall thickness of 2 in., and it is 5000 ft long on a uniform slope. A pipe transports water from a reservoir and discharges it to the atmosphere at an elevation of 150 ft below the reservoir's free surface. There is a valve installed at the downstream end of the pipe, which allows for a flow rate of 25 ft^3/s. Figure 7.9 illustrates a schematic of the valve and the sloped pipe that discharges the water into the environment. What is the maximum water hammer pressure generated at the valve if the valve closes in 1.4 seconds? Is the valve suddenly slammed and closed? (*Note*: there is no longitudinal stress in the pipe, and the pipe is free to move longitudinally throughout.)
A 857 psi due to sudden closure of the valve
B 450 psi due to slow closure of the valve
C 790 psi due to sudden closure of the valve
D 400 psi due to slow closure of the valve

Answer
The following is a summary of the given data in the question:

$$\text{Steel pipe length} : L = 5000\text{ ft}$$

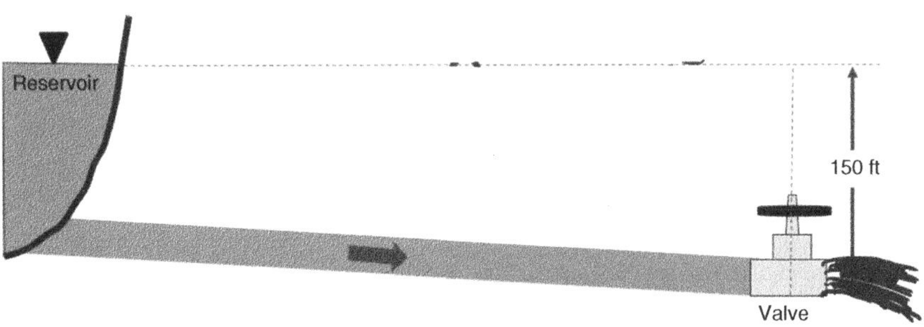

Figure 7.9 An illustration of the valve and sloped valve that discharges water into the environment.

Pipe diameter: $D = 18\text{ in.} = 1.5\text{ ft}$

Pipe thickness: $t = 2\text{ in.} = 0.166667\text{ ft}$

Flow rate: $Q = 25\text{ cfs}$

Valve closing time: $t = 1.4$

Firstly, the acoustic wave velocity, parameter C, must be calculated. Due to the fact that the pipe is steel and not plastic, the following equations apply:

$$C = \sqrt{\frac{E_c}{\rho}}$$

where:

$$\frac{1}{E_c} = \frac{1}{E_b} + \frac{Dk}{E_p t}$$

E_b represents the bulk modulus of water, which is 3.0×10^5 psi. According to Table 7.2, the modulus elasticity E_p of pipe material in steel equals 28×10^6. Assuming a Poisson's ratio of 0.25, the following equation can be used to determine the value of k:

$$k = \left(\frac{5}{4} - \varepsilon\right) = 1.25 - 0.25 = 1$$

$$\frac{1}{E_c} = \frac{1}{3.0 \times 10^5} + \frac{18\text{ in.} \times 1}{(28 \times 10^6) \times 2\text{ in.}} \rightarrow E_c = 2.74 \times 10^5\text{ psi}$$

$$C = \sqrt{\frac{E_c}{\rho}} = \sqrt{\frac{2.74 \times 10^5\text{ psi}}{0.01347}} = 4510\text{ ft/s}$$

Equation (7.9) is used to calculate the maximum amount of time it takes for a sound wave to cause valves to slam and water hammer

$$t = 2 \times \frac{L}{c} = 2 \times \frac{5000}{4510} = 2.2\text{ second}$$

The valve closes in 1.4 seconds, resulting in the sudden closure of the valve and the hammering of water in the pipe due to the pressure wave. In order to calculate the fluid velocity, the next step is to calculate the area of the pipe.

$$A = \pi\frac{d^2}{4} = \pi\frac{1.5^2}{4} = 1.7670\text{ ft}^2$$

The following equation expresses the relationship between the volumetric flow rate, area, and velocity of fluid inside the pipe:

$$Q = V \times A \rightarrow V = \frac{Q}{A} = \frac{25}{1.7670} = 14.1\ \text{ft/s}$$

By using Eq. (7.7), it is possible to calculate the pressure increase in the pipe as a result of water hammering:

$$\Delta P_{\text{max}} = \rho \times C \times V = 1000\,\text{kg/m}^3 \times 4510\,\text{ft/s} \times 0.3048\,\text{m/ft} \times 14.1\,\text{ft/s} \times 0.3048\,\text{m/ft} = 5,907,797\,\text{Pa} = 857\,\text{psi}$$

Thus, option A is the correct answer.

Further Reading

American Petroleum Institute (API) 594 (2004). *Check Valves: Flanged, Lug, Wafer and Butt-welding*, 6e. Washington, DC: American Petroleum Institute (API).

American Society of Mechanical Engineers (ASME) B16.34 (2017). *Valves – Flanged, Threaded, and Welding End*. New York: American Society of Mechanical Engineers (ASME).

Ballun, J.V. (2007). *A Methodology for Predicting Check Valve Slam*. American Water Works Association (AWWA).

Ford, R. (2014). Power industry applications: A valve selection overview. *Valve World Magazine* 19 (8): 96–103.

Gustorf, H. and Root, P. (2016). Developments in axial valve design. *Valve World Magazine* 21 (10): 49–53.

ITT Water & Wastewater AB (2019). *Hydraulic Transient Analysis: Preventing Water Hammer*. ITT Water & Wastewater AB: Sundbyberg.

Kruisbrink, A. (2010). The need for dynamic characteristics of check valves. *Valve World Magazine* 15 (9): 65–66.

Nesbitt, B. (2007). *Handbook of Valves and Actuators: Valves Manual International*, 1e. Oxford: Elsevier.

NORSOK L-001 (2017). *Piping and Valves*. Revision 4. Lysaker: NORSOK.

Norwegian Oil Industry Association (2013). *Valve Technology*. 2nd revision. Norsk Olje & Gass.

Oxler, G. (2009). Non-return valve and/or check valve for pump system – a new approach. *Valve World Magazine* 14 (4): 75–77.

Provoost, G.A. (1982). The dynamic characteristic of non-return valves. *Conference Paper Submitted to 11th Symposium of the Section of Hydraulic Machinery, Equipment and Cavitation*, Netherlands (13–17 September 1982).

Schmitz, H. (2018). What is water hammer and how do I fix it? *Livintanor.* https://livinator.com/whats-water-hammer-and-how-do-i-fix-it/ (accessed 23 March 2022).

Skousen, P.L. (2011). *Valve Handbook*, 3e. New York: McGraw-Hill.

Smit, P. and Zappe, R.W. (2004). *Valve Selection Handbook*, 5e. New York: Elsevier.

Sotoodeh, K. (2015). Axial flow nozzle check valves for pumps and compressors protection. *Valve World Magazine* 20 (1): 84–87.

Sotoodeh, K. (2018). Comparing dual plate and swing check valves and the importance of minimum flow for dual plate check valves. *American Journal of Industrial Engineering.* 5 (1): 31–35. https://doi.org/10.12691/ajie-5-1-5.

Sotoodeh, K. (2021). Analysis and failure prevention of nozzle check valves used for protection of rotating equipment due to wear and tear in the oil and gas industry. *Journal of Failure Analysis and Prevention* https://doi.org/10.1007/s11668-021-01162-2.

Sotoodeh, K. (2021). *Subsea Valves and Actuators for the Oil and Gas Industry*, 1e. Austin, TX: Elsevier (Gulf Professional Publishing).

VALMETALIC (2018). Design and selection of check valves. https://www.valmatic.com/Portals/0/pdfs/DesignSelectionCheckValves.pdf (accessed 22 March 2022).

8

Safety Valves

8.1 Introduction

Pressure safety valves (PSVs) and pressure relief valves (PRVs) are collectively referred to as safety valves, which are used in all pressurized equipment and facilities such as piping, pumps, compressors, turbines, and boilers when overpressurizing can occur. The principles of protection from overpressurizing are briefly presented in Chapter 2. In addition, the sizing of safety valves is covered in full detail in Chapter 2. The objective of this chapter is to provide an overview of essential design considerations and calculations for PSVs and PRVs, including the relief of pressure and capacity as well as generating reaction forces. There is a noticeable sound made by these valves during operation, which can be measured using the methods and calculations described in Chapter 6. For whatever reason, if something goes wrong with the pressure piping or equipment which leads to overpressure scenarios, the excessive pressure can damage or burst the expensive equipment and piping. PRVs and PSVs are installed over pipes and equipment to prevent undesirable events and enhance safety and reliability. If the pressure inside the equipment or piping exceeds the allowable limit, the safety valve automatically opens and releases the extra pressure. The result is a reduction in pressure across the protected equipment or piping. Both PSVs and PRVs are sometimes referred to as spring valves. There are two main differences between a PSV and a PRV; the first is that PSVs are typically used on gas services as well as steam services. However, PRVs are primarily used to provide overpressure protection for liquid handling systems. Another main difference between these two types of valves is that a PRV opens proportionally to the increase in fluid pressure, while a PSV opens suddenly. As shown in Figure 8.1, three PSVs are installed on the piping system to protect it in case of an overpressure situation. Three PSVs are used in lieu of one for safety and reliability reasons. In fact, if one or two PSVs malfunction and require maintenance, the third one may be utilized. Three ball valves are

Industrial Valves: Calculations for Design, Manufacturing, Operation, and Safety Decisions, First Edition. Karan Sotoodeh.

Figure 8.1 Three PSVs installed on piping for overpressure protection. *Source:* Pichitstocker/Adobe Stock.

located upstream of PSVs to isolate the piping from the downstream PSV when it is removed for maintenance.

8.2 Safety Valve Parts

Figure 8.2 depicts a conventional safety valve, which features the following essential parts: The main pressure-containment element is a body made of carbon steel or another corrosion-resistant alloy such as stainless steel. In valves, pressure-containing components are those components whose malfunction could cause leakage from the valve into the environment. It is important to note that the fluid enters the valve through a port at the bottom, and it exits through the left port following a 90° rotation.

8.3 Safety Valve Design and Operation

8.3.1 Design and Operation Parameters

Three primary standards are relevant to the design and operation of safety valves. The first is American Petroleum Institute (API) 520, *Sizing, Selection, and Installation of Pressure-Relieving Devices Part 1 – Sizing and Selection*, the second is

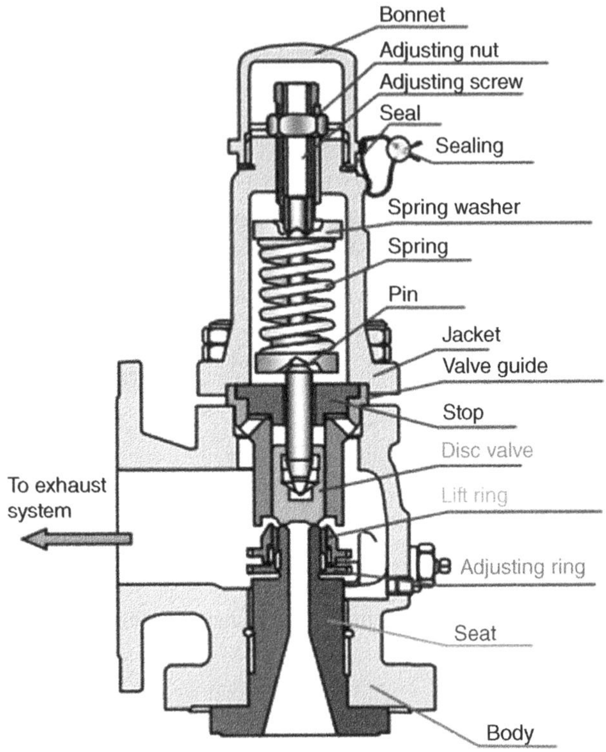

Figure 8.2 An essential component of a conventional safety valve.

API 526, *Flanged Steel Pressure-Relief Valves*, and the third is API 521, *Pressure-Relieving and Depressuring Systems*. In determining how safety valves should be designed, it is critical to determine what factors can increase pressure inside protected piping and facilities. It is also critical to determine how much each factor can increase pressure inside protected components and equipment. Table 8.1 summarizes the most relevant parameters for the design of a safety valve.

The following are some terminologies associated with safety valve design parameters:

Relieving pressure: It is also referred to as the inlet flow pressure as indicated by parameter P_1 and can be calculated using Eq. (8.1). Using Eq. (8.2), this pressure can be expressed as absolute relieving pressure. It is equal to the set pressure plus the allowable overpressure plus atmospheric pressure minutes loss pressure. The relieving pressure is equal to the *test pressure*.

Table 8.1 Factors that should be considered in the design of safety valves.

1) Fluid properties	3) Relieving conditions
A) Type of fluid service and its state	A) Required relieving capacity
B) Molecular weight	B) Set pressure
C) Viscosity	C) Allowable overpressure
D) Specific gravity	D) Superimposed backpressure
E) Ratio of specific heats	E) Buildup backpressure
F) Compressibility factor	F) Relieving temperature
2) Operating conditions	
A) Operating pressure	
B) Operating temperature	
C) Maximum allowable working pressure (MAWP)	

Calculation of Safety Device Inlet, Upstream, or Relieving Pressure

$$P_1 = P_{set} + P_{over} - P_{loss} \tag{8.1}$$

Calculation of Safety Device Inlet, Upstream, or Absolute Relieving Pressure

$$P_1 = P_{set} + P_{over} + P_{atmosphere} - P_{loss} \tag{8.2}$$

The *maximum allowable working pressure* (MAWP) is the highest pressure at which a device or equipment can be safely operated without any change in overpressure scenarios. The MAWP can be considered equal to the design pressure of the equipment, according to some valve literature.

A *set pressure* refers to the level of increasing static inlet pressure at which a PRV opens or leaks. In general, the set pressure is equal to the valve opening pressure, the popping pressure, or the start of leakage pressure. A *bench test* is conducted on a PSV to determine set pressure and seat tightness at ambient temperature by adjusting the valve spring. The first step in designing a safety valve is determining a suitable set pressure, which is typically expressed as a percentage of MAWP. The set pressure could be less, equal, or greater than the MAWP.

Overpressure or *allowable overpressure* is the pressure increase over the set pressure of a safety valve, which is usually expressed as a percentage.

Table 8.2 Set pressure and maximum accumulated pressure limits for safety valves.

	Single device installation		Multiple device installation	
	Maximum set pressure (%)	Maximum accumulated pressure (%)	Maximum set pressure (%)	Maximum accumulated pressure (%)
Nonfire case				
First relief device	100	110	100	116
Additional device	—	—	105	116
Fire case				
First relief device	100	121	100	121
Additional device	—	—	105	121
Supplement device	—	—	110	121

All values are expressed as a percentage of the MAWP.

The overpressure scenario may occur in both the protected pressure equipment and the safety valve.

The *maximum accumulated pressure* is calculated using Eq. (8.3) by adding the set pressure and the allowable overpressure.

Calculation of Maximum Accumulated Pressure

$$\text{Maximum accumulated pressure} = \text{Set pressure} + \text{Allowable overpressure} \tag{8.3}$$

In accordance with ASME codes, Table 8.2 summarizes the maximum accumulation and set pressure for PRVs. Based on Table 8.2, two conditions are considered for overpressure protection of pressurized equipment and piping: fire conditions and nonfire conditions. Furthermore, one or more safety devices may be installed on the pressurized equipment.

Example 8.1 For a nonfire case, a single safety valve is installed on a piping system with a MAWP of 100 psi. Given the maximum values of set pressure and maximum accumulated pressure, what would be the values of allowable overpressure and absolute relieving pressure assuming that there is no pressure loss in the valve?

Answer

The maximum set pressure for a single safety valve in nonfire conditions is MAWP = 100 psi (see Table 8.2). According to the same table, the maximum accumulated pressure for this case is approximately 1.1 times the MAWP of 110 psi. Knowing the set pressure and maximum accumulated pressure, it is possible to calculate overpressure as per Eq. (8.3) as follows:

Maximum accumulated pressure = Set pressure + Allowable overpressure
→Allowable overpressure = Maximum accumulated pressure − Set pressure = 110−100 = 10 psi

Thus, the maximum allowable overpressure is 10% of the set pressure. The next step is to calculate the absolute relieving pressure using Eq. (8.2) as follows:

$$P_1 = P_{set} + P_{over} + P_{atmosphere} - P_{over} + P_{atmosphere} - P_{loss}$$
$$= 100 + 10 + 14.7 = 124.7 \text{ psi}$$

Example 8.2 The top of a pressure vessel is equipped with two safety valves in the event of a fire. If MAWP = 100 psi, and the set and accumulated pressure values are at their maximum, calculate the absolute relieving pressure for both valves. (*Note*: no pressure loss inside both valves.)

Answer

The maximum set pressure for the first valve in the fire case = MAWP = 100 psi

The maximum set pressure for the second valve in the fire case = 105% MAWP = 105 psi

The maximum accumulated pressure for the first and second valve in the fire case = 121% MAWP = 121 psi

→Allowable overpressure (first valve) = Maximum accumulated pressure − Set pressure = 121 − 100 = 21 psi

→Allowable overpressure (second valve) = Maximum accumulated pressure − Set pressure = 121 − 105 = 16 psi

$$\text{Absolute relieving pressure (first valve)} = P_{set} + P_{over} + P_{atmosphere} - P_{loss}$$
$$= 100 + 21 + 14.7 = 135.7 \text{ psi}$$

$$\text{Absolute relieving pressure (second valve)} = P_{set} + P_{over} + P_{atmosphere} - P_{loss}$$
$$= 105 + 16 + 14.7 = 135.7 \text{ psi}$$

Backpressure refers to the statistical outlet pressure of the safety valve, which can either be fixed or variable. In plain English, backpressure is defined as a pressure that exists at the outlet of a safety valve that affects the opening pressure and the valve capacity. Furthermore, for a conventional safety valve subjected to constant backpressure, the set pressure is effectively reduced by an amount proportional to the backpressure. Due to this condition, a pressure increase equal to the backpressure should be applied to the required set pressure in order to compensate. In this case, the initial set pressure plus the constant backpressure is considered the adjusted set pressure. In the event of variable backpressure, the effective set pressure will be dependent on the fluctuations in the backpressure. The effects of backpressure on a safety valve depend on many factors such as the type of backpressure (fixed or variable), the valve position (open or closed), the fluid phase (gases, vapors, or liquids), and the type of valve construction (balanced bellows or conventional). Backpressure is equal to the sum of superimposed and accumulated backpressures.

Superimposed backpressure is a static pressure that exists at the outlet of a pressure relief device before and during the PSV operation, which is caused by other pressures in the discharge system. As an example, if a PSV outlet is attached to a closed system with a pressure of 2 psi, the superimposed backpressure is 2 psi. If the discharge outlet of the PSV is connected to the atmosphere, the superimposed backpressure would be 0 psi.

Buildup backpressure refers to the existing pressure at the outlet of a pressure relief device resulting from the flow passing through the safety device into a discharge system. Buildup backpressure is affected by the flow rate and the flow phase and varies from one safety valve to another as well as in different situations. Later in this chapter, more information about methods for calculating buildup pressure will be discussed.

Backpressure correction factor K_b or K_B is the ratio of the safety valve capacity with backpressure included to the rated valve capacity without backpressure that is normally applicable for valves that handle vapors and gases calculated in accordance with Eq. (8.4).

Backpressure correction factor K_b or K_B is the ratio of the safety valve capacity with backpressure included C_1 to the rated valve capacity without backpressure C_2 that is typically applicable for valves handling vapors and gases, based on Eq. (8.4). It is not possible to provide a standard value for the backpressure correction factor since this depends on the design of the valve. The responsibility lies with the manufacturers to provide their own data about this parameter, which is backed up by testing.

Calculation of Backpressure Correction Factor

$$K_b = \frac{C_1}{C_2} = \frac{\text{Safety valve capacity with backpressure included}}{\text{Safety valve capacity without backpressure}} \tag{8.4}$$

Calculation of the backpressure correction factor for balanced bellows safety valves is based on the *percentage of gauge backpressure* calculated according to Eq. (8.5).

Calculation of Gauge Backpressure for Bellows Safety Valves

$$\text{Percentage of gauge backpressure} = \frac{P_b}{P_s} \times 100 = \frac{\text{Backpressure}}{\text{Set pressure}} \times 100 \tag{8.5}$$

It is possible to determine the value of K_b using the diagrams in Figures 8.3 and 8.4 for bellows and conventional safety valves, respectively. Bellows or *balanced bellows safety valves* are those which are designed to minimize the effect of back-pressure on the operational characteristics of the valve. As a result, the bellows surround an area equal to the inlet of the orifice to protect it from backpressure resulting from the discharge of the safety valve. However, it should be noted that

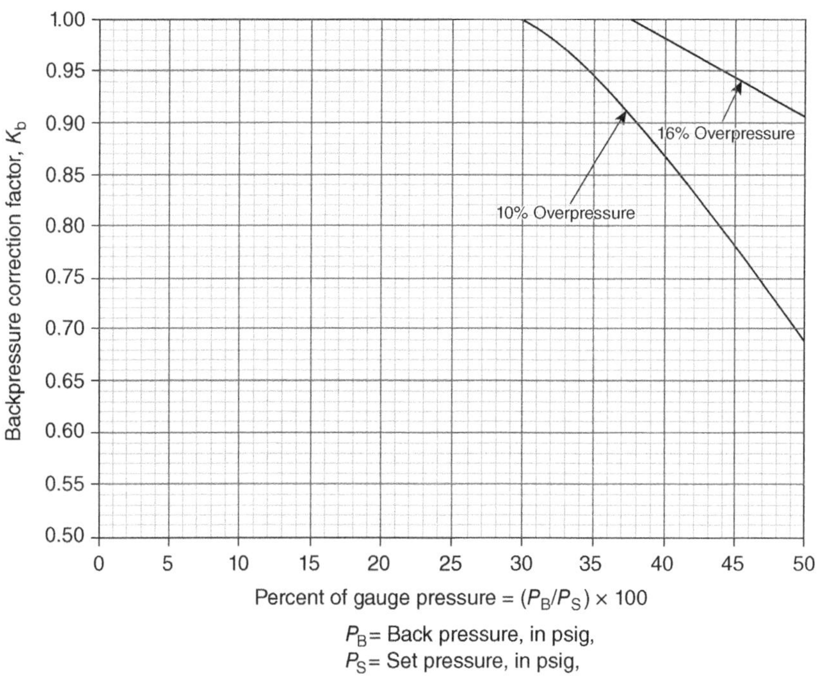

Figure 8.3 Backpressure correction factor for gas and vapors based on API 520 for balanced bellows valves.

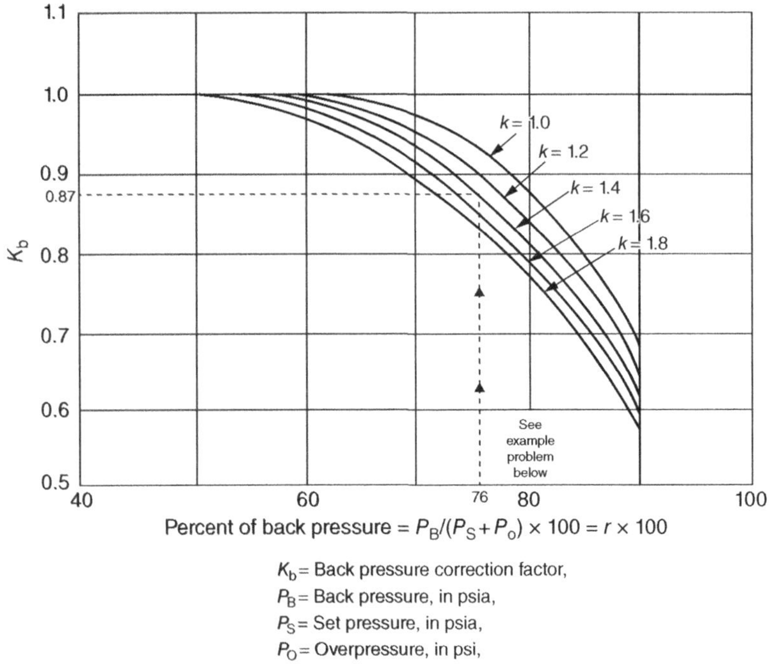

Figure 8.4 Backpressure correction factor for gas and vapors based on API 520 for conventional relief valves.

the main benefit of using a bellows safety valve is that there is no effect of backpressure on the relieving pressure.

In Figure 8.4, it is shown that the percentage of backpressure and the ratio of specific heat k are used to calculate the value of K_b for conventional relief valves handling gases or vapors. Equation (8.6) can be used to calculate the *percentage of backpressure* for conventional safety valves. The ratio of specific heat is defined as the ratio of heat capacity at constant pressure to heat capacity at constant volume. Heat or thermal capacity is defined as the amount of heat required to change the temperature of an object by one degree. Ratio of specific heat k for different gases can be extracted from Table 8.3.

Calculation of Backpressure Percentage for Conventional Safety Valves

$$\text{Backpressure percentage} = \frac{P_b}{P_s + P_o} \times 100$$
$$= \frac{\text{Absolute backpressure}}{\text{Absolute set pressure} + \text{Overpressure}} \times 100 \tag{8.6}$$

Table 8.3 Ratio of specific heats "*k*" for different gases.

Gas	Molecular weight	Ratio of specific heat (*k*)
Air	29.96	1.40
Ammonia	17.03	1.40
Argon	40	1.31
Carbon dioxide	44.01	1.40
Carbon monoxide	28	1.40
Ethane	30.07	1.19
Ethylene	28.03	1.24
Helium	4	1.66
Hexane	86.18	1.06
Hydrogen	2.02	1.41
Hydrogen sulfide	34	1.32
Methane	16.4	1.31
N-Butane	58.12	1.09
Natural gas (specific gravity = 0.60)	18.9	1.27
Nitrogen	28	1.40
Oxygen	32	1.40
Pentane	72.15	1.07
Propane	44.09	1.13
Propylene	47.08	1.15
Steam	18.01	1.31
Sulfur dioxide	64.04	1.29

There is also a backpressure correction factor K_w used for balanced bellows pressure relief or safety valves in liquid services. This value can be derived directly from Figure 8.5 based on the percentage of gauge backpressure referred to as API 520.

Example 8.3 The set pressure in a balanced bellows safety valve is 100 psig and the backpressure is 35 psig. What is the value of backpressure correction factor if the maximum allowable overpressure is 10% of the set pressure?

Answer

In this case, the percentage of gauge pressure is 35, and a 10% overpressure curve should be applied. According to Figure 8.3, the backpressure correction factor is approximately 0.94, which indicates that the backpressure reduces the valve capacity by 6%.

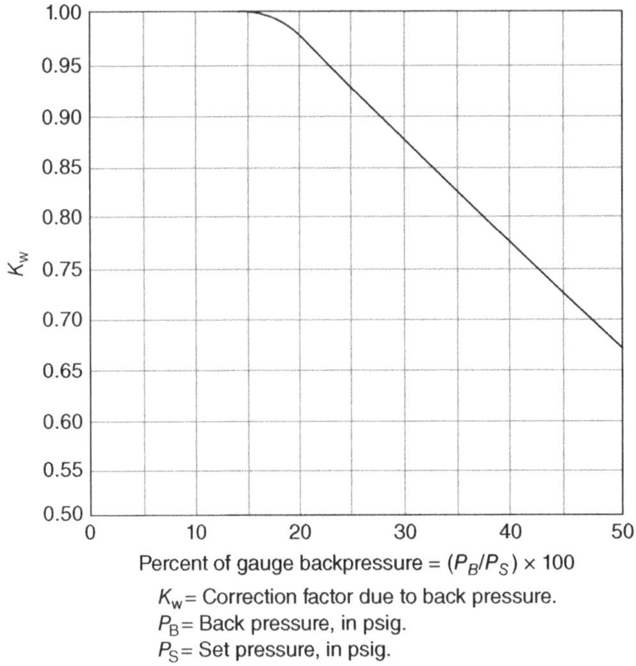

Figure 8.5 The capacity correction factor K_w of balanced bellows PRVs in liquid services caused by backpressure.

Example 8.4 The set pressure or overpressure of a conventional safety valve are 100 psi and 10 psi, respectively. It is estimated that the superimposed and buildup backpressures are 70 psig and 10 psi, respectively. In light of the fact that the valve deals with carbon dioxide, what is the percentage reduction in valve capacity caused by backpressure?

Answer

Backpressure = Superimposed backpressure + Buildup backpressure = 70 + 10 = 80 psig

Absolute backpressure = Backpressure + 14.7 = 80 + 14.7 = 94.7 psia

$$\text{Backpressure percentage for the conventional safety valve} = \frac{P_b}{P_s + P_o} \times 100$$

$$= \frac{\text{Absolute backpressure}}{\text{Absolute set pressure} + \text{Overpressure}} \times 100 = \frac{94.7}{100 + 14.7 + 10} \times 100 = 76$$

According to Table 8.3, carbon dioxide has a specific heat ratio k of 1.4. As shown in Figure 8.4, 0.87 represents the backpressure correction factor K_b. Due to the backpressure, this valve's capacity has been reduced by 13%.

Conventional relief valves do not perform satisfactorily when excessive backpressure develops after the safety valve opens to discharge gas or liquid. In this instance, the backpressure tends to reduce the valve disk lifting forces, resulting in the valve remaining open. In addition, excessive backpressure may cause the valve to operate in an unstable manner, resulting in the valve vibrating and chattering. In this case, the disk of the valve can move up and down and cause contact with the seat of the valve, which damages its internals. In addition to changing the set pressure, backpressure can introduce corrosive conditions into the inner chamber of the safety valve. In the case of a conventional safety valve, backpressure must not exceed 10% of the set pressure when the valve's allowable overpressure is 10%. When the backpressure is expected to exceed the provided limit, bellows balanced safety valves must be used instead. A balanced bellows safety valve is required in the second circumstance when the superimposed backpressure varies significantly from the set pressure. When using a balanced bellows safety valve, the total backpressure can reach a maximum of 50% of the set pressure.

Coefficient of discharge K_d is a unitless valve which is typically supplied by the valve manufacturer and it is typically used for calculation of flow passing through the safety valve. Based on Eq. (8.7), this factor is simply calculated by dividing the actual flow capacity of the safety valve by the theoretical flow capacity of the valve, and it is assumed to be equal to 0.9.

Calculation of Coefficient of Discharge

$$K_d = \frac{\text{Actual flow capacity of the safety valve}}{\text{Theoritical flow capacity of the safety valve}} \tag{8.7}$$

There is a very rough estimation of buildup backpressure (see Eq. 8.8) that does not fully consider important factors such as fluid properties and pressure losses due to the discharge piping arrangement.

An Estimation of the Buildup of Backpressure

$$P_{bb} = P_1 \times \frac{A_1}{A_2} \tag{8.8}$$

where:

P_{bb}: Buildup backpressure (psia);
P_1: Pressure at the inlet or the relieved pressure (psia) calculated from Eq. (8.2);
A_1: Cross-sectional area of the valve inlet $= \frac{\pi \times \text{Valve inlet size}^2}{4}$ (in.^{2});
A_2: Cross-sectional area of the valve outlet $= \frac{\pi \times \text{Valve outlet size}^2}{4}$ (in.^{2}).

Example 8.5 What is a rough estimate of backpressure buildup at a PSV 2″ × 3″ when the set pressure is 150 psig and 10% overpressure is permitted? (Note that there is no pressure loss within the valve.)

Answer

According to Eq. (8.2), the inlet or relieving pressure is calculated as follows:

$$P_1 = P_{\text{set}} + P_{\text{over}} + P_{\text{atmosphere}} - P_{\text{loss}} = 150 + 15 + 14.7 = 179.7\text{ psia}$$

$$A_1 = \text{The cross-sectional area of the valve inlet} = \frac{\pi \times \text{Valve inlet size}^2}{4}$$

$$= \frac{\pi \times 2^2}{4} = 3.14\text{ in.}^2$$

$$A_2\text{: The cross-sectional area of the valve outlet} = \frac{\pi \times \text{Valve outlet size}^2}{4}$$

$$= \frac{\pi \times 3^2}{4} = 7.07\text{ in.}^2$$

$$P_{\text{bb}} = 179.7 \times \frac{3.14}{7.07} = 79.81\text{ psia} = 65.1\text{ psig}$$

There is a more accurate way to calculate the buildup backpressure at any terminal point (t) located on the safety valve outlet piping or at the discharge point shown in Figure 8.6 with number 2.

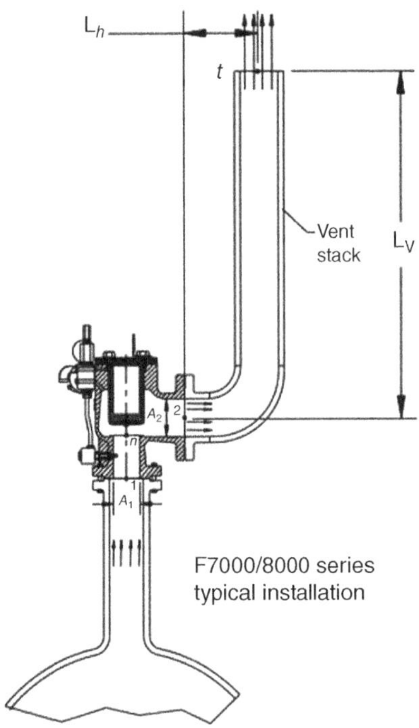

Figure 8.6 Schematic of a safety valve for calculating backpressure (flow safe).

Equation (8.9) is employed to calculate the buildup of backpressure at terminal pressure, which is point "*t*."

Calculation of the Terminal Pressure at Point "*t*"

$$P_t = \frac{K_d \times K_b \times A_1}{A_t \times P_1 \times (2/(k+1))^{k/(k-1)}(1/Z)^{1/2}} \tag{8.9}$$

where:

P_t: Terminal pressure (psi);
K_d: Coefficient of discharge calculated from Eq. (8.7) (dimensionless);
K_b: Backpressure correction factor (dimensionless);
A_1: Cross-sectional area of the valve inlet $\frac{\pi \times \text{Valve inlet size}^2}{4}$ (in.^{2});
A_t: Internal area of the outlet pipe at the point "*t*" (in.^{2});
Z: Gas compressibility factor calculated according to Eq. (8.10) (dimensionless);
k: Ratio of specific heats obtained from Table 8.3 for different gases (dimensionless);
P_1: Inlet pressure (psi) calculated according to Eq. (8.10).

Note 1

In thermal physics and thermodynamics, the ratio of specific heat is the ratio of heat capacity at constant pressure to heat capacity at constant volume. Heat capacity can be defined as the amount of heat necessary to change an object's temperature in one unit.

Note 2

In thermodynamics, the gas compressibility factor indicates how much the real gas differs from the ideal gas at given pressure and temperature. The *Z*-factor of gas is close to 1 at low pressures and high temperatures, which means that the behavior of real gases is similar to that of ideal gases. A gaseous mixture typically consists of billions of energetic gas molecules that interact with one another. As a result, it is extremely complex and difficult to model a real gas. Thus, the concept of an ideal gas was developed in order to simplify the behavior of gases. Scientists and engineers can model and predict the behavior of gases using the theory of ideal gases. There are two main rules associated with ideal gases: the first rule states that the molecules of ideal gases do not attract or repel one another, but are constantly in motion. In ideal gases, intermolecular forces are negligible. The only type of interaction between molecules in an ideal gas model is an elastic collision, which occurs when molecules strike against each other and kinetic energy is exchanged. It should be noted, however, that the total molecular kinetic energy inside the gas remains constant, and the kinetic energy does not change before or after the collision. Furthermore, the molecules occupy very little volume inside the gas. Helium is known as a type of gas that shares many characteristics with ideal gases. In contrast to ideal gases, real gases do not follow the laws of kinetic molecular theory. Nonideal gases are also known as real gases. Carbon dioxide, carbon monoxide, helium, oxygen, and nitrogen are all examples of real gases. When the pressure and temperature are the same, the volume of a real gas is typically less than the volume of an ideal gas. A real gas can therefore be considered supercompressible. The ratio of the real volume to the ideal volume is known as the *Z*-factor or compressibility factor, which is determined by the parameter *Z* and measured using Eq. (8.10). Gases possess a *Z*-factor that is close to 1 at low pressures and high temperatures, meaning that they behave very similarly to ideal gases. The volume of most gases at a low pressure and high temperature is so great that the volume of the molecules can be ignored, and the distance between the molecules is so large that any attractive or repulsive forces do not affect the behavior of the gas. However, as pressure increases, the volume of a gas decreases. So, the volume of molecules should also be taken into account. Furthermore, the attractive or repellent forces between molecules become relevant.

Gas *Z*-Factor or Compressibility Factor Calculation

$$Z = \frac{\text{Actual or real volume of the gas at specified pressure and temperature}}{\text{Ideal volume of the gas at specifiedf pressure and temperature}} \tag{8.10}$$

Example 8.6 There is a set pressure of 150 psig for a PSV in a size of 2″ × 3″ with an allowable overpressure of 10%. In this valve, air is received and connected to a line coming from the outlet, which has a size of 3″ and a schedule of 40. There is a 1.5-ft horizontal run of the outlet pipe, which is followed by a 6-ft vertical run of the vent stack. How much pressure is present at the outlet pipe taking the following assumptions into consideration?

A The actual flow capacity of the valve is equal to 0.915 times that of the theoretical value.

B The backpressure has no effect on the flow capacity of the PSV.

Answer

Due to the fact that the ratio of the valve's actual capacity to its theoretical capacity is 0.915, then $K_d = 0.915$. Moreover, since the backpressure has no effect on the PSV flow capacity, $K_b = 1$. As air is an ideal gas, $Z = 1$, and according to Table 8.3, the specific heat ratio of air is equal to 1.4. The cross-sectional area of the valve inlet and outlet pipes at point "t" is calculated as follows:

$$A_1 = \text{Cross-sectional area of the valve inlet} = \frac{\pi \times \text{Valve inlet size}^2}{4} = 3.14\,\text{in.}^2$$

P_1: Inlet pressure (psi) calculated according to Eq. (8.10):

$$P_1 = P_{set} + P_{over} + P_{atmosphere} - P_{loss} = 150 + 15 + 14.7 = 179.7$$

$$\begin{aligned}\text{Outlet pipe internal diameter} &= \text{Outlet pipe outside diameter} - 2 \times \text{Thickness}\\ &= 3.5\ \text{in.} - 2 \times 0.216\ \text{in.} = 3.068\ \text{in.}\end{aligned}$$

The values for the pipe's external diameter and thickness are taken from ASME B36.10 standard.

$$\begin{aligned}A_t &: \text{Internal area of the outlet pipe at the point"}t\text{"}\\ &= \frac{\pi \times \text{Outlet pipe inside diamter}^2}{4} = \frac{\pi \times 3.068^2}{4} = 7.39\ \text{in.}^2\end{aligned}$$

$$\begin{aligned}P_t &= \frac{K_d \times K_b \times A_1}{A_t \times P_1 \times (2/(k+1))^{k/(k-1)}(1/Z)^{1/2}}\\ &= \frac{0.915 \times 1 \times 3.14}{7.39 \times 179.7 \times (2/(1.4+1))^{1.4/(1.4-1)}(1/1)^{1/2}}\\ &= 0.004\ \text{psia}\end{aligned}$$

Equation (8.11) is used to calculate the backpressure at valve discharge, which is point "2."

Calculation of the Backpressure at Point "2"

$$P_2 = \left[P_t^2 + \frac{[2\ln(P_2/P_1) + \sum r + fL]}{\text{ID}} \times \left(\frac{K_d \times K_b \times A_n \times P_1}{A_t}\right)^2 \times \left(\frac{k}{Z}\right) \times \left(\frac{2}{k+1}\right)^{2k/(k-1)}\right]^{1/2} \tag{8.11}$$

where:

P_2: Outlet or discharge pressure at point "2" (psi) calculated according to Eq. (8.11);
P_t: Terminal pressure (psi);
P_1: Inlet pressure (psi) calculated according to Eq. (8.1) or (8.2);
K_d: Coefficient of discharge calculated from Eq. (8.7) (dimensionless);
K_b: Backpressure correction factor (dimensionless);
A_n: Cross-sectional area of the valve outlet $= \frac{\pi \times \text{Valve outlet size}^2}{4}$ (in.^{2});
A_t: Internal area of the outlet pipe at the point "t" (in.^{2});
r: Sum of the outlet pipe resistance coefficient (dimensionless);
f: Outlet pipe friction factor (dimensionless);
ID: Outlet pipe internal diameter (in.);
L: Output pipe length equal to the sum of the horizontal and vertical lengths (in.);
Z: Gas compressibility factor (dimensionless);
k: Ratio of specific heats obtained from Table 8.3 for different gases (dimensionless).

Opening or cracking pressure refers to the increasing inlet static pressure of a safety valve at which there is a measurable lift or at which the discharge becomes continuous, as determined by sight, feel, or hearing. Accordingly, the *opening pressure*, sometimes known as *cracking pressure*, is a pressure value that is lower than the set pressure at which leakage begins from the safety valve.

Closing or reseat pressure refers to the pressure required for a valve to reseat after it has been opened. Actually, the main point is that the safety valve is usually reseated at a pressure that is slightly lower than the set pressure, as shown in Figures 8.7 and 8.8. The difference between these two pressure values calculated using Eq. (8.12) is known as *blowdown*.

Calculation of Blowdown

$$\text{Blowdown} = \text{Set pressure} - \text{Closing pressure (reseat pressure)} \quad (8.12)$$

The reseat or closing pressure is calculated by knowing the set pressure and the inlet line pressure loss to the safety valve as per Eq. (8.13). According to ASME Section VIII Div. 01 and API 520 part II, the nonrecoverable losses in the inlet pipe to the safety valve shall not exceed 3% of the set pressure. Nonrecoverable losses are primarily due to friction rather than static losses. As an example, if the design pressure for a pressure vessel is 10 bars, and the PSV is installed 10 m above the vessel, the static pressure loss for the water supply is approximately 1 bar, and the valve must not be set at more than 9 bars ($10 - 1 = 9$ bar).

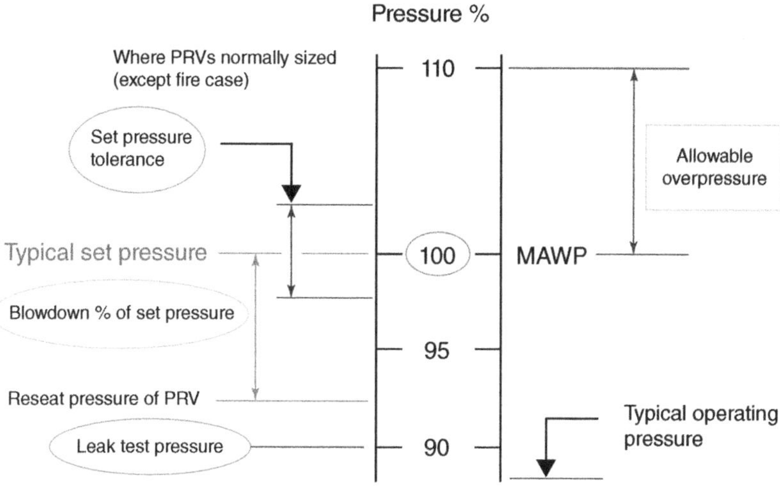

Figure 8.7 A chart illustrating the relationship between reseat pressure, set pressure, and blowdown.

Calculation of the Closing or Reseat Pressure

$$\text{Closing or reseat pressure} = \text{Set pressure} - \text{Non-recoverable losses at the valve inlet line} = \text{Set pressure} - 3\% \times \text{Maximum set pressure} \quad (8.13)$$

It should be noted that most safety valve manufacturers and codes and standards recommend a difference between reseat and operating pressure of 3–5% in order to ensure proper reseating of the valve and the restoration of proper tightness.

Example 8.7 Approximately, 10 m above a water storage tank handling water with a density of 1000 kg/m^3 is a PSV. If the design pressure of the storage tank is 10 bars and the allowable overpressure is 10%, what are the values of the maximum set pressure, fixed set pressure, relieving pressure, and reseat pressure? Moreover, the fixed set pressure must be a round number and as close to the maximum set pressure as possible, and there should not be any head loss in the valve.

Answer

The static pressure loss is approximately 1 bar due to the location of the safety valve, which is 10 m above the storage tank. Consequently, the maximum and set pressure are almost equal to 9 bar.

Static pressure loss = 1 bar

Maximum set pressure of the PSV = Fixed set pressure = 9 bar

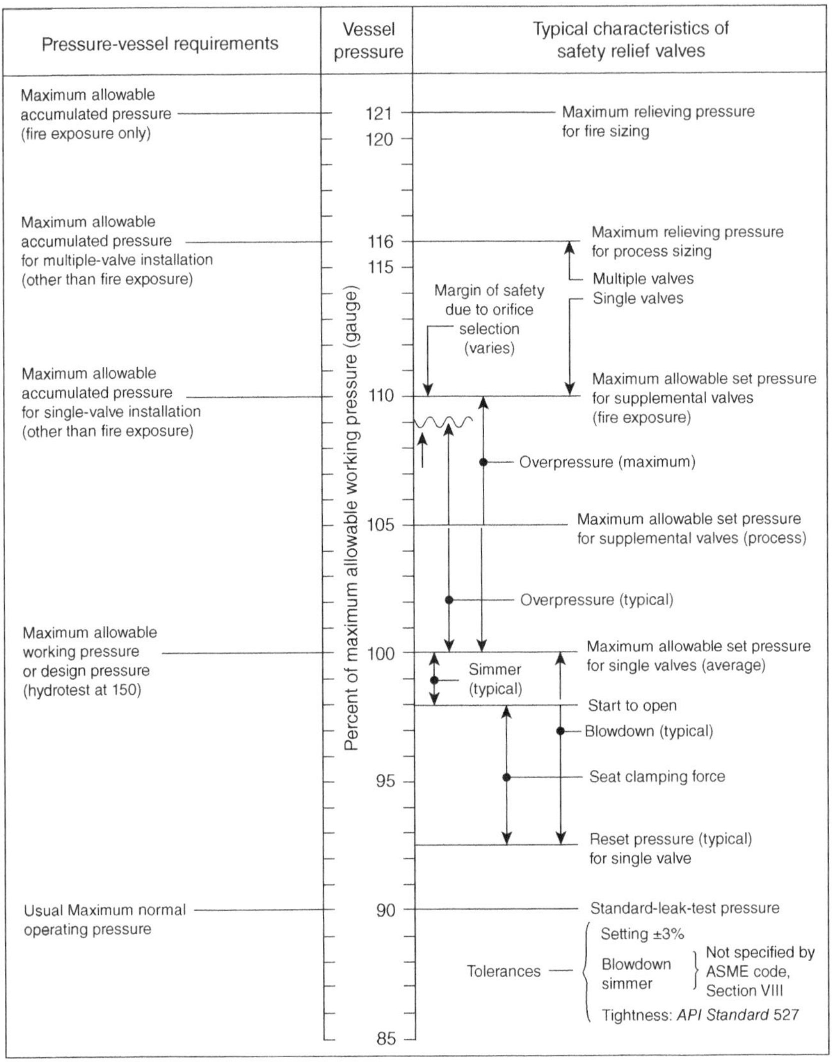

Notes:
1. The operating pressure may be any lower pressure required.
2. The set pressure and all other values related to it may be moved downward if the operating pressure permits.
3. This figure conforms with the requirements of Section VIII. Division I. of the ASME Code.
4. The pressure conditions shown are for safety relief valves installed on a pressure vessel (vapor phase).

Figure 8.8 Allowable operating, working, relief, set, and blow-down pressures in a safety valve.

Thus, the PSV opens at a pressure less than the fixed set pressure of 9 bar when the pressure in the storage tank is 10 bar.

$$\text{PSV relieving pressure} = 9 + 10\% \times 9 + 1 = 10.9\,\text{bara} = 9.9\,\text{barg}$$

$$\text{Maximum safety valve inlet line loss} = 3\% \times 9\,\text{bar} = 0.27\,\text{bar}$$

$$\text{Equation (8.13)} \rightarrow \text{Reseat pressure} = \text{Set pressure} - 3\% \times \text{Maximum set pressure} = 9 - 0.27 = 8.73\,\text{bar}$$

Simmer refers to a situation where the inlet pressure approaches the set pressure without still opening or popping the safety relief valve. In fact, simmer makes a small opening between the disk and the seat, which is called the startup opening. As a general rule, PRVs begin simmering or opening at a pressure of 90% of the set pressure. According to Eq. (8.14), simmer may be calculated as follows:

Simmer Calculation

$$\text{Simmer} = \text{Maximum allowable set pressure} - \text{Start to open pressure} \tag{8.14}$$

Figure 8.8 illustrates allowable operating, working, relief, set, and blow-down pressures in a safety valve.

The following paragraph summarizes the pressure changes during the operation of a safety valve. A pressure less than the set pressure causes the PRV to open initially. Gradually, the pressure reaches the set pressure and then exceeds the set pressure based on allowable overpressure. Once the pressure reaches the relieving pressure, which is considered a maximum pressure scenario, the pressure gradually decreases until the valve is closed.

8.3.1.1 Overpressure Criteria

It is imperative that all piping and equipment be protected when internal or external pressure exceeds the design condition of the system due to a variety of reasons, such as an emergency, operational error, instrument malfunction, or fire. Pressure-relieving devices were designed to ensure that a system or any of its components would not be subject to or at risk of a pressure that exceeded the allowable pressure specified in the relevant design code. According to ASME B31.3, process piping code, overpressure is referred to as occasional variations in pressure that may occur in piping systems, including safety valves. There is no doubt that the allowances for pressure increases are correlated with temperature. Actually, the most severe coincident pressure and temperature condition should be taken into consideration as the design pressure can be equal to the *pressure rating*. There is, however, the possibility of exceeding the design pressure in some cases as an overpressure scenario if all the following conditions are met:

a) Pipes and valves shall not contain pressure-containing components made from materials such as cast iron or other nonductile or brittle materials.
b) The stress generated from overpressure cannot exceed the yield strength of the piping and valve material at temperature.
c) In the case of an overpressure scenario, the longitudinal stress generated in piping and valves must be contained within limits stipulated by the relevant ASME code.
d) During the life of the piping system, the total number of pressure-temperature variations above the design conditions is not to exceed 1000 times.
e) The increased pressure should, in no case, exceed the test pressure used for the piping system, which is equal to 1.5 times the design pressure.
f) There may be occasional variations above design conditions, provided that they do not exceed one of the following limits:

Under certain circumstances and with the approval of the client or end user, it is permitted to exceed the pressure rating or the allowable stress for pressure design at the temperature of the increased condition not more than:

a) 33% for more than 10 hours at any one time and no more than 100 h/year, or
b) 20% for no more than 50 hours at any one time and no more than 500 h/year.

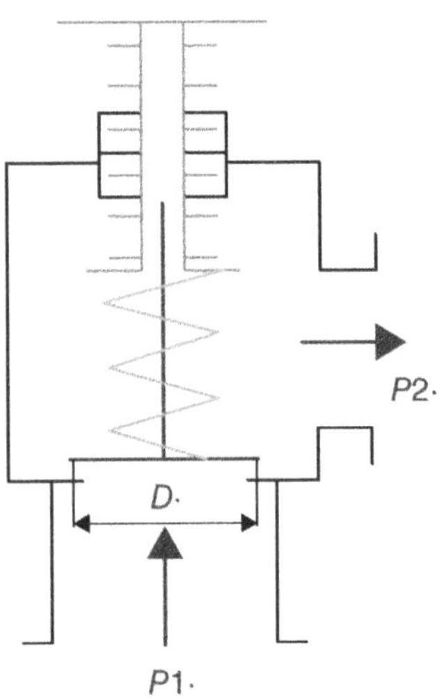

Figure 8.9 A safety valve schematic.

8.3.2 Principle of Operation

As illustrated in Figure 8.9, a safety valve is activated as a result of two dynamic forces caused by the velocity and pressure of gas or liquid. In the *closing* or *idle* position, the internal pressure of the protected equipment, such as a pressure vessel, is less than the set pressure of the safety valve; therefore, the spring retains the disk of the valve against the seat and keeps the valve in the closed position. When Eq. (8.15) is satisfied, the valve is closed.

Safety Valve Closing Condition

$$P_1 \times \frac{\pi D^2}{4} < P_2 \times \frac{\pi D^2}{4} + Kx \tag{8.15}$$

where:

P_1: Inlet pressure (Pa or N/m^2);
P_2: Outlet pressure (Pa or N/m^2);
D: *Seat diameter* refers to the smallest diameter of contact between the fixed and moving portions of the valve's pressure-containing elements (m);
Kx: Action of spring (N).

Conversely, the valve is *open*, or in the *working position*, when the internal pressure of the protected equipment exceeds the valve's set pressure, causing the spring to be compressed. As the fluid is released, the internal pressure is reduced. As soon as the internal pressure falls below the valve's set pressure, the spring closes the valve against the seat. If the condition in Eq. (8.16) is satisfied, the valve is open.

Safety Valve Opening Condition

$$P_1 \times \frac{\pi D^2}{4} > P_2 \times \frac{\pi D^2}{4} + Kx \tag{8.16}$$

Example 8.8 Flanges are used to connect a 6″ pipe with a standard thickness to an inlet of a PSV attached to a pressure vessel. Accordingly, the internal diameter of the PSV flange end is equal to the pipe's internal diameter, but the diameter of the seat where the disk sits on the seat is 80% of the flange diameter. If the inlet pressure is 10 bar and the outlet pressure is 2 bar, what is the minimum spring force necessary to maintain the closed position of the valve?

Answer
Following ASME B36.19, the standard for steel pipes, the outside diameter of a 6″ pipe is 6.625″ equal to 168.3 mm, and the standard wall thickness is 0.280″ equal to 7.11 mm. Equation (8.17) is used to calculate the internal diameter of the pipe based on the given outside diameter and the wall thickness:

Pipe Internal Diameter Calculation

$$\text{ID} = \text{OD} - 2 \times T \tag{8.17}$$

where:

ID: Internal diameter of the pipe (in., mm);
OD: Outside diameter of the pipe (in., mm);
T: Pipe thickness (in., mm);
$\text{ID} = \text{OD} - 2 \times T = 168.3 - 2 \times 7.11 = 154.08\ \text{mm} = 0.154\ \text{m}$;
PSV seat diameter $= 80\,\% \times 0.154 = 0.1232\ \text{m}$;

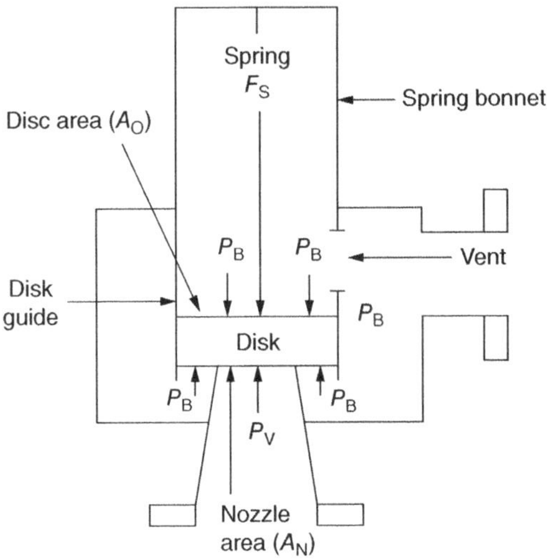

Figure 8.10 Schematic representation of a safety valve with the bonnet vented to the valve discharge.

P_1 : Inlet pressure $= 10$ bar $= 10^6$;
P_2 : Inlet pressure $= 2$ bar $= 2 \times 10^5$.

The condition in Eq. (8.15) must be met in order for the valve to remain in the closed position:

$$P_1 \times \frac{\pi D^2}{4} < P_2 \times \frac{\pi D^2}{4} + Kx \rightarrow \text{Minimum“}Kx\text{” or spring force to keep the valve}$$

$$\text{in closed position} = P_1 \times \frac{\pi D^2}{4} - P_2 \times \frac{\pi D^2}{4} = \frac{\pi D^2}{4}(P_1 - P_2)$$

$$= \frac{\pi \times 0.1232^2}{4}\left(10^6 - 2 \times 10^5\right) = 9536.49\,\text{N}$$

As illustrated in Figure 8.10, a safety valve schematic diagram is illustrated in which the pressure inside the valve bonnet is released to the valve discharge or outlet line. The force acting on the disk with the area of A_D is equal to the valve inlet pressure multiplied by the area of the nozzle A_N, which is equal to the sum of the spring force F_S and the force caused by backpressure P_B acting on the top of the disk A_N. In this case, the minimum amount of force required to open the valve is determined by Eq. (8.18).

The Minimum Force Necessary to Open the Safety Valve

$$P_v A_N = F_S + P_B A_N \tag{8.18}$$

where:

P_v: Fluid inlet pressure;
A_N: Nozzle area;
F_S: Spring force;
P_B: Backpressure.

In order to select a PRV, three factors must be taken into consideration: minimum inlet diameter D, the flow rate through the PSV, also known as the relieving flow, and time for the valve to open. Equations (8.19), (8.20), and (8.21) are used to calculate these parameters.

Calculation of the Minimum Inlet Diameter of a Safety Valve

$$D = 28.6 \times \frac{\sqrt{Q_{max}}}{H_{max}^{0.23}} \tag{8.19}$$

where:

D: Minimum diameter of a safety valve's inlet (mm);
Q_{max}: Discharged flow rate at the maximum pressure and opening (l/s);
H_{max}: Pressure head at maximum valve opening (m).

Example 8.9 Calculate the minimum diameter of the inlet of a safety valve that discharges a flow of 400 l/s at the maximum opening with a head pressure of 234 m.

Answer

$$D = 28.6 \times \frac{\sqrt{Q_{max}}}{H_{max}^{0.23}} = 28.6 \times \frac{\sqrt{400}}{234^{0.23}} = 28.6 \times \frac{20}{3.5} = 163\text{ mm}$$

Calculation of the Relieving Flow Through a Safety Valve

$$Q_{relv} = Q_{max} \times \frac{\sqrt{\text{Head}}}{\sqrt{\text{Head at the maximum openning}}} \tag{8.20}$$

Example 8.10 Calculate the relieving flow for the safety valve in the previous example if the head is 160 m.

Answer

$$Q_{relv} = Q_{max} \times \frac{\sqrt{\text{Head}}}{\sqrt{\text{Head at the maximum openning}}} = 400 \times \frac{\sqrt{160}}{\sqrt{234}} = 330.76\,\text{l/s}$$

Calculation of the Time Required for Valve to Open

$$t = K \times \left(\frac{D}{50}\right) \times \sqrt{\frac{100}{1.25 \times H_{\text{open}}}} \tag{8.21}$$

where:

t: Time required for the valve to open (second);
K: Constant that equals 0.4 for fast opening and 0.9 for slow opening valves (dimensionless);
D: Minimum diameter of a safety valve's inlet (mm);
H_{open}: The head at which the valve opens (m).

Example 8.11 Estimate the time required to open a safety valve if the minimum diameter of the safety valve's inlet is 200 mm and the head at which it opens is 100 m.

Answer
We will consider two cases of fast opening and slow opening for the valve in this example.

$$\text{Fast opening} \rightarrow t = K \times \left(\frac{D}{50}\right) \times \sqrt{\frac{100}{1.25 \times H_{\text{open}}}}$$

$$= 0.4 \times \left(\frac{200}{50}\right) \times \sqrt{\frac{100}{1.25 \times 100}} = 1.43 \text{ seconds}$$

$$\text{Slow opening} \rightarrow t = K \times \left(\frac{D}{50}\right) \times \sqrt{\frac{100}{1.25 \times H_{\text{open}}}}$$

$$= 0.9 \times \left(\frac{200}{50}\right) \times \sqrt{\frac{100}{1.25 \times 100}} = 3.2 \text{ seconds}$$

8.3.3 Safety Valve Reaction Forces

Reaction forces can be generated by PSVs, slug flow forces, and water hammering. Water hammering due to changes in flow velocity and sudden valve closure is discussed in Chapter 7. Slug flow forces are not covered since they are not related to industrial valves. When the pressure inside the apparatus exceeds the set pressure or allowable pressure level, the PSV opens, the excess pressure is released to the flare lines, and the extremely high pressure returns to its normal, safe level. As the safety valve is open and blowing, reaction forces are generated by a combination of backpressure and sudden impulses. While reaction forces at the inlet are relatively small, the main reaction force is generated at the outlet, especially in the case of gas fluid service. This is because high velocity and an increase in outlet pressure are expected. The impact of the PSV load should be taken into account so that the pipe-work is adequately supported.

The PSV load can be calculated in two ways. The first is based on ISO 4126-9, Safety devices for protection against excessive pressure – Application and installation of safety devices other than standalone bursting disc safety devices, based on Eq. (8.22). A steady-state flow should be considered for the calculations of the PSV reaction forces according to ISO 4126-9. Flow in a steady state refers to a process in which the fluid properties at any point in the system do not change over time. These fluid properties include temperature, pressure, velocity, and mass flow rate.

The Reaction Forces for Pressure Safety Valves or Relief Valves as Defined by ISO 4126-9 in Steady-State Flow

$$F = \frac{Q_m V_{out}}{3600} + (P_b - P_u)\frac{A_{out}}{10} \tag{8.22}$$

where:

F: Reaction force (N);
Q_m: Mass flow (kg/h);
V_{out}: Velocity of the fluid in the outlet pipe (m/s);
P_b: Backpressure (MPa);
P_u: Superimposed backpressure (MPa);
A_{out}: Flow area of outlet pipe (mm^2).

Example 8.12 With an inlet of 4″ and an outlet of 6″, a PSV has a capacity of 53,484 lb/h. A gas with a density of 0.68 kg/m^3 passes through the safety valve. The static pressure at the outlet of the PSV during the opening process is 2 bars, whereas the buildup backpressure is 3 bars. Calculate the PSV reaction forces according to ISO 4126-9, taking into account that the wall thickness of the outlet pipe is 0.280 in. (7.112).

Answer
The first step is to calculate the mass flow rate in kilograms per hour in SI units as follows:

$$Q_m = 53{,}484\ (\text{lb/h}) = 53{,}484\ (\text{lb/h}) \times 0.453592\ (\text{kg/lb}) = 24{,}260\ (\text{kg/h})$$
$$= 6.739\ (\text{kg/s})$$

In order to calculate the outlet piping flow area, the following formula must be used:

According to ASME B36.10, the piping standard, a 6″ pipe has an outside diameter of 6.625 in. (168.275). Based on the given outside diameter and the wall thickness, the following Eq. (8.6) is used to calculate the internal diameter of the pipe:

$$\mathrm{ID} = \mathrm{OD} - 2 \times T = 168.275 - 2 \times (7.112) = 154.051\ \mathrm{mm} = 0.154051\ \mathrm{m}$$

From the pipe's internal diameter, the flow area of the pipe can be calculated as follows:

$$A_{\mathrm{out}} = \pi \frac{\mathrm{ID}^2}{4} = \pi \frac{154.051^2}{4} = 18{,}638.29\ \mathrm{mm}^2$$

The following equation gives the relationship between mass flow rate, flow area, the density of fluid at the outlet pipe, and fluid density. Using the mass flow rate, pipe flow area, and density of the fluid inside the valve and the outlet pipe, one can obtain the fluid velocity as follows:

$$Q_{\mathrm{m}} = V_{\mathrm{out}} \times A_{\mathrm{out}} \times \rho \rightarrow V_{\mathrm{out}} = \frac{Q_{\mathrm{m}}}{A_{\mathrm{out}}\rho} = \frac{6.739}{0.154051 \times 0.68} = 66.33\ \mathrm{m/s}$$

While the valve is opening, the static backpressure at the outlet of the PSV is 2 bar or 0.2 MPa. In addition, the buildup backpressure is 3 bars or 0.3 MPa. Calculating backpressure is as follows:

$$\text{Backpressure} = \text{superimposed back} - \text{pressure} + \text{buildup backpressure} = 0.2 + 0.3 = 0.5\ \mathrm{MPa}$$

$$F = \frac{Q_{\mathrm{m}} V_{\mathrm{out}}}{3600} + (P_{\mathrm{b}} - P_{\mathrm{u}}) \frac{A_{\mathrm{out}}}{10} = \frac{24{,}260 \times 66.33}{3600} + (0.5 - 0.2) \frac{18{,}638.29}{10} = 447 + 559.15 = 1006.15\ \mathrm{N}$$

According to API 520 Part II contains sizing, selection, and installation of pressure-relief devices in refineries. According to Eq. (8.23) extracted from API 520 for calculation of reaction forces from PSVs, the standard assumes that the discharge from the PSV is discharged into the atmosphere (open discharge), and the flow condition is steady-state. In a closed system under steady-state conditions, the PSV reaction forces are not so large that they can be ignored.

Pressure Safety or Relief Valve Reaction Forces as Per API 520 Part II in Open Discharge Condition for Gas Service

In imperial units, the following equation should be used:

$$F = \frac{W}{366} \times \sqrt{\frac{KT}{(K + 1)M}} + \mathrm{AP} \tag{8.23a}$$

Table 8.4 Definitions of parameters used for PRV load calculation.

Used symbol	Designation	Unit (metric or imperial)	
F	Reaction force at the point of discharge to the atmosphere	N (newton)	Pound
W	Flow of any gas or vapor	kg/s	Pound mass per hour (lbm/hour)
K	Ratio of specific heats at the outlet conditions calculated from C_p/C_v equation	—	—
C_p	Specific heat at constant pressure	—	—
C_v	Specific heat at constant volume	—	—
T	Temperature at the outlet	°K	°R (Rankine same as Fahrenheit scale)
M	Molecular weight of the process fluid	—	
A	Area of the outlet at the point of discharge	mm^2	in.^{2}
P	Static pressure within the outlet at the point of discharge	Barg	psig

In metric units, the following equation should be used:

$$F = 129\ \mathrm{W} \times \sqrt{\frac{KT}{(K+1)M}} + 0.1\ \mathrm{AP} \tag{8.23b}$$

A description of each parameter and its unit of measurement is provided in Table 8.4.

Example 8.13 A PSV is installed for a gas service with a flow rate of 81, 804 kg/h. The ratio of specific heat at the outlet of PSV is 1.001 and the temperature at the outlet of the PSV is 515.6 °C. The molecular weight of the process fluid is 18.02 and the area of the outlet at the point of discharge is 345.0424 cm^2. The static pressure within the outlet at the point of discharge is 1.05 kg/cm^2. Calculate the PSV reaction force in Newton.

Answer

All the given data can be converted to metric units as per Table 8.5, as follows:

Using all the given data in Eq. (8.23), the PSV reaction load is calculated as follows:

$$F = 129\ \mathrm{W}\sqrt{\frac{KT}{(K+1)M}} + 0.1\ \mathrm{AP}$$

$$= 129 \times 22.72\ (\mathrm{kg/s})\sqrt{\frac{1.001 \times 788.75}{2.001 \times 18.02}} + 0.1 \times 34504.2378 \times 1.03$$

$$= 2930.88 \times 4.68 + 3554 = 17{,}270\ \mathrm{N}$$

Table 8.5 Provided data for PSV load calculation in gas service and the conversion to metric units.

Provided data	Given data value	Given data value in metric system
Flow of gas or vapor (*W*)	81,804 kg/h	22.72 kg/s
Ratio of specific heat (*K*)	1.001	1.001
Temperature at inlet (*T*)	515.6 °C	788.75 K
Modular weight of process fluid (*M*)	18.02	18.02
Area of outlet at the point of discharge (*A*)	345.0424 cm^2	34,504.2378 mm^2
Static pressure within the outlet at the point of discharge (*P*)	1.05 kg/cm^2	1.03 bar

Note: Some of the literature recommends applying a dynamic load factor of 2 to the PSV reaction forces. Thus, in this case, the reaction force that is calculated through this equation should be doubled to 34,540 N.

Using Eqs. (8.24) and (8.25), there is an alternative method to calculate reaction forces for a PRV discharging gas, vapor, or steam directly to the atmosphere without any discharge piping, using API 520.

PSV Reaction Force for Open Discharge of Gas, Vapor, or Steam to the Atmosphere (Open Discharge) in Accordance with API 520

Gas/Vapor

$$F = \frac{CKAP_1\sqrt{k/(k+1)}}{332.7} + F_g$$
$$F_g = \left(\frac{KAP_1k_r}{1.383A_0} - P_a\right) \times A_o \tag{8.24}$$

If F_g is less than or equal to zero, use $F_g = 0$

Steam

$$F = \frac{AP_1}{1.335} + F_s$$
$$F_s = \left(\frac{AP_1}{1.886A_0} - P_a\right) \times A_oK_n \tag{8.25a}$$

If F_s is less than or equal to zero, use $F_s = 0$

where:

F: The total reaction force at the point of discharge to the atmosphere in pounds-force;

F_g: The reaction force due to static pressure at the valve outlet for gas/vapor applications, expressed in pounds;

F_s: The reaction force due to static pressure at the valve outlet for steam applications in pounds;

C: The coefficient is obtained from the ratio of the specific heats of the gas or vapor at standard conditions in Table 8.7 (dimensionless);

K: Effective coefficient of discharge equal to 0.975 for safety valves handling gas, vapor, or steam according to API 520. According to API 520, the effective coefficient of discharge for safety valves in liquid service is 0.65 and for two-phase flows, it is 0.85;

A: An effective discharge area in square inches is selected from the ASME or API standards according to Table 8.6;

Table 8.6 ASME and API standard orifice sizes.

Serial number	Orifice designation	API effective area (in.^{2})	ASME effective area (in.^{2})	PSV inlet × outlet sizes (in.)
1	D	0.110	0.124	1″ × 2″ 1.5″ × 2″ 1.5″ × 2.5″
2	E	0.196	0.221	1″ × 2″ 1.5″ × 2″ 1.5″ × 2.5″
3	F	0.307	0.347	1″ × 2″ 1.5″ × 2″ 1.5″ × 2.5″
4	G	0.503	0.567	1.5″ × 2.5″ 1.5″ × 3″ 2″ × 3″
5	H	0.785	0.887	1.5″ × 3″ 2″ × 3″
6	J	1.287	1.453	2″ × 3″ 2.5″ × 4″ 3″ × 4″
7	K	1.838	2.076	3″ × 4″
8	L	2.853	3.221	3″ × 4″ 4″ × 6″
9	M	3.600	4.065	4″ × 6″
10	N	4.340	4.900	4″ × 6″
11	P	6.380	7.205	4″ × 6″
12	Q	11.05	12.47	6″ × 8″
13	R	16.00	18.06	6″ × 8″ 6″ × 10″
14	T	26.00	29.35	8″ × 10″

Table 8.7 Values of the gas constant (*C*) based on the ratio of specific heat (*k*) according to API 520.

K	*C*	*k*	*C*	*k*	*C*	*k*	*C*
1.00	315	1.31	347	1.60	372	1.90	394
1.01	317	1.31	348	1.61	373	1.91	395
1.02	318	1.32	349	1.62	374	1.92	395
1.03	319	1.33	350	1.63	375	1.93	396
1.04	320	1.34	351	1.64	376	1.94	397
1.05	321	1.35	352	1.65	376	1.95	397
1.06	322	1.36	353	1.66	377	1.96	398
1.07	323	1.37	353	1.67	378	1.97	398
1.08	325	1.38	354	1.68	379	1.98	399
1.09	326	1.39	355	1.69	379	1.99	400
1.10	327	1.41	356	1.70	380	2.00	400
1.11	328	1.41	357	1.71	381	—	—
1.12	329	1.42	358	1.72	382	—	—
1.13	330	1.43	359	1.73	382	—	—
1.14	331	1.44	360	1.74	383	—	—
1.15	332	1.45	360	1.75	384	—	—
1.16	333	1.46	361	1.76	384	—	—
1.17	334	1.47	362	1.77	385	—	—
1.18	335	1.48	363	1.78	386	—	—
1.19	336	1.49	364	1.79	386	—	—
1.20	337	1.50	365	1.80	387	—	—
1.21	338	1.51	365	1.81	388	—	—
1.22	339	1.52	366	1.82	389	—	—
1.23	340	1.53	367	1.83	389	—	—
1.24	341	1.54	368	1.84	390	—	—
1.25	342	1.55	369	1.85	391	—	—
1.26	343	1.56	369	1.86	391	—	—
1.27	344	1.57	370	1.87	392	—	—
1.28	345	1.58	371	1.88	393	—	—
1.29	346	1.59	372	1.89	393	—	—
1.30	347	1.60	373	1.90	394	—	—

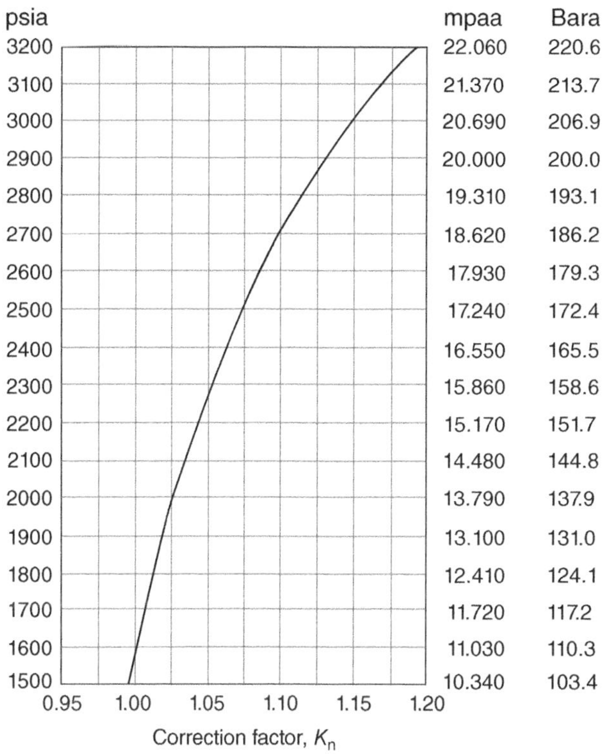

Figure 8.11 Correction factor for high pressure steam, K_n.

A_o: Valve outlet cross-sectional area (in.^{2}) calculated from Eq. (8.25);

P_1: The absolute relieving pressure is calculated using Eq. (8.2) (psia);

P_a: Atmospheric pressure equal to 14.7 psi;

k: Ratio of specific heats of the fluid as per Table 8.3 (dimensionless);

K_n: According to Figure 8.11, the high-pressure steam correction factor is used when the steam absolute relieving pressure is greater than 1400 psia. As a result of the ASME standard, this factor is used to compensate for the deviation between the steam flow as determined by Napier's equation and the actual saturated steam flow at high pressures. It can be calculated from Eq. (8.26) rather than from the figure to obtain the high-pressure steam correction factor;

K_r: The correction factor for ratios of specific heat of other than 1.4 according to Table 8.8.

Calculation of the Valve Outlet Cross-Sectional Area

$$A_o = \pi \times \frac{\text{Valve outlet size}^2}{4} \tag{8.25b}$$

Calculation of the Correction Factor for High-Pressure Stream K_n

$$K_{\mathrm{n}} = \frac{0.1906P_1 - 1000}{0.2292P_1 - 1061} \tag{8.26}$$

Table 8.8 Correction factors for the specific heat.

K	K_r
1.01	1.15
1.05	1.13
1.10	1.11
1.15	1.09
1.20	1.07
1.25	1.05
1.30	1.03
1.35	1.02
1.40	1.00
1.45	0.98
1.50	0.97
1.55	0.95
1.60	0.94
1.65	0.93
1.70	0.91
1.75	0.90
1.80	0.89
1.85	0.87
1.90	0.86
1.95	0.85
2.00	0.84

Example 8.14 In a subflare system, a safety valve with dimensions of 1 1/2″ × 2 1/2″ is managing natural gas with a specific gravity of 0.6. The valve has a capacity of 5900 lb/h, and the valve is designed to have an orifice area of 0.397 in.^{2}. If the valve set pressure is set at 210 psig with only 10% overpressure, calculate the reaction forces generated by the safety valve during operation if there is no pressure loss within the valve. (*Note*: API standard is used for the design of this valve.)

Answer

According to Eq. (8.2), the relieving pressure is calculated as follows:

$$P_1 = P_{set} + P_{over} + P_{atmosphere} - P_{loss} = 210 + 21 + 14.7 = 245.7\,\text{psia}$$

Table 8.3 states that the ratio of specific heat (k) for natural gas is 1.27. As a result, the correction factor for ratios of specific heat K_r according to Table 8.8 is 1.04. The next step is to calculate the cross-sectional area of the valve outlet based on the valve outlet size as determined by Eq. (8.25) as follows:

$$A_o = \pi \times \frac{\text{Valve outlet size}^2}{4} = \pi \times \frac{2.5^2}{4} = 4.909\,\text{in.}^2$$

There is a calculated orifice area of 0.397 in.^{2}, which is not a standard orifice size. According to the API standard, the standard orifice size that meets the estimated orifice area in Table 8.6 is orifice designation G, equal to 0.503 in.^{2}. According to Table 8.7, the gas constant is obtained as 344 based on the value of specific heat ratio. All given and calculated data are summarized as follows:

$$K = 0.975, A = 0.503\,\text{in.}^2, P_1 = 245.7\,\text{psia}, K_r = 1.04, P_a = 14.7\,\text{psia},$$
$$A_o = 4.909\,\text{in.}^2, k = 1.27, C = 344$$

In the next step, the reaction force due to the static pressure at the safety valve outlet will be calculated based on Eq. (8.24) as follows:

$$F_g = \left(\frac{KAP_1k_r}{1.383A_0} - P_a\right) \times A_o = \left(\frac{0.975 \times 0.503 \times 245.7 \times 1.04}{1.383 \times 4.909} - 14.7\right)$$
$$\times 4.909 = 18.45\,\text{lb}$$

The total force generated by the safety valve can now be calculated as follows:

$$F = \frac{CKAP_1\sqrt{k/(k+1)}}{332.7} + F_g = \frac{344 \times 0.975 \times 0.503 \times 245.7\sqrt{1.27/(1.27+1)}}{332.7}$$
$$+ 18.45 = 111.64\,\text{lb}$$

Equation (8.27) is used for the calculation of reaction forces for a PRV discharging liquid into the atmosphere.

PSV Reaction Force for Open Discharge of Liquid Service to the Atmosphere (Open Discharge)

$$F = \frac{(3.44 \times 10^{-7}) \times W^2}{\rho A_0} \tag{8.27}$$

where:

F: Reaction force at the point of discharge to the atmosphere (lb);
W: Actual relieving capacity (lb/h);
A_0: Area at discharge (in.^{2});
ρ: Density of the fluid (lb/ft^3).

Equation (8.28) is used for the calculation of reaction forces for a PRV discharging a two-phase gas and liquid flow into the atmosphere, according to API 520-part II recommended practice.

PSV Reaction Force for Open Discharge of Gas–Liquid Service to the Atmosphere (Two-Phase Discharge)

$$F = \frac{W^2}{(2.898 \times 10^6)A_0}\left[\frac{X}{\rho_g} + \frac{1-X}{\rho_L}\right] + A_o(P_E - P_A) \tag{8.28}$$

where:

F: Reaction force at the point of discharge to the atmosphere (lb);
W: Actual relieving capacity (lb/h);
A_o: Area at discharge outlet to the atmosphere (in.^{2});
x: Mass fraction of gas and vapor portion calculated as W_G/W;
W_G: Actual relieving capacity of the gas (lb/h);
ρ_g: Vapor density at exit (lb/ft^3);
ρ_l: Liquid density at exit (lb/ft^3);
P_E: Pressure at pipe exit (psia);
P_A: Ambient pressure (psia).

Example 8.15 A PSV has a 1″ CL150 raised face flange inlet and a 2″ CL150 raised face flange outlet. The design pressure and temperature are 285 psi gauge and 86 °F and the material of the valve and connected pipe are in carbon steel. The piping material grade is ASTM A106 Gr. B. The overpressure crude oil in the liquid phase is released from the valve to the atmosphere. The density of the crude oil in this case is 800 kg/m^3. The actual capacity of the valve is 26,896 lb/h. The area of discharge from the PSV is 0.196 in.^{2}. What is the reaction load of the PSV?

Answer
Equation (8.27) should be used for the calculation of the reaction forces. All the required data are in imperial units, except for the density of the fluid.

$$\rho = 800\ \text{kg/m}^3 = 49.94\ \text{lb/ft}^3$$

$$F = \frac{(3.44\times10^{-7})\times W^2}{\rho A_0} = \frac{3.44\times10^{-7}\times(26{,}896)^2}{49.94\times0.196} = \frac{248.85}{9.79} = 25.41\,\text{lb} = 113\,\text{N}$$

Applying the design load factor of 2 provides a 51-lb equal to 226 N reaction force for the PSV.

The main question is, what is the limit of reaction forces generated by the PSV? It should be noted that the PSV reaction force is an occasional load. The total of the sustained loads S_L plus occasional loads S_o should be at maximum equal to 1.33 of the allowable stress at hot temperature. ($S_L + S_o \leq 1.33 \times S_h$). The term "sustained loads" refers to a combination of internal pressure and weight loading that are constant during the lifetime of the pipe. In this context, weight refers to both the pipe itself and related components such as valves, flanges, strainers, etc. It is also important to consider the weight of the fluid during the operation or during the hydrotest. The weight of some pipes is increased by the use of thermal insulation or internal cladding. In addition, in cold regions, snow can accumulate on the piping. So, the weight of the snow is considered in calculating the sustained load on the piping. Sustained load analysis takes into account only longitudinal or axial principal stress (δ_a), which is produced by the fluid pressure. The sustained loads may result in the collapse of the piping if the load is not appropriately managed through adequate pipe supports at the correct location. The ASME B31.3 code requires that the sustained load or stress always be less than, or at a maximum equal to, the hot allowable stress, parameter S_h. Sustained load is equal to weight plus longitudinal stress. Equation (8.29) shows the limitation and calculation of sustained loads:

Limitation and Calculation of Sustained Loads

$$S_L \leq S_h \rightarrow S_a + S_w \leq S_h \tag{8.29}$$

where:

S_L: Sustained load equal to the weight loads plus longitudinal (axial) loads (psi);
S_h: Allowable stress at temperature or hot temperature (can be extracted from ASME B31.3; psi);
S_a: Axial or longitudinal stress as a result of piping inside pressure, which can also be shown as δ_a (psi);
S_w: Stress due to weight (psi).

Example 8.16 A PSV is estimated to produce a load of 51 lb. It is assumed that none of the occasional loads occurs at the same time. The total sustained load is at its maximum and equal to the allowable stress at hot temperatures. The PSV and its connecting pipe are constructed from carbon steel, and the pipe is constructed of A106 Gr.B. Assuming the operating temperature of the PSV is ambient, is the produced load by the PSV acceptable?

Answer
If the sustained load is equal to the allowable stress at hot temperature ($S_L = S_h$), then the maximum allowable occasional stress would be $0.33 \times S_h$.

$$S_L + S_o \leq 1.33 \times S_h \rightarrow S_h + S_o \leq 1.33 \times S_h \rightarrow S_o \leq 0.33 \times S_h$$

The piping material connected to the PSV is in ASME A106 Gr. B and the temperature is 86 °F. Accordingly, the maximum allowable stress on a pipe at a given temperature is 20,000 psi. There is an allowable PSV load of $0.33 \times 20,000$ psi, which equals 6600 psi. The pressure produced by the PSV, 51 lb, is much less than 6600 psi. So, the PSV load is within acceptable limits.

8.3.4 Safety Valve Capacity Conversion

It is possible to determine the capacity of a safety or relief valve in terms of a gas or vapor other than the medium for which the valve is rated and designed using ASME Section VIII Div. 01 and appendix 11. Equations (8.30), (8.31), and (8.32) are used to calculate the capacity of a relief valve for steam, air, gas or vapor, respectively. This formula can also be used when the required flow rate of any gas or vapor is known and it is required to determine the rated capacity of a valve for steam or air service.

Safety Valve Rated Capacity for Steam

$$W_s = 51.5KAP \tag{8.30}$$

Safety Valve Rated Capacity for Air

$$W_a = CKAP\sqrt{\frac{M}{T}} \tag{8.31}$$

$$C = 356,\ M = 28.97$$

Safety Valve Rated Capacity for Gas or Vapor

$$W = CKAP\sqrt{\frac{M}{T}} \tag{8.32}$$

In addition, if the compressibility factor Z should be included in Eq. (8.32) for the light hydrocarbons, then the formula would be adjusted as follows:

$$W = CKAP\sqrt{\frac{M}{ZT}}$$

where:

W_s: Rated capacity lb/h of steam;
W_a: Rated capacity, converted to lb/h of air at 60 °F inlet temperature;
W: Flow of any gas or vapor in lb/h;
C: The constant for gas or vapor that depends on the ratio of specific heats from Table 8.7;
K: Coefficient of discharge calculated from Eq. (8.7);
A: Actual discharge area of the safety valve (in.^{2});
P: Discharge pressure (psia) calculated according to Eq. (8.2);
M: Molecular weight obtained from Table 8.3;
T: Absolute temperature at the valve inlet (°F + 460).

Example 8.17 The capacity of a safety valve is 3500 lb/h under a set pressure of 190 psi. What is the capacity of the valve to relieve air at 100 °F for the same set pressure?

Answer
Using Eq. (8.30) to calculate the safety valve's capacity to handle steam:

$$W_s = 51.5KAP \rightarrow 3500 = 51.5KAP \rightarrow KAP = \frac{3500}{51.5} = 67.96$$

Using Eq. (8.31) to calculate the safety valve's capacity to handle air:

$$W_a = CKAP\sqrt{\frac{M}{T}} = 356 \times 67.96 \times \sqrt{\frac{28.97}{460 + 100}} = 5503 \text{ lb/h}$$

Example 8.18 A safety valve installed on a pressure vessel must release 5200 lb of propane per hour considering the valve inlet temperature is 125 °F. What is the overall capacity of this valve in pounds of steam per hour if it has been designed for steam service?

Answer
In accordance with Table 8.3, the ratio of the specific heat k for propane is 1.13. The molecular weight of propane as taken from the same table of data is found to be 44.09. In accordance with Table 8.7, the gas constant value C is found to be 330. Using Eq. (8.32), the safety valve rated capacity for a gas or vapor can be calculated as follows:

$$W = CKAP\sqrt{\frac{M}{T}} \rightarrow 5200 = 330KAP\sqrt{\frac{44.09}{460 + 125}} \rightarrow KAP = 57.395$$

It is now necessary to apply Eq. (8.30) in order to determine the safety valve rating based on the steam service as follows:

$$W_s = 51.5KAP = 51.5 \times 57.395 = 2955\ \text{lb/h}$$

Example 8.19 It is necessary to discharge 1000 lb of ammonia per hour through a PSV at a temperature of 150 °F. Is the total capacity of this valve the same as the set pressure that is used for steam?

Answer

It is known that ammonia has a molecular weight of 17.03 and its specific heat ratio is 1.4. Based on Table 8.7, we can determine that the value of constant C based on Table 8.7 should be 355.5. Equation (8.32) is used to calculate the safety valve rated capacity for ammonia by the following formula:

$$W = CKAP\sqrt{\frac{M}{T}} \rightarrow 1000 = 355.5KAP\sqrt{\frac{17.03}{460 + 150}}$$

$$\rightarrow KAP = \frac{1000}{64.96} = 15.394$$

In order to determine the safety valve rating for steam service, the following Eq. (8.30) must be applied:

$$W_s = 51.5KAP = 51.5 \times 15.394 = 792.80\ \text{lb/h}$$

Questions and Answers

8.1 In the context of nonfire conditions, two PSVs are installed above pressure equipment. For the second safety valve, the MAWP is 100 psi and the values of set pressure and accumulated pressure are at their maximum. Identify the incorrect statements from the following list:

- **A** The set pressure for the valve is 105 psi.
- **B** The allowable overpressure is 10 psi.
- **C** The maximum accumulated pressure is 116 psi.
- **D** Based on no head loss in the valve, the absolute relieving pressure is 130.7 psi.

Answer

Based on Table 8.2, the maximum set pressure and maximum accumulated pressure values for the second safety valve in the nonfire case are 1.05 times and 1.16 times the MAWP, respectively.

Set pressure $= 1.05 \times 100 = 105$ psi

Maximum accumulated pressure $= 1.16 \times 100 = 116$ psi

Equation (8.2) $\rightarrow$ Allowable overpressure = Maximum accumulated pressure – Set pressure $= 116 - 105 = 11$ psi

Equation (8.2) $\rightarrow$ Absolute relieving pressure = Set pressure + overpressure + atmospheric pressure – pressure loss $= 105 + 11 + 14.7 = 130.7$ psi

Therefore, the incorrect answer is B since the allowable overpressure is 11 psi.

8.2 The MAWP for a PSV made of carbon steel material is 20 bar, which is equal to the design pressure. During normal operation, the safety valve experiences overpressure conditions. Under which conditions is it acceptable to experience an overpressure condition? Carbon steel material has a tensile strength of 70 ksi and a yield strength of 36 ksi.

A The total number of pressure-temperature variations above the design conditions over the design life of the piping system is 2000 times.
B As a result of the overpressure scenario, the safety valve is subjected to a stress of more than 40 ksi.
C Overpressure reached 35 bars only once in the overpressure scenario.
D In all three cases described here, an overpressure scenario is not acceptable.

Answer
The condition described in option A is not acceptable because the total number of overpressure scenarios shall not exceed 1000 times during the design life of the piping system. It is also incorrect to choose Option B since the stress generated during the overpressure scenario must not exceed the yield strength of the carbon steel material, which is 36 ksi. Option C is not appropriate because it is not acceptable to have an overpressure scenario of more than 1.5 times the design pressure equal to 30 bar. Option D is the right answer.

8.3 Which sentence is correct regarding the PRV installed on the pressure vessel illustrated in Figure 8.12?

A The amount of force produced by a PRV during opening is negligible.
B The PRV is used to drain liquid.
C Overpressurized fluid in the pressure vessel is released into the atmosphere through the PRV.
D None of these.

Answer
Option A is incorrect as the generated relief valve load to the atmosphere, also known as open discharge, is not negligible. In regard to option B, it is

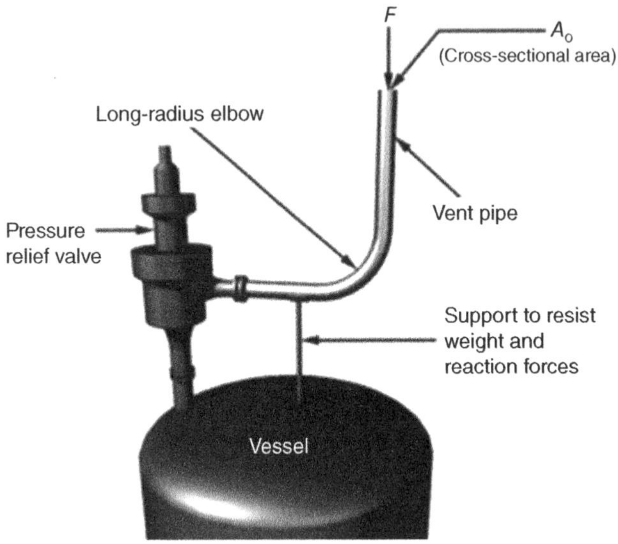

Figure 8.12 PSV on a pressure vessel.

not possible to determine whether gas or liquid is released into the environment from the figure. Therefore, option B is not entirely accurate. Accordingly, option C is correct, as the overpressure fluid in the vessel is released into the environment through the PRV. Therefore, option C is the correct answer, and option D is incorrect.

8.4 Which statement regarding the maximum relieving pressure and set pressure relationship to MAWP is accurate, assuming that the valve's atmospheric pressure and pressure loss are both zero?

A The set pressure of a pressure vessel should not exceed 10% of the MAWP of the vessel when a single pressure relief device protects the vessel.

B The relief pressure shall not exceed 16% of the maximum operating pressure of a pressure vessel protected by multiple pressure relief devices.

C The pressure relief shall not exceed 21% of the MAWP of the vessel when the vessel is protected from fire.

D All three options are correct.

Answer

As shown in Table 8.2, the maximum set pressure of a safety valve when a single device is installed on a pressure vessel is 10% of MAWP. Therefore, option A is correct. As both atmospheric and pressure losses occur within the valve, the relieving pressure and accumulated pressure are equal to the set pressure plus overpressure. Accordingly, options B and C are correct. Therefore, option D is the correct answer.

8.5 There is a PSV installed 10 m above pressure equipment with a design pressure of 20 bar. When the maximum set pressure and the fixed set pressure are equal, the PSV opens at 18 bar, and the allowable overpressure is 10%, which of the following statements is incorrect?

A There is a relieving pressure of 20.9 bar.
B A simmer is equal to 1 bar.
C The maximum head loss in the inlet line to the PSV is 0.7 bar.
D Closing or reseat pressure is 18.43 barg.

Answer

The first step is to determine the fixed set pressure of the PSV. The static pressure loss associated with PSV's location at a distance of 10 m is 1 bar. Therefore, it is possible to calculate the maximum and fixed set pressure as follows:

Maximum set pressure = Fixed set pressure = Design pressure − Static pressure loss $= 20 - 1 = 19$ bar

$$\text{Relieving pressure} = \text{Set pressure} + \text{Overpressure}$$
$$= 19 + 19 \times 10\% = 20.9 \text{ barg}$$

Referring to Eq. (8.14), simmer is the difference between the set pressure and the start to open pressure as calculated as follows:

Simmer $= 19 - 18 = 1$ bar.

The maximum head loss according to both API 521 and ASME Sec. VIII Div.01 is $3\,\% \times 19$ bar $= 0.57$ bars.

In relation to Eq. (8.13), the closing pressure or reseat pressure is the difference between the set pressure and the maximum head loss, $19 - 0.57 = 18.43$ barg.

Therefore, all options except C are correct.

8.6 According to the manufacturer of a pressure vessel, the maximum allowable working pressure is 6 bar, and the maximum allowable accumulated pressure is 6.3 barg. In the case of a PSV used to protect a pressure vessel from overpressure scenarios, what is the set pressure of the PSV if the overpressure amounts to 10%? In this case, a single PSV is used to protect the equipment.

A 5.7 bar
B 6 bar
C 6.3 bar
D 6.5 bar

Answer

Options C and D are not valid, since if a single PSV is used, its set pressure cannot exceed MAWP, which is 6 bar, as shown in Table 8.2. Using Eq. (8.2), the maximum allowable accumulated pressure can be calculated as follows:

$$\text{Maximum accumulated pressure} = \text{Set pressure}$$

$$+ \text{Allowable overpressure (10\% of the set pressure)}$$

$$= 1.1 \times \text{Set pressure} \rightarrow \text{Set pressure} = \frac{\text{Maximum accumulated pressure}}{1.1}$$

$$= \frac{6.3}{1.1} = 5.7\,\text{bar} \rightarrow \text{Option A is the correct answer.}$$

8.7 In determining the set pressure of a safety valve, which of the following statements is correct?

A The set pressure of the PSV should never exceed the maximum allowable working pressure (MAWP).

B The set pressure of the PSV should never exceed the maximum allowable accumulation pressure (MAAP).

C The set pressure of the PSV should never exceed the normal working pressure (NWP) of the protected equipment.

Answer

Option B is the correct answer since MAAP equals the sum of set pressure and allowable overpressure.

8.8 According to the standards for conventional safety valves, the required set pressure is 8.5 barg, unless, of course, the valve experiences a contact backpressure of 1.0 barg. In light of the effects of backpressure on the set pressure, how much should the set pressure be adjusted?

A 7.5 bar

B 8.5 bar

C 9.5 bar

D 10.5 bar

Answer

The set pressure, also known as cold set pressure is reduced due to the effect of backpressure. This has to be compensated for by adding the backpressure value to the cold set pressure so as to get the adjusted set value, which is equal to $8.5 + 1 = 9.5$ bar. Therefore, option C is the correct answer.

8.9 There is a PSV installed on the pressure equipment that handles the natural gas. The valve has an inlet and an outlet size of 1.5″ and 2.5″, respectively, and the orifice size selected is G. Considering the fact that 10% overpressure is allowed, the required capacity of the valve is 6500 lb/h and the set pressure is 150 psi. During the operation of a valve, what is the total reaction forces that are produced?

A 60.25 lb

B 64.67 lb

C 68.16 lb

D 72.55 lb

Answer

According to Eq. (8.2), the relieving pressure is calculated as follows:

$$P_1 = P_{set} + P_{over} + P_{atmosphere} - P_{loss} = 150 + 15 + 14.7 = 179.7\ \text{psia}$$

Table 8.3 states that the ratio of specific heat (k) for natural gas is 1.27. As a result, the correction factor for ratios of specific heat K_r according to Table 8.8 is 1.04. The next step is to calculate the cross-sectional area of the valve outlet based on the valve outlet size as determined by Eq. (8.25) as follows:

$$A_o = \pi \times \frac{\text{Valve outlet size}^2}{4} = \pi \times \frac{2.5^2}{4} = 4.909\ \text{in.}^2$$

According to the API standard, the standard orifice size that meets the estimated orifice area in Table 8.6 is orifice designation G, equal to 0.503 in.^{2}. According to Table 8.7, the gas constant is obtained as 344 based on the value of specific heat ratio. All given and calculated data are summarized as follows:

$$K = 0.975, A = 0.503\ \text{in.}^2, P_1 = 179.7\ \text{psia}, K_r = 1.04, P_a = 14.7\ \text{psia},$$
$$A_o = 4.909\ \text{in.}^2, k = 1.27, C = 344$$

In the next step, the reaction force due to the static pressure at the safety valve outlet will be calculated based on Eq. (8.24) as follows:

$$F_g = \left(\frac{KAP_1k_r}{1.383A_0} - P_a\right) \times A_o = \left(\frac{0.975 \times 0.503 \times 179.7 \times 1.04}{1.383 \times 4.909} - 14.7\right) \times 4.909 = -5.89\ \text{lb}$$

Since F_g is less than zero, it should be considered equal to zero.
The total force generated by the safety valve can now be calculated as follows:

$$F = \frac{CKAP_1\sqrt{k/(k+1)}}{332.7} + F_g = \frac{344 \times 0.975 \times 0.503 \times 179.7\sqrt{1.27/(1.27+1)}}{332.7} + 0 = 68.16\ \text{lb}$$

Thus, option C is the correct choice.

8.10 In order for a safety valve to function properly, it should be able to release 10,000 ft^3/min of air at 60 °F, and its atmospheric pressure should be equal to 14.7 psia. If the set pressure of the safety valve for both fluid services is equal, what is the flow capacity of this safety valve in pounds per hour for saturated steam?

A 25,460 lb/h
B 28,168 lb/h
C 26,750 lb/h
D 23,233 lb/h

Answer

The first step is to convert the air flow capacity from cubic feet per minute to pounds per hour as follows:

$$W_a = 10{,}000 \times 0.0766 \times 60 = 45{,}960\ \text{lb/h}$$

Now, Eq. (8.31) is used to calculate the rated capacity of the valve for air as follows:

$$W_a = CKAP\sqrt{\frac{M}{T}}(C = 356, M = 28.97) \rightarrow 45,960$$

$$= 356 \times KAP \times \sqrt{\frac{28.97}{460 + 60}} \rightarrow KAP = \frac{45{,}960}{84.028} = 546.96$$

Now, Eq. (8.30) is used to calculate the rated capacity of the valve for steam as follows:

$$W_s = 51.5KAP = 51.5 \times 546.96 = 28{,}168\ \text{lb/h}$$

Thus, option B is the correct answer.

Further Reading

American Petroleum Institute (API) 520 (2020). *Sizing, Selection, and Installation of Pressure-Relieving Devices Part 1 – Sizing and Selection*. Washington, DC: American Petroleum Institute (API).

American Petroleum Institute (API) 521 (2007). *Pressure-Relieving and Depressuring Systems*, 5e. Washington DC: American Petroleum Institute (API).

American Petroleum Institute (API) 526 (2017). *Flanged Steel Pressure-Relief Valves*, 7e. Washington, DC: American Petroleum Institute (API).

American Society of Mechanical Engineers (ASME) (2004). *Carbon, Alloy and Stainless-Steel Pipes. ASME B36.10/19*. New York: American Society of Mechanical Engineers (ASME).

American Society of Mechanical Engineers (ASME) (2012). *Design and Fabrication of Pressure Vessels. Boiler and Pressure Vessel Code. ASME Section VIII Div.02.* New York: American Society of Mechanical Engineers (ASME).

Crosby Valve Inc (1997). *Crosby Pressure Relief Valve Engineering Handbook.* Technical document number. TP-V300. Crosby Valve Inc.

Hellemans, M. (2009). *The Safety Relief Valve Handbook: Design and Use of Process Safety Valves to ASME and International Codes and Standards*, 1e. Oxford: Butterworth-Heinemann, and imprint of Elsevier.

Nesbitt, B. (2007). *Handbook of Valves and Actuators: Valves Manual International*, 1e. Oxford: Elsevier.

Skousen, P.L. (2011). *Valve Handbook*, 3e. New York: McGraw-Hill.

Smit, P. and Zappe, R.W. (2004). *Valve Selection Handbook*, 5e. New York: Elsevier.

Sotoodeh, K. (2022). *Piping Engineering: Preventing Fugitive Emission in the Oil and Gas Industry*, 1e. New York: Wiley.

Sotoodeh, K. (2022). *Cryogenic Valves for Liquified Natural Gas Plants*, 1e. Austin, TX: Elsevier (Gulf Professional Publishing).

Spirax Sarco (2009). Module 9.3, safety valve selection.

9

Safety and Reliability

9.1 Introduction

In different industries, including oil and gas, valves are essential components of piping systems that regulate and control fluid flow, prevent backflow, and maintain safety. In the offshore industry, various types of valves are used, including ball valves, gate valves, butterfly valves, globe valves, and check valves. Various reasons such as corrosion and mechanical malfunction can result in industrial valves failing to function. This can lead to adverse consequences such as environmental pollution, asset loss, loss of production, or even death. It is true that valves are connected to pipes in order to produce or transport goods. These valves are an essential part of the process, as their failure could halt the entire process. Safety and reliability assurance of industrial valves is such an important topic that has been addressed in past studies. In addition, a subsurface safety valve (SSV) is also installed in the well. In the process of oil and gas production, they monitor the well for abnormal conditions such as an overpressure scenario. This is to prevent blowouts and make sure the well is closed in case of any abnormal conditions. The reliability analysis of SSVs has been a popular topic among researchers. In a separate study, pressure relief valves that are installed on piping and mechanical equipment to relieve the overpressure fluid or gas from the piping and equipment have been subjected to a reliability analysis. In the offshore oil and gas industry, there have recently been a couple of studies that focus on online monitoring and strict factory acceptance tests (FATs) to enhance the safety and reliability of valves. For example, the effect of partial stroke testing on the reliability of safety-critical valves is analyzed in previous research. The safety-critical valves are those that are connected to either a process shutdown (PSD) system or an emergency shutdown system (ESD). PSDs and ESDs are two types of shutdown systems for process systems. As part of the plant and facility safeguarding systems, PSD systems minimize or

Industrial Valves: Calculations for Design, Manufacturing, Operation, and Safety Decisions, First Edition. Karan Sotoodeh.

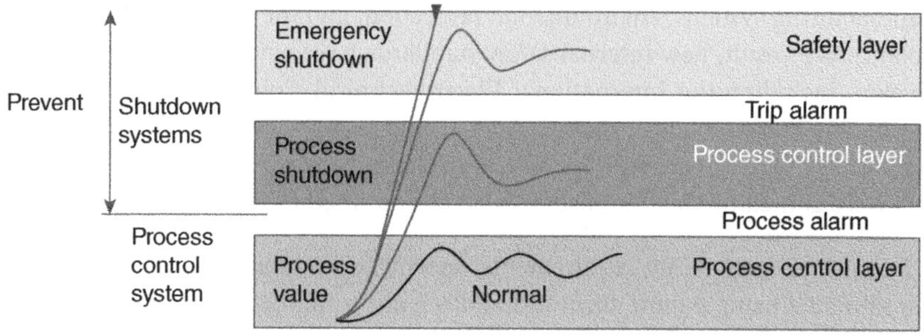

Figure 9.1 Plant safety implementation and protection with PSD and ESD layers.

prevent the occurrence and consequences of process parameters like pressure, temperature, and level exceeding operating limits. PSD systems are capable of stopping a part or the entire operation or of depressurizing or blowing down parts of the process. Process shutdown valves are actuated valves connected to the basic process control systems that ensure that the process is managed and controlled within acceptable limits. The use of an ESD system minimizes or prevents emergency conditions such as extreme or abnormally high values of process parameters (e.g. high fluid flow in piping) or leakage or emission of hydrocarbons to the environment. In this sense, ESD differs from PSD in that it is a safety layer rather than a process control layer, as illustrated in Figure 9.1. In general, PSD and ESD valves in oil and gas projects are listed in safety analysis reports (SARs) and must meet safety integrity level (SIL) 2 or 3. SIL, an important parameter in evaluating the safety and reliability of industrial valves, is discussed in more detail further in this chapter.

9.2 Safety Standards

A safety-related system protects equipment and industrial processes where danger or risk may arise as a result of a failure or malfunction. It is important to remember that in the old days, industrial safety was focused on areas such as safe work practices, the control of hazardous materials, and the protection of personnel and equipment. The safety issues and strategies that are addressed today, however, are far deeper and more complex than those that were addressed in the past. Therefore, safety systems of today reduce risk while operating in a manner that continuously improves productivity and profitability by reducing risk. The safety management process was largely self-regulated until the 1980s. Following the advent of electronic control devices, more complex issues arose concerning manufacturing systems, environmental protection, and the assets and facilities in a plant. As a result, new internal safety standards have emerged and continue to develop, including the International Electrotechnical Commission (IEC) 61508 and 61511. The number of manufacturers integrating their products with these standards is increasing. With the advent of these standards, many essential benefits were gained by the final user, which are outlined as follows:

- The risk and reliability requirements, specifications, and design processes are formulated using a more technical and scientific approach;
- Definition of risk that is more accurate;
- The ability to demonstrate the effectiveness of safety-related systems more easily and widely;
- Implementing safety-related issues in a more cost-effective manner;
- Maintenance operations can be evaluated more effectively.

IEC 61508 is an international standard published by the International Electrotechnical Commission addressing the functional safety of electrical, electronic, and programmable electronic safety-related systems. The standard is divided into seven parts which can be summed up as follows:

- **Part 1:** General requirements (Normative);
- **Part 2:** Requirements for electrical/electronic/programmable electronic safety-related systems (Normative);
- **Part 3:** Software requirements (Normative);
- **Part 4:** Definitions and abbreviations (Informative);
- **Part 5:** Examples of methods for determination of SILs (Informative);
- **Part 6:** Guidelines on the application of parts 2 and 3 (Informative);
- **Part 7:** Overview of techniques and measures (Informative).

Part 6 of IEC 61508 provides some essential definitions that will be discussed in the following section.

Hazard: The term refers to a possible cause of harm, recognizing that the term includes both threats that occur within a short period of time such as fires and explosions and those that have a long-term effect on a person's health, such as the release of toxic substances such as hydrogen sulfide.

Hazardous situation: The term refers to situations in which a person is exposed to certain hazards.

Hazardous event: Generally, it refers to a hazardous situation that leads to a negative result.

Risk: It is a combination of the probability of occurrence of harm and the severity of that harm.

Tolerable risk: According to the current values of society, it is an acceptable risk based on the given context.

Residual risk: The residual risk is the risk that remains even after protective measures have been taken.

IEC 61511 is an international standard titled "Safety Instrumented Systems for Process Industry Sector: Functional Safety" that was developed in order to integrate safety instruments into safety systems in the process industry sector as an extension of IEC61508 for the process industry. Standard IEC 61508 governs the *safety instrumented system (SIS)* for process industries, including sensors, logic solvers, and final elements in particular. Figure 9.2 depicts a SIS architecture that consists of three pressure transmitters, a logic solver, and finally two actuated valves as the last two components.

The pipeline pressure transmitters monitor the pipeline pressure against the predefined limit and send signals to the logic solver where it is determined what action is appropriate based on the logic solver output in the form of an electrical signal. This enables the final element to perform an action, such as shutting down the processing system through a valve, after receiving the electrical signal from the logic solver. As a result of this configuration, overpressure incidents are avoided by removing the source of overpressure or reducing the likelihood of an overpressure event to such a low level that it is no longer considered likely. SISs are widely used

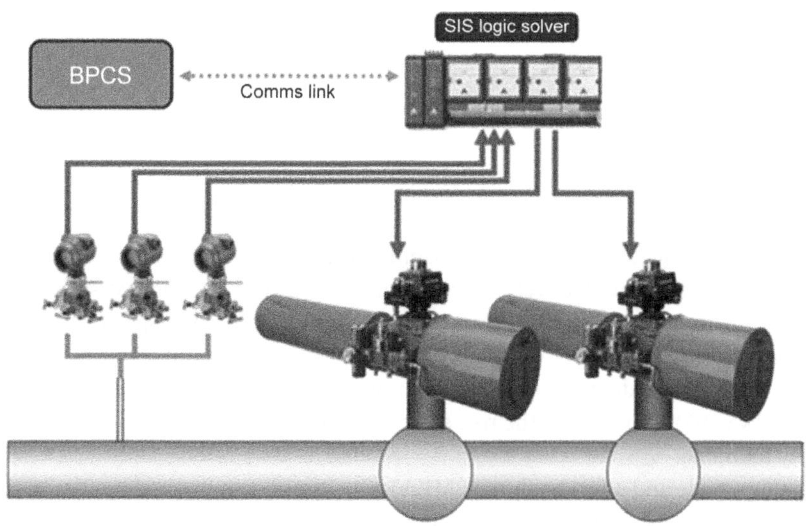

Figure 9.2 A safety instrumented system (SIS) architecture. *Source:* Courtesy of Emerson Electric Co.

in oil and gas industries, for example, with a variety of design and application concepts. The main benefits of SIS can be summed up as the protection of equipment, personnel, and the environment, as well as the reduction of capital expenditures and operations costs. In addition to this, IEC 61511 deals with the interface between SIS and other safety-related systems relating to process hazards and risks. This standard includes three parts that are listed as follows:

- **Part 1:** Framework, definitions, and requirements for systems, hardware, and software;
- **Part 2:** Guidelines for the application of the IEC 61511 standard;
- **Part 3:** An overview of the examples and methods for determining safety integrity in the application of hazard and risk analysis.

9.3 Risk Analysis

It is the purpose of all safety standards to minimize the risk that is inherent to the manufacturing and processing processes. To achieve this goal, it is necessary to either eliminate the risk or reduce it to a level that is acceptable. In a real-world scenario, the risk could be negligible, tolerable, or unacceptable. Thus, more specifically, the purpose of safety systems is to reduce the risk to a level that is acceptable or tolerable. Hence, safety can be defined as a state of being free from unacceptable risks. According to Eq. (9.1), the risk is given a score or calculated.

Calculation of Risk Score

$$\text{Risk Score or risk priority number (RPN)} = \text{Hazard frequency or probability} \times \text{Hazard severity} \tag{9.1}$$

It is possible to determine the probability and severity of a hazard based on Tables 9.1 and 9.2, respectively. Accordingly, the lowest and highest scores for the hazard frequency (probability) and severity are 1 and 10, respectively. This makes the risk score range from a minimum of one to a maximum of one hundred. There must be a reduction or elimination of those risks that have higher scores (e.g. RPN = 95). However, if the risk score is as low as five, the risk factor is most likely tolerable.

Failure mode and effect analysis (FMEA) is an assessment method used to identify and eliminate known and potential problems and errors from a system before it is delivered to the customer. This tool is widely used for determining the reliability of systems. Among the major advantages of this system is that it enhances

Table 9.1 Determination of the risk probability score.

Rating	Occurrence probability	Minimum probability percentage (%)
10	Extremely high	50
9	Very high	33
8	Very high	10 – 15
7	High	5
6	Marginal	1
5	Marginal	0.25
4	Unlikely	0.05
3	Low	0.007
2	Very low	0.0007
1	Remote	0.000007

Table 9.2 Determination of the severity of the risk.

Rating	Effect	Severity effect
10	Hazardous without warning	Very high severity without warning
9	Hazardous with warning	Very high severity with warning
8	Very high	Destructive and unsafe failure
7	High	System is inoperable due to major equipment damage
6	Moderate	System is inoperable due to minor damage
5	Low	System is inoperable without damage
4	Very low	High degradation of performance
3	Minor	Less performance degradation
2	Very minor	Minimal system operability
1	None	No effect

safety and reliability, and reduces the possibility of failure, which, in turn, increases customer satisfaction. Injury to humans and adverse environmental effects can be minimized through the use of failure reduction opportunities to improve health, safety, and environment (HSE). At the moment, this is the most common risk analysis approach being used. In accordance with historical records,

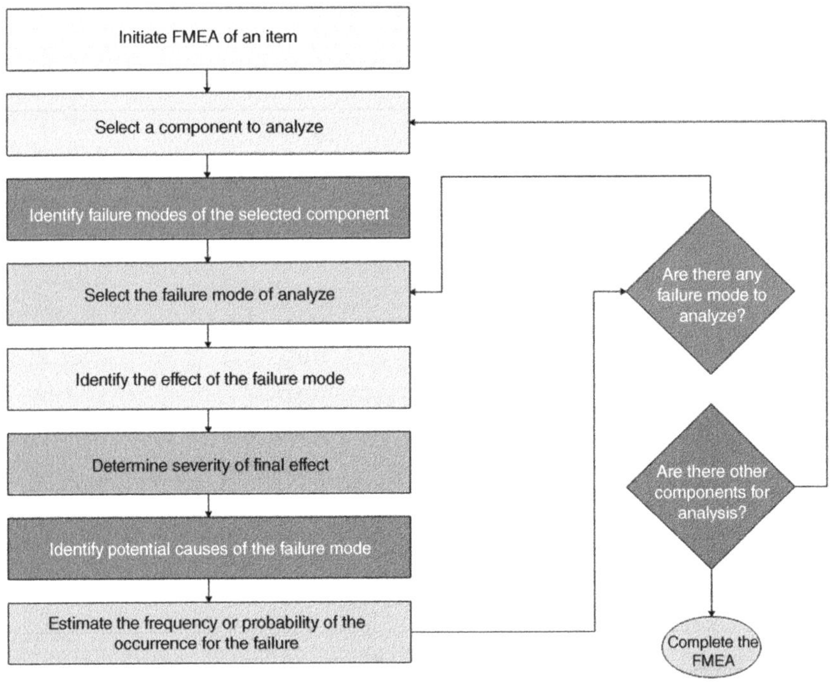

Figure 9.3 FMEA procedure chart.

FMEA was first developed by the American military in the 1940s and implemented in the 1950s for the design of flight control systems. Figure 9.3 is an illustration of steps and a procedure for implementing FMEA in practice.

Example 9.1 In the offshore oil and gas industry, pipeline valves are known for being the most expensive and critical valves. An example of a pipeline valve is discussed in this example. The valve is connected to the emergency shutdown system and it is considered to be a part of SIS. This valve was found to have four specific types of risk that are listed in Table 9.3. The table also includes estimates of the probability and severity of the listed risks. Which of the risks has the highest priority that needs to be reduced or eliminated? In addition, which risk item is acceptable as it is?

Answer

In order to calculate the RPN, each risk item is assigned a severity and occurrence score as well as a risk severity-occurrence score. These values are entered into Table 9.4. On the right side of the table, in the last column, we can see RPNs

Table 9.3 Identified risks for the pipeline valve.

Risk item	Potential failure mode or risk item	Severity (S)	Occurrence (%)
A	Damage to the valve body as a result of pipeline loads	Hazardous with warning	1
B	Ingress of dense oil in the seat arrangement and malfunction of the valve seat	Very high	5
C	Damage to the seat and ball during the commissioning and hydrotest as a result of dirt and welding debris	Hazardous without warning	0.007
D	Failure in shutting down the emergency shutdown valve during the subsea pipeline failure	Hazardous with warning	0.000007

Table 9.4 Scores of risks associated with pipeline valves.

Risk item	Potential failure mode or risk item	Severity (S)	Severity score	Occurrence (%)	Occurrence score	Risk priority number (RPN)
A	Damage to the valve body as a result of pipeline loads	Hazardous with warning	9	1	6	54
B	Ingress of dense oil in the seat arrangement and malfunction of the valve seat	Very high	8	5	7	56
C	Damage to the seat and ball during the commissioning and hydrotest as a result of dirt and welding debris	Hazardous without warning	10	0.007	3	30
D	Failure in shutting down the emergency shutdown valve during the subsea pipeline failure	Hazardous with warning	9	0.000007	1	9

are calculated and risk item B has the highest score equal to 56. This makes it the top priority to reduce or eliminate this risk. Risk item A has been assigned a score of 54 and has been given the second highest priority for risk reduction measures, having a score of 54. In terms of risk number, item D has the lowest risk score. So it can be accepted as is.

9.4 Basic Safety and Reliability Concepts

In this section, some fundamental concepts that are used in the IEC 61508 standard are explained in order to provide a deeper understanding of safety issues.

9.4.1 System Incidents and Failures

It is very important to understand the concept of system incidents and failures before we go into the details of basic safety and reliability concepts as well as how these concepts are calculated. Figure 9.4 shows the most relevant system incidents and failures, those incidents and failures that are all connected to the system's safety and reliability.

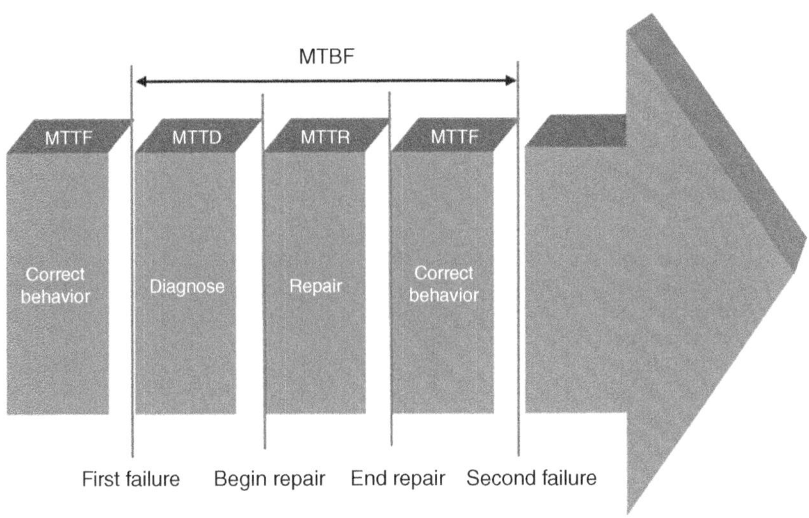

Figure 9.4 A system incidents and failures chart.

9.4.1.1 Failure Rate

The *failure rate* refers to the frequency of failure for a component per unit of time, which is typically shown as a Greek letter λ. Several safety parameters are calculated based on failure rates, such as reliability and incidents that are explained in more detail later in this chapter. There is a general rule of thumb that the failure rate of a component or system is related to time. This means that the failure rate and its effect will vary over the system's or the component's lifespan. In the case of a valve, a failure occurring five years after its installation might have a greater impact on the cost and risk of the operation than if the valve failed after only one year of operation.

There is a method to estimate the failure rate of a valve by analyzing its performance history over the years of operation. In order to calculate the failure rate of a valve, it is simply necessary to divide the number of failures by the total time the valve has been in operation (see Eq. (9.2)). It is usually the valve manufacturer or the end user who provides a history of the valves' failures during operation.

Calculation of the Failure Rate Based on the Valve's Performance During Operation

$$\lambda = \frac{\text{Number of failures}}{\text{Total time of the valve in operation}} = \frac{n}{\tau} \tag{9.2}$$

Example 9.2 For the past seven years, 300 industrial valves have been operating in a refinery. Since the valves have been in operation for seven years, only five valves have failed. In this example, what is the average failure rate for the group of valves?

Answer

The average failure rate per hour for the group of valves is calculated as follows:

$$\lambda = \frac{\text{Number of failures}}{\text{Total time of the valve in operation}} = \frac{n}{\tau} = \frac{5}{300 \times 7 \times 365 \times 24}$$

$$= 2.7 \times 10^{-7} \text{ failures/h}$$

It is helpful to know that some people and engineers prefer to use the failure rate in the unit of the number of failures per year. This is because the number of failures per hour in this instance is so low. As a result, the average failure rate per year is calculated as follows:

$$\lambda = \frac{\text{Number of failures}}{\text{Total time of the valve in operation}} = \frac{n}{\tau} = \frac{5}{300 \times 7}$$

$$= 0.00238 \text{ failures/year}$$

Specifically, IEC 61511 states that the estimated rates of failure of a subsystem can be determined through a quantified failure mode analysis of the design using failure data from a recognized industry source or from previous experience of using the system in the same environment as it is intended to be used for its intended application. This needs to be done in a way in which the experience is sufficient to prove the claimed mean time to failure (MTTF) on a statistical basis up to a single-sided confidence interval of at least 70%. The calculation of the failure rate at the upper, (single-sided) confidence interval of 70% can be performed by applying Eq. (9.3).

Calculation of Failure Rate Representing a 70% Single-Side Upper Confidence Interval

$$\lambda = \frac{1}{2\tau} Z_{0.3\ 2(n+1)} \tag{9.3}$$

where:

τ: Accumulated exposure time;
n: Number of reported failures; and
$Z_{0.3\ 2(n+1)}$: 30% percentage of the chi-square distribution with 2 (n+1) degrees of freedom.

In this analysis, the failure rate is calculated as per Eq. (9.2) in order to determine the final result. In the study, we took into account the total percentage of failures, the distribution of failures into operation modes, such as dangerous, safe, detected, and undetected, and finally, the 70% single-sided upper confidence interval.

Final Failure Rate Calculation

$$\begin{aligned}\text{Final failure rate } \left(\text{h}^{-1}\right) = {} & \text{Failure rate} \times \text{Percentage of failure} \\ & \times \text{Distribution percentage into operational modes} \\ & \text{(open or closed or both positions)} \\ & \times \text{Failure distribution percentage into dangerous,} \\ & \text{safe, detected, and undetected modes.}\end{aligned} \tag{9.4}$$

Example 9.3 As can be seen in Table 9.5, we have forecasted failure rates for butterfly valves during the period 2014–2020. All valve failures reported by an end user to the valve manufacturer over the first three years of valve operation when the valve was covered by the manufacturer's warranty. How reliable are butterfly valves according to your estimation?

Table 9.5 The number of butterfly valve failures during operations reported by end users to valve manufacturers from 2014 to 2020.

Year	Number of valves	Number of reported failures	Total hours in the service	Estimated failure rate (h^{-1})
2014	1,673	—	—	—
2015	1,961	—	—	—
2016	3,169	—	—	—
2017	4,648	5	6,803	8.4×10^{-8}
2018	3,243	12	9,778	1.4×10^{-7}
2019	3,351	13	11,060	1.3×10^{-7}
2020	3,243	1	11,242	1.0×10^{-8}
Total	21,288	31	38,883	9.1×10^{-8}

Answer

τ: accumulated exposure time = 38,883 hours;
n: number of reported failures = 31;
$Z_{0.3\ 2(n+1)}$: 30% of chi-square distribution with 2 (n+1) degrees of freedom.
Using the 70% single-sided upper confidence interval for the calculation of failure rate as per Eq. (9.3) is provided as follows:

$$\lambda = \frac{1}{2\tau} Z_{0.3\ 2(n+1)} = 10^{-7}\,\text{h}^{-1}$$

Among the most important things to keep in mind is that there is a relationship between probability of failure rate and failure rate that can be expressed in Eq. (9.5).

Relationship Between Probability of Failure and Failure Rate

$$\text{Probability of failure } (P) = \lambda \times t \tag{9.5}$$

Example 9.4 There is a failure rate of 500 FIT for safety valves. What is the probability of the valve failing during one year of operation?

Answer

A FIT of 500 indicates 500 failures in one billion hours of operation

$$\lambda = \frac{\text{Number of failures}}{\text{Total time of the valve in operation}} = \frac{500}{10^9} = 5 \times 10^{-7}\ \text{failures/h}$$

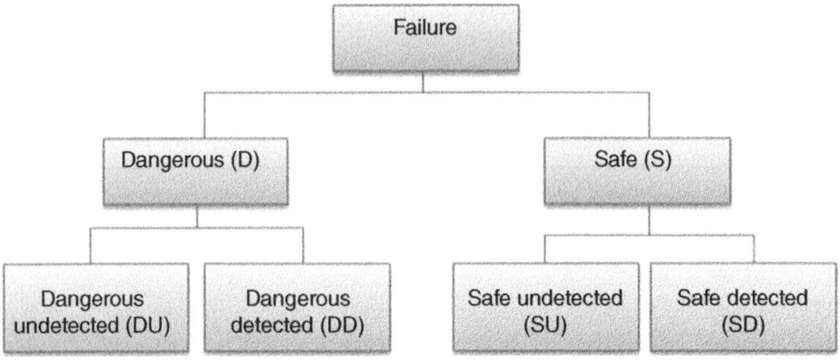

Figure 9.5 Failure mode classification.

The duration of operation is one year which is equal to $365 \times 24 = 8760$ hours.

Using Eq. (9.5) → Probability of failure (P) $= \lambda \times t = 5 \times 10^{-7} \times 8760 = 4.38 \times 10^{-3}$

Failure rates (probabilities) are categorized into two types: safe failure (λ_S) or dangerous failure (λ_D) rates. A dangerous failure rate relates to failures that will prevent achieving the required SIL. Safe failures put the safety function in its safe state, such as the emergency shutdown of a valve in case of failure. It is noticeable that SIL addresses only dangerous failures (λ_D) The dangerous failures can be either detected or undetected. The possibility of detected dangerous failures and undetected dangerous failures are shown as λ_{DD} and λ_{DU}. Equation (9.6) shows the relationship between λ_D, λ_{DD} and λ_{DU}. Failure modes' classification is shown in Figure 9.5.

Relationship Between Dangerous Failure Rate and Dangerous Detected and Dangerous Undetected Failure Rates According to IEC 61508

$$\lambda_D = \lambda_{DU} + \lambda_{DD} \tag{9.6}$$

Safe failure rates are divided into safe detected and safe undetected failure rates. The safe failure rates can be divided into two categories: safe detected failure rates and safe undetected failure rates. Furthermore, it is important to note that the safe failure rate is equal to the sum of the safe detected and undetected failure rates, as indicated in Eq. (9.7).

Relationship Between Safe Failure Rates

$$\lambda_S = \lambda_{SU} + \lambda_{SD} \tag{9.7}$$

where:

λ_S: Safe failure rate;
λ_{SU}: Safe undetected failure rate;
λ_{SD}: Safe detected failure rate;
SD: Estimated percentage of failures belonging to the "safe detected" category;
SU: Estimated percentage of failures belonging to the "safe undetected" category;
DD: Estimated percentage of failures belonging to the "dangerous detected" category;
DU: Estimated percentage of failures belonging to the "dangerous undetected" category.

9.4.1.2 Repair Rate

Repair rate is an evaluation of the frequency of successful repairs performed per unit of time on a failed component. In general, it is signified by the Greek letter μ (Mu) and is used to calculate the metrics that are specified in the following paragraphs. In mathematic terms, the repair rate is defined as follows:

Calculation of Repair Rate

$$\mu = \frac{1}{\text{MTTR}} \tag{9.8}$$

$$\mu = \frac{\text{Number of repairs}}{\text{Total maintenance time}} \tag{9.9}$$

where:

The mean time to recovery (MTTR) is discussed later in this chapter.

9.4.1.3 Mean Time to Failure (MTTF)

During the operation, an industrial valve is installed, and it will function until the first failure occurs. The average time it takes from the time an operation begins until the first failure occurs is known as the *mean time to failure (MTTF)*. In order to determine MTTF, Eqs. (9.10) to (9.11) can be used.

Calculation of MTTF

$$\text{MTTF} = \frac{\text{Total hours of opeartion}}{\text{Total number of units}} \tag{9.10}$$

$$\begin{aligned}\text{MTTF} = {} & \text{The time of the first failure} - \text{The time when the system} \\ & \text{or component starts operating}\end{aligned} \tag{9.11}$$

Calculation of MTTF Based on the Failure Rate

$$\text{MTTF} = E(t) = \int_0^\infty tf(t)\text{d}t \tag{9.12}$$

$$\text{MTTF} = \int_0^\infty R(t)\text{d}t \tag{9.13}$$

$$\text{MTTF} = \text{limit} \lim_{s \to 0} R(s) = \frac{1}{\lambda} \tag{9.14}$$

where:

MTTF: Mean time to failure;
$E(t)$: Expected value of MTTF;
$R(t)$: An indicator of system or component reliability that indicates the probability that the system or component will not fail in the time period from 0 to "t." In other words, it indicates the probability that the given item will survive at least until time "t";
s: In math, the Laplace transform variable can be defined as a linear transformation which takes a parameter like "x" and transforms it into a new and more complex variable called "s." As a result of this factor, we may also be able to convert some of the differential equations into pure algebraic equations;
$R(s)$: It refers to the Laplace transform for the reliability function, $R(t)$;
λ: Failure rate.

9.4.1.4 Mean Time Between Failure (MTBF)

Mean time between failure (MTBF) is defined as the average time duration between two repairable failures that is calculated in accordance with Eq. (9.15) as follows:

Calculation of Mean Time Between Failure (MTBF)

$$\text{MTBF} = \frac{\text{Total hours of operation between two failures}}{\text{Total number of incidents}} \tag{9.15}$$

Refer to Figure 9.4, MTBF can be calculated as per Eq. (9.16) alternatively.

An Alternative Way to Calculate Mean Between Failure (MTBF)

$$\begin{aligned}\text{MTBF} = {} & \text{Mean time to detection (MTTD)} + \text{Mean time to recovery (MTTR)} \\ & + \text{Mean time tofailure (MTTF)}\end{aligned} \tag{9.16}$$

9.4.1.5 Mean Time to Repair and Recovery (MTTR)

An MTTR is the average time required to repair a failed component and restore it to operational status. This metric includes the time spent during the alert and diagnostic processes, prior to the initiation of repair activities. The average time spent solely on the repair process is called the mean time to repair and it is also shown with MTTR and it is calculated according to Eq. (9.17).

Calculation of Mean Time to Repair (MTTR)

$$\text{MTTR} = \frac{\text{Total hours of maintennace}}{\text{Total number of repairs}} \tag{9.17}$$

9.4.1.6 Mean Time to Detection (MTTD)

Mean time to detection (MTTD) is the average time elapsed between the occurrence of a component failure and its detection, which is calculated according to Eq. (9.18).

Calculation of Mean Time to Detection (MTTD)

$$\text{MTTD} = \frac{\text{Total hours of incident detection}}{\text{Total number of incidents}} \tag{9.18}$$

Example 9.5 In the case of an industrial valve, the MTBF value is 3,000,000, MTTF = 2,999,992, and the repair time is eight hours. Calculate the mean time to detection (MTTD).

Answer

Using Eq. (9.16): MTBF = Mean time to detection (MTTD) + Mean time to recovery (MTTR) + Mean time to failure (MTTF) $\rightarrow$ 3,000,000 = MTTD + 2,999,992 + 8 $\rightarrow$ 3,000,000 = MTTD + 3,000,000 $\rightarrow$ MTTD = 0

Thus, the mean time to detect the failure is zero.

9.4.2 Reliability and Unreliability

The reliability of a system such as an industrial valve is the possibility that it will perform correctly during a specific period of time. The valve does not need to be repaired or maintained during the period of operation. Additionally, the system adheres to defined performance according to protocols and specifications. The more comprehensive definition of the term reliability can be expressed as follows: The probability that a component, equipment, or system will be able to satisfactorily perform its intended function when required to do so, under given conditions, such as design limits and environmental conditions, operating time, and frequency and accuracy of maintenance for a specific period of time. Following are four important aspects of this definition:

- The component, equipment, or system's intended function must be understood. This aspect refers to the explicit definition of what the device is expected to do in order to accomplish the task.
- It is necessary to specify the time when the component or system must be operated. Secondly, this aspect relates to the requested operability of a system, such as an industrial valve.
- A satisfactory performance of the component or system must be established. Lastly, there is the issue of evaluating what the device is capable of with regard to a good performance.
- There must be a clear understanding of the specific design limits. Lastly, there is the issue of the operating conditions under which the device operates, such as pressure and temperature.

In the mathematical sense, reliability is the probability that a device will work correctly in the interval starting from time 0 to time "*t*" in the time interval. An example of the reliability chart as a function of time for an industrial valve is illustrated in Figure 9.6. When the operating time is increased from "*t*" to TTF (Time to Failure), where it is estimated that the device will fail with a probability of almost 100%, the reliability is reduced from 100% to zero. It is predicted that at time "*t*," the probability of failure will be 73%. In other words, the probability that the system will operate without failure from 0 to "*t*" is 0.73. It is not possible to calculate

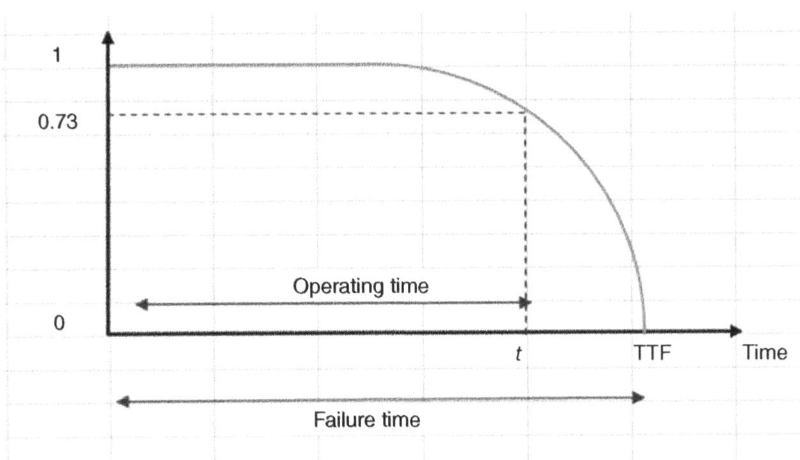

Figure 9.6 Reliability chart as a function of time for an industrial valve.

reliability for time periods longer than "t," as this is outside the normal operating time period of the system. It is important to note that the reliability of the system depends on the failure rate, which will be discussed in more detail later in this chapter.

Example 9.6 In the case of a newly manufactured and tested industrial valve, it operates correctly when it is put into service at time $t = 0$. What is the reliability of the valve when it is put into service? What is the reliability of the valve over an indefinite period of time? Is it true that the valve has a 95% reliability rate?

Answer

The reliability of the valve may be almost 100% once it is placed into operation. Since the valve will eventually fail, its likelihood of success and reliability for an infinite period will be zero. A statement such as a valve has a 95% reliability rating is not accurate because reliability should be expressed over a period of time. It is accurate to state, for example, that the valve has a reliability of 98% over a period of 10,000 hours.

As an alternative method of calculating the reliability, there is an Eq. (9.19) which can be used to perform the calculation.

General Reliability Function

$$\frac{1}{R(t)} \times \mathrm{d}R(t) = -\lambda(t)\mathrm{d}t \tag{9.19}$$

where:

$R(t)$: Reliability function;
λ: Failure rate (1/h).

Equation (9.23) is obtained by integrating both sides of Eq. (9.20) over the time interval $[0, t]$ as follows:

$$\int_1^{R(t)} \frac{1}{R(t)} \times \mathrm{d}R(t) = -\int_0^t \lambda(t)\mathrm{d}t \tag{9.20}$$

$$\rightarrow \ln R(t) = -\int_0^t \lambda(t)\mathrm{d}t \tag{9.21}$$

$$\rightarrow R(t) = e^{-\int_0^t \lambda(t)\mathrm{d}t} \tag{9.22}$$

$$\rightarrow R(t) = e^{-\lambda t} \tag{9.23}$$

Example 9.7 The failure rate for a valve in an industrial setting can be assumed to be 0.00002 failures/h. In this case, calculate the valve mean time to failure.

Answer

Based on the failure rate λ, MTTF can be calculated by using Eq. (9.14) as follows:

$$\text{MTTF} = \text{limit} \lim_{s \to 0} R(s) = \frac{1}{\lambda} = \frac{1}{0.00002} = 50{,}000 \text{ hours}$$

This means that the valve is expected to fail after 50,000 hours of operation, a very long time after it has been installed.

Example 9.8 It can be assumed that a valve that is used for an industrial application will fail on average 0.0004 times/h. Calculate the reliability and mean time to failures for 15 hours of continuous operation.

Answer

Equations (9.22) and (9.23) are used to calculate the reliability of the valve in the following way:

$$R(t) = e^{-\int_0^t \lambda(t)\mathrm{d}t} = e^{-\lambda t} = e^{-0.0004 \times 15} = 2.7183^{-0.006} = 0.9940$$

$$\text{MTTF} = \int_0^\infty R(t)\mathrm{d}t = \int_0^\infty e^{-\lambda t}\mathrm{d}t = \int_0^\infty e^{-0.0004 \times t}\mathrm{d}t = \frac{1}{0.0004} = 2500 \text{ hours}$$

The reliability of the industrial valve has been calculated to be 0.9940 and the mean time to failure is estimated to be 2500 hours.

It is not unusual to find engineering systems, such as an industrial valve, composed of a variety of components and configurations that are subject to reliability analysis. In the case of safety-critical valves, for example, these consist of three components, which are a valve, an actuator, and a control panel, as will be explained in greater detail later on in this chapter. As a result of this, it is important to take into account the overall reliability of the system as a whole. As a matter of fact, the failure of one component in a system can result in the dysfunction of the system as a whole. In order to determine the reliability of the whole system, the combinatorial rules of probability are applied based on the type of subsystems and elements which are connected in order to comprise the whole. For safety critical valves, it is very common to use a series network, which consists of three blocks or units, including a valve, actuator, and control panel. This means that a safety critical valve is typically comprised of three units acting in series. So in the event of one component failing, the entire system will also fail. In other words,

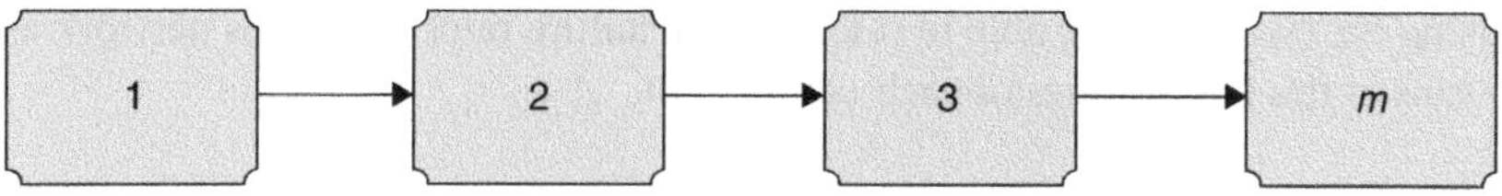

Figure 9.7 A series of network diagram including "*m*" components.

the entire system must function normally for the system to be successful. Figure 9.7 illustrates a series of "*m*" components that are located in a series.

The reliability of the series systems network is expressed by Eq. (9.24).

Calculation of the Reliability Series

$$R_s = P\,(X_1X_2X_3 ... \,X_m) \tag{9.24}$$

where:

R_s: Series system reliability or probability of success;
X_i: An event donating the success of unit i, for $i = 1, 2, 3, ...,$ and m;
$P(X_1X_2X_3 ... \,X_m)$: Probability of occurrence of events $X_1, X_2, X_3, ..., X_m$.

For a system that can fail due to the failure of each individual component, the series system reliability can be calculated using Eqs. (9.25) and (9.26):

$$R_s = P(X_1) \times P(X_2) \times P(X_3) \times \cdots \times P\,(X_m) \tag{9.25}$$

where:

$P(X)$ is the occurrence probability of event X_i, for $i = 1, 2, 3, ...,$ and m;
Considering $R_i = P(x_i)$.

Calculation of the Reliability Series (Main Equation)

$$R_s = \prod_{i=1}^{m} R_i \tag{9.26}$$

where:

R_i is the unit i reliability for $i = 1, 2, 3, ...,$ and m.

For identical units where $R_i = R$, it is possible to simplify the calculation of the reliability series by using Eq. (9.27) as follows:

Calculation of the Reliability Series for Identical Units

$$R_s = 1 - m(1 - R) \tag{9.27}$$

Using Eq. (9.28), we are able to calculate the failure rates of a series network as the sum of the failure rates of each component.

Calculation of the Failure Rate Series for *N* Components

$$\lambda_S = \sum_{i=1}^{N} \lambda_i \tag{9.28}$$

where:

λ_S: The failure rate of the entire system including all N components in series;
λ_i: The failure rate of component "i" that could be a number from one to N.

Example 9.9 Imagine that four identical valves are installed in a pipeline in such a way that failure of any one of them as a single unit can cause problems during the transportation of crude oil from the plant to the customer. Calculate the reliability of the crude oil transportation system including these four valves. All four valves have a reliability of 0.97.

Answer

Using Eq. (9.26), it is possible to calculate the reliability of the entire system by following the steps as follows:

$$R_s = \prod_{i=1}^{4} R_i = R_1 \times R_2 \times R_3 \times R_4 = (0.97) \times (0.97) \times (0.97) \times (0.97)$$
$$= 0.8853$$

Since all units have the same reliability, it is possible to use Eq. (9.27) in lieu of Eq. (9.26) to calculate the total reliability of the crude oil transportation system as follows:

$$R_s = 1 - m(1 - R) = 1 - 4 \times (1 - 0.97) = 0.88$$

It is important to know about the *failure rate allocation method*. In this method, failure rates are allocated to components of a system based on the knowledge of the required failure rate for a given system. The assumptions that are associated with this method are as follows:

- System components for a series configuration
- System components fail independently
- Time to component failure is exponentially disturbed

Accordingly, the calculation of the system failure rate is based on Eq. (9.28). If the system failure rate is λ_{rs}, then the allocation component failure rate must meet the condition in Eq. (9.29).

$$\sum_{i=1}^{N} \lambda_i^* \leq \lambda_{\mathrm{sr}} \tag{9.29}$$

where:

λ_i^* is the failure rate allocated to component $i = 1, 2, 3, n$ according to the required failure rate λ_{rs}.

The following steps are required in this method:

1) Estimate the component failure rates λ_i for $i = 1, 2, 3, \ldots, n$ using the past data.
2) Calculate the relative weight θ_i of component i using Eq. (9.30).

Calculation the Relative Weight

$$\theta_i = \frac{\lambda_i}{\sum_{i=1}^{N} \lambda_i} \tag{9.30}$$

It should be borne in mind that θ_i indicates the relative failure rate of component i, and

$$\sum_{i=1}^{N} \theta_i = 1 \tag{9.31}$$

By using the relationship given in Eq. (9.32), it will be possible to assign the required failure rate to component i.

Calculation of Required or Allocated Failure Rate

$$\lambda_i^* = \theta_i \times \lambda_{\mathrm{rs}} \tag{9.32}$$

Example 9.10 Let us suppose that there are five independent control subsystems in series and that the failure rate of the whole system is expected to be 0.0006 failures per hour. Using the experience from the past, the failure rate for these five subsystems are estimated at $\lambda_1 = 0.0001$ failures/h, $\lambda_2 = 0.0002$ failures/h, $\lambda_3 = 0.0003$, $\lambda_4 = 0.0004$, and $\lambda_5 = 0.0005$. Assign the required system failure rate to five subsystems.

Answer

To begin with, the first step would be to calculate the failure rate for the entire system according to the past experience as per Eq. (9.28).

$$\lambda_S = \sum_{i=1}^{N} \lambda_i = 0.0001 + 0.0002 + 0.0003 + 0.0004 + 0.0005$$
$$= 0.015 \text{ failures/h}$$

Using Eq. (9.30), the relative weights for five subsystems are calculated as follows:

$$\theta_1 = \frac{0.0001}{0.015} = 0.00667 \rightarrow \lambda_1^* = \theta_1 \times \lambda_{rs} = (0.00667)\,(0.0006)$$
$$= 0.00004 \text{ failures/h}$$

$$\theta_2 = \frac{0.0002}{0.015} = 0.1333 \rightarrow \lambda_2^* = \theta_2 \times \lambda_{rs} = (0.1333)\,(0.0006)$$
$$= 0.00007 \text{ failures/h}$$

$$\theta_3 = \frac{0.0003}{0.015} = 0.2 \rightarrow \lambda_3^* = \theta_3 \times \lambda_{rs} = (0.2)\,(0.0006) = 0.00012 \text{ failures/h}$$

$$\theta_4 = \frac{0.0004}{0.015} = 0.2667 \rightarrow \lambda_4^* = \theta_4 \times \lambda_{rs} = (0.2667)\,(0.0006)$$
$$= 0.00016 \text{ failures/h}$$

$$\theta_5 = \frac{0.0005}{0.015} = 0.333 \rightarrow \lambda_5^* = \theta_5 \times \lambda_{rs} = (0.333)\,(0.0006)$$
$$= 0.00019 \text{ failures/h}$$

Besides series networks, parallel networks are also widely used and it represents a system with units as illustrated in Figure 9.8 that work at the same time as an alternative to series networks. The success of the system depends on the functioning of at least one of the units.

A block represents a unit of the system. According to Eq. (9.33), the failure probability of the parallel system or network can be calculated.

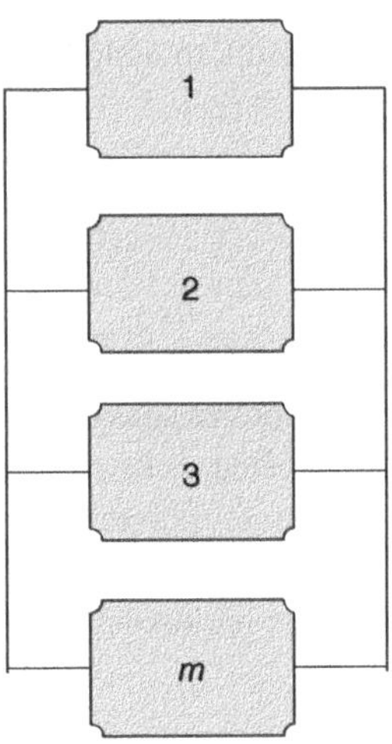

Figure 9.8 A parallel network diagram including "*m*" components.

Failure Probability of a Parallel System

$$F_P = P\,(X_1 X_2 X_3 \ldots X_m) \tag{9.33}$$

where:

F_P: Failure probability of the parallel system;

X_i: An event donating the failure of unit i, for $i = 1, 2, 3, \ldots,$ and m;

$P(X_1X_2X_3\ldots X_m)$: Probability of occurrence of events $X_1, X_2, X_3, \ldots, X_m$.

In addition, the failure probability of a parallel system can also be calculated according to Eq. (9.34).

An Alternative Method for Calculating the Failure Probability of a Parallel System

$$F_P = \prod_{i=1}^{m} F_i \tag{9.34}$$

where:

F_i: The failure probability of each unit for unit numbers $i = 1, 2, 3, \ldots, m$.

The reliability of a parallel network can now be calculated right away with the help of Eq. (9.35).

Calculation of the Reliability for a Parallel System

$$R_P = 1 - F_P = 1 - \prod_{i=1}^{m} F_i \tag{9.35}$$

where:

R_P: Parallel system reliability.

For identical units with similar failure probability, the calculation of reliability for a parallel system can be simplified according to Eqs. (9.36) and (9.37).

Calculation of the Reliability of a Parallel System Consisting of Identical Units

$$R_P = 1 - F^m \tag{9.36}$$

$$\text{Since} \quad R + F = 1 \rightarrow R_P = 1 - (1 - R)^m \tag{9.37}$$

Figure 9.9 depicts the reliability plot for parallel systems with m's of 1, 2, 3, and 4. Based on the picture here, it can be clearly seen that the parallel system reliability increases in proportion to the unit reliability and number of units.

Example 9.11 Two identical pipeline valves are installed and operated on a pipeline; one of them is the replacement of the other. Ideally, at least one of the valves should function properly to ensure that the oil is successfully transported from the

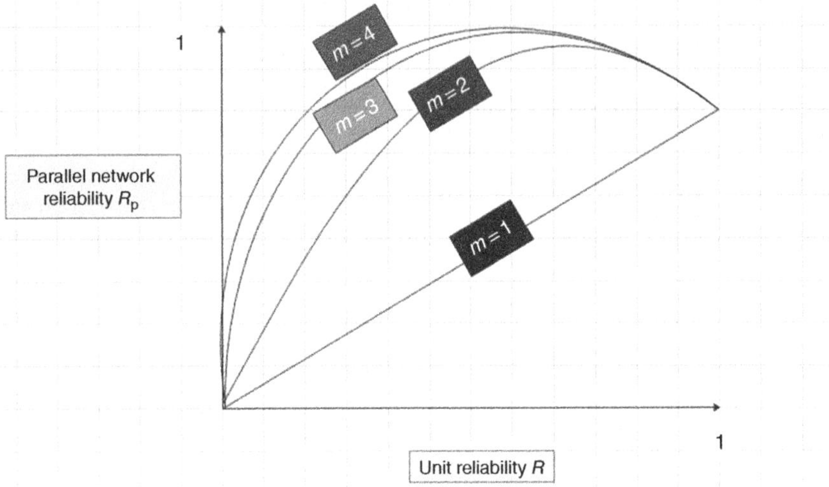

Figure 9.9 A reliability plot of a parallel system based on m = 1, 2, 3, and 4 units.

platform to the end user. Assuming that each of these two valves has a reliability of 0.96, calculate the reliability of the system with respect to these two valves.

Answer

In order to calculate the reliability of a parallel system consisting of identical units, we can turn to Eq. (9.37). The answer is as follows:

$$R_P = 1 - (1 - R)^m = 1 - (1 - 0.96)^2 = 0.9984$$

Accordingly, the reliability of the pipeline system is equal to 0.9984 with regard to these two pipeline valves.

Alternatively, a reliability block diagram as illustrated in Figure 9.10 may be used for a complex system listing all the units or subsystems as well as their failure rate values. In this scenario, the analytical methods must be revised as shown above to take into consideration the large-scale and complex nature of the network. It should be noted that in the illustrated RBD, both parallel and series connections between various components are present, and this is what is known as the *hybrid* model. This can be seen in the figure below, where in this example, the reliability or any other relevant required data can first be retrieved from subsystem B by using parallel reliability equations. After that, this complex network will be simplified to one that has four subsystems in series.

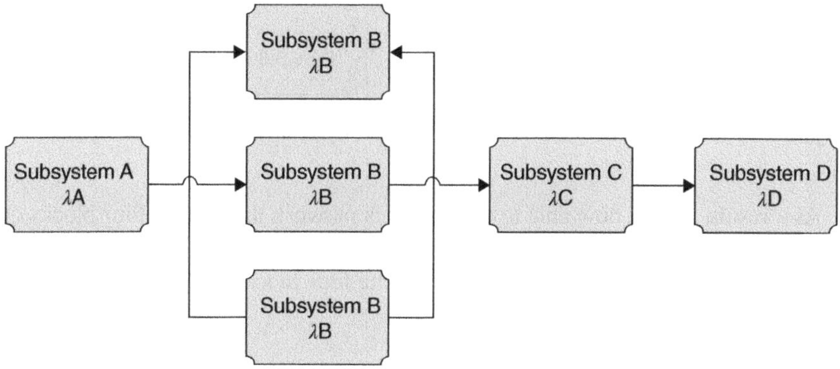

Figure 9.10 An example of a network diagram that includes series and parallel blocks along with their failure rates.

Example 9.12 An illustration of the arrangement of safety units in an entire system is given in Figure 9.11. Estimate the reliability of the entire system including all seven units.

Answer

First, we need to calculate the level of reliability of subsystem B, which includes four units in parallel so that we have B1, B2, B3, and B4. As a result, it is possible to calculate the reliability of the parallel network B with the help of Eq. (9.35), as follows:

$$R_B = R_P = 1 - F_P = 1 - \prod_{i=B1}^{B4} F_i = 1 - (1-0.2)(1-0.8)(1-0.6)(1-0.7)$$
$$= 1 - 0.0192 = 0.9808$$

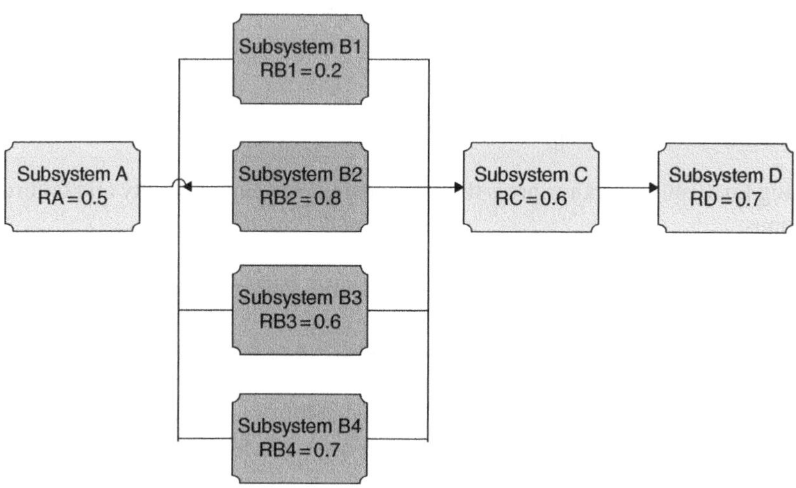

Figure 9.11 An example of a network diagram that contains seven series and parallel blocks, along with their failure rates.

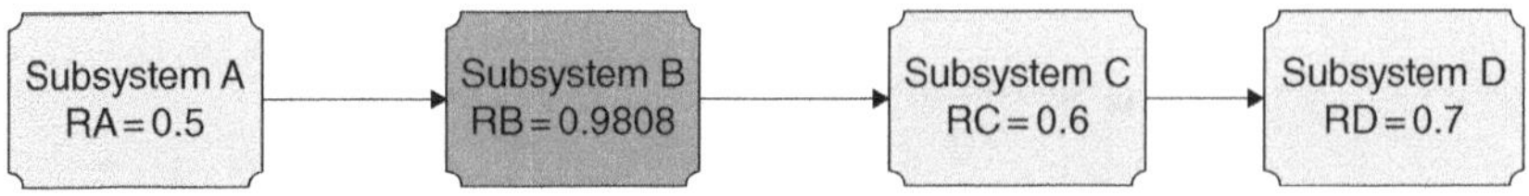

Figure 9.12 An illustration of a simplified network is shown in the previous figure.

As a result, we are now able to simplify this network diagram to four blocks of A, B, C, and D as illustrated in Figure 9.12.

The reliability of this network including the four blocks in series is calculated using Eq. (9.26) which is as follows:

$$R_S = \prod_{i=A}^{D} R_i = R_A \times R_B \times R_C \times R_D$$
$$= (0.5) \times (0.9808) \times (0.6) \times (0.7) = 0.2060$$

Thus, the overall reliability of the entire system, including seven units, is 20.60%.

We need to know about unreliability, which can be considered a measure of failure. The term unreliability is defined as the probability of a system or component failing within a time interval from 0 to "t." Unreliability is calculated according to Eq. (9.38).

Calculation of Unreliability

$$\text{Unreliability } U(t) = 1 - \text{Reliability } (t) \tag{9.38}$$

It is possible to measure the degree of unreliability of an item by considering the *failure function* ($F(t)$) and calculate this according to Eq. (9.39).

Calculation of the Failure Function

$$F(t) = \Pr(T \leq t) = \int_0^t f(u)\mathrm{d}u \quad \text{for } t > 0 \tag{9.39}$$

Using Eq. (9.40), we can express the relationship between reliability and failure functions.

The Relationship Between the Reliability and Failure Functions

$$R(t) = \Pr(T > t) = 1 - F(t) = \int_t^\infty f(u)\mathrm{d}u \tag{9.40}$$

Example 9.13 An industrial valve has a reliability of 0.98 over a period of 1000 hours of use. What is the exact unreliability value for the valve during the given time period?

Answer

Unreliability $U(t) = 1 - \text{Reliability}\ (t) = 1 - 0.98 = 0.02$

9.4.3 Availability and Unavailability

The concept of reliability, as it was mentioned earlier, refers to the period when a system or component can operate successfully and during the period when it cannot be repaired. Availability is a term introduced to describe an operational condition of a device, a system, or its component parts, even if it was corrected either by maintenance or repair. This means that if a system or device works, it is available for use regardless of the fact that it has already been repaired or if it has been running successfully without requiring any kind of repair from the start of operation. Figure 9.13 is a representation of the relationship between reliability, availability, unreliability, and unavailability of a device or component in order to provide successful or unsuccessful operation.

Availability is the function of failure rate, repair time, and operating time that is calculated in accordance with Eq. (9.41). A good system or component is expected to have an availability close to one. In order to calculate the unavailability, Eq. (9.41) is used.

Calculation of Availability

$$\text{Availability} = \frac{\text{Operating time}}{\text{Operating time} + \text{Repair time}} = \frac{\text{MTTF}}{\text{MTTF} + \text{MTTR}} = \frac{\text{MTBF}}{\text{MTBF} + \text{MTTR}} \tag{9.41}$$

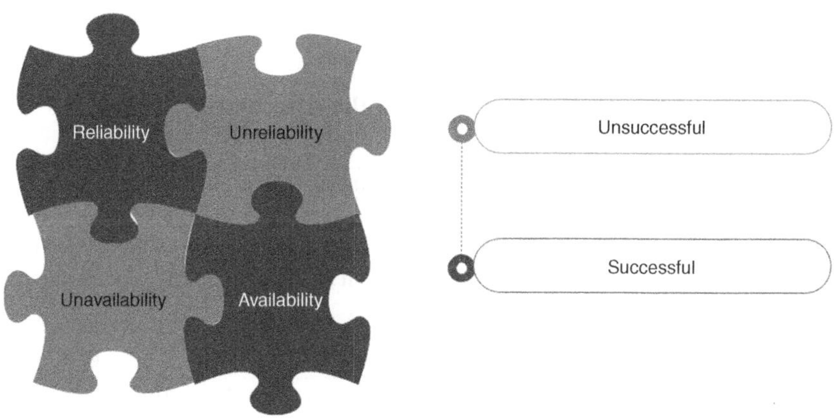

Figure 9.13 Relationship between reliability, availability, unreliability, and unavailability.

Example 9.14 There is an average repair time of 8 hours for a safety device with an operating life of 50,000 hours without failure. For a period of 20 hours, calculate the reliability and unreliability of the device, as well as the availability of the device.

Answer

The first step is to calculate MTTF by using Eq. (9.10):

$$\text{MTTF} = \frac{\text{Total hours of opeartion}}{\text{Total number of units}} = \frac{50{,}000}{1} = 50{,}000\ \text{hours}$$

Now, it is possible to calculate the failure rate by using Eq. (9.14):

$$\lambda = \frac{1}{\text{MTTF}} = \frac{1}{50{,}000} = 0.00002\ \text{failures/h}$$

Equations (9.22) and (9.23) are used to calculate the reliability of the valve in the following way:

$$R(t) = e^{-\int_0^t \lambda(t)dt} = e^{-\lambda t} = e^{-0.00002 \times 20}$$

$$= 2.7183^{-0.0004} = 0.99960 = 99.960\%$$

Thus, the reliability of the safety device is 99.96%.

Using Eq. (9.38): Unreliability $U(t) = 1 - \text{Reliability}\ (t) = 1 - 0.99960 = 0.0004$

Equation (9.41):

$$\text{Availability} = \frac{\text{Operating time}}{\text{Operating time} + \text{Repair time}} =$$

$$\frac{\text{MTTF}}{\text{MTTF} + \text{MTTR}} = \frac{50{,}000}{50{,}000 + 8} = 0.99984 = 99.984\%$$

Alternatively, a calculation of the availability can be done using the failure rate and repair rate provided in Eq. (9.42) as follows:

Calculation of the Availability Based on the Alternative Method

$$\text{Availability} = \frac{\text{Repair rate}}{\text{Failure rate} + \text{Repair rate}} = \frac{\mu}{\lambda + \mu} \tag{9.42}$$

Example 9.15 A safety-critical system will have a failure rate of 0.0000005 failures/h with a repair rate of 0.1 failures/h. What is the availability of the system?

Answer

Using Eq. (9.42) $\rightarrow$ $\text{Availability} = \frac{\text{Repair rate}}{\text{Failure rate} + \text{Repair rate}} = \frac{0.1}{0.1 + 0.0000005}$

$= 0.999995 = 99.9995\%$

The availability of a series systems network is expressed by Eq. (9.43).

Calculation of the Availability Series

$$A_s = P(X_1X_2X_3 \ldots X_m) \tag{9.43}$$

where:

A_s: Series system availability connected to probability of success;
X_i: An event donating the success of unit i, for $i = 1, 2, 3, \ldots,$ and m;
$P(X_1X_2X_3 \ldots X_m)$: Probability of occurrence of events $X_1, X_2, X_3, \ldots, X_m$.

For a system that can fail due to the failure of each individual component, the series system availability can be calculated using Eqs. (9.44) and (9.45):

$$A_s = P(X_1) \times P(X_2) \times P(X_3) \times \cdots \times P(X_m) \tag{9.44}$$

where:

$P(X)$ is the occurrence or availability probability of event X_i, for $i = 1, 2, 3, \ldots,$ and m;
Considering $A_i = P(x_i)$.

Calculation of the Availability Series (Main Equation)

$$A_s = \prod_{i=1}^{m} A_i \tag{9.45}$$

where:

A_i is the unit i availability for $i = 1, 2, 3, \ldots,$ and m.

For identical units where $A_i = A$, it is possible to simplify the calculation of the availability series by using Eq. (9.46) as follows:

Calculation of the Availability Series for Identical Units

$$A_s = 1 - m\,(1 - A) \tag{9.46}$$

Example 9.16 As a result of the intertwining of two safety systems, failure of any one of them results in the failure of the whole system as a whole. In the first system, the failure rate is 0.0002 failures/h, and the rate of repair is 0.1 failures/h. The second system has a failure rate of 0.0003 failures/h and a repair rate of 0.2 failures/h. What is the reliability of this system as a whole?

Answer
The first thing we need to do is calculate the availability of both systems by applying Eq. (9.42) as follows:

$$A_1 = \text{Availability of the first system} = \frac{\text{Repair rate}}{\text{Failure rate} + \text{Repair rate}}$$
$$= \frac{0.1}{0.0002 + 0.1} = 0.9980$$

$$A_2 = \text{Availability of the second system} = \frac{\text{Repair rate}}{\text{Failure rate} + \text{Repair rate}}$$
$$= \frac{0.1}{0.0003 + 0.2} = 0.49925$$

Equation (9.45) is used to calculate the availability of the entire system as follows:

$$A_S = \prod_{i=1}^{2} A_i = A_1 \times A_2 = 0.9980 \times 0.49925 = 0.49825$$

In order to calculate the unavailability of a system as a function of the operating and repair times, we can use Eq. (9.47).

Calculation of Unavailability

$$\text{Unavailability} = 1 - \frac{\text{Operating time}}{\text{Operating time} + \text{Repair time}} \tag{9.47}$$

Example 9.17 In an industrial application, a valve has been operated for 60,000 hours without any failures. After 60,000 hours of operation for eight hours, the valve has already been repaired once. What is the value of the availability and unavailability of this valve?

Answer

$$\text{Availability} = \frac{\text{Operating time}}{\text{Operating time} + \text{Repair time}} = \frac{60{,}000}{60{,}000 + 8}$$
$$= 0.99986 = 99.986\%$$

$$\text{Unavailability} = 1 - \text{Availability} = 1 - 0.99986 = 0.00014 = 0.014\%$$

According to Eq. (9.48), the unavailability of the parallel system or network can be calculated.

Unavailability of a Parallel System

$$U_P = P\,(X_1 X_2 X_3 \ldots X_m) \tag{9.48}$$

where:

U_P: Unavailability of the parallel system;

X_i: An event donating the failure of unit i, for $i = 1, 2, 3, ...,$ and m;
$P(X_1X_2X_3 ... X_m)$: Probability of occurrence of events $X_1, X_2, X_3, ..., X_m$.

In addition, the unavailability of a parallel system can also be calculated according to Eq. (9.49).

An Alternative Method for Calculating the Unavailability of a Parallel System

$$U_P = \prod_{i=1}^{m} U_i \tag{9.49}$$

where:

U_i: The unavailability of each unit for unit numbers $i = 1, 2, 3, ..., m$.

The availability of a parallel network can now be calculated right away with the help of Eq. (9.50).

Calculation of the Availability for a Parallel System

$$A_P = 1 - U_P = 1 - \prod_{i=1}^{m} U_i \tag{9.50}$$

where:

A_P: Parallel system availability.

Example 9.18 In parallel with each other, four safety systems are working. The operating and repair times of these four systems can be found in Table 9.6. How available is the entire system as a whole?

Table 9.6 The operating and repair times for four subsystems that operate in parallel.

Subsystem number	Operating time (hours)	Repair time (hours)
1	50,000	10
2	60,000	12
3	65,000	15
4	70,000	18

Answer

As a first step, according to Eq. (9.41), we need to calculate the availability for each subsystem as follows:

$$\text{Availability 1} = \frac{\text{Operating time}}{\text{Operating time} + \text{Repair time}} = \frac{50{,}000}{50{,}000 + 10}$$

$$= 0.9980 = 99.80\%$$

$$\rightarrow \text{Unavailability } (U1) = 1 - A1 = 1 - 0.9980 = 0.002$$

$$\text{Availability 2} = \frac{\text{Operating time}}{\text{Operating time} + \text{Repair time}} = \frac{60{,}000}{60{,}000 + 12}$$

$$= 0.9998 = 99.98\%$$

$$\rightarrow \text{Unavailability } (U2) = 1 - A2 = 1 - 0.9998 = 0.0002$$

$$\text{Availability 3} = \frac{\text{Operating time}}{\text{Operating time} + \text{Repair time}} = \frac{65{,}000}{65{,}000 + 15}$$

$$= 0.99977 = 99.977\%$$

$$\rightarrow \text{Unavailability } (U2) = 1 - A3 = 1 - 0.99977 = 0.00023$$

$$\text{Availability 4} = \frac{\text{Operating time}}{\text{Operating time} + \text{Repair time}} = \frac{70{,}000}{70{,}000 + 15}$$

$$= 0.99979 = 99.979\%$$

$$\rightarrow \text{Unavailability } (U2) = 1 - A4 = 1 - 0.99979 = 0.00021$$

Equation (9.50) can be used to determine the availability of a parallel network immediately as follows:

$$A_P = 1 - U_P = 1 - \prod_{i=1}^{4} U_i = 1 - (0.002)(0.0002)(0.00023)(0.00021)$$

$$= 1 - 0.00000000000001932 = 0.99999999$$

9.5 Safety Integrity Level (SIL) Calculations

9.5.1 SIL

SIL is a term attached to international standards such as IEC 61508 which provides suppliers and end users with a common framework to design products and systems when it comes to safety-related applications. The IEC has published IEC 61508 as an international standard for electrical, electronic, and programmable electronic safety-related systems. With the help of SIL, it is possible to specify and design safety systems in a scientific, numerical manner, in order to allow for a quantitative evaluation of failure risk. A SIL is a scientific and numerical method for

Table 9.7 SIL Levels according to IEC 61508 for low-demand conditions.

SIL	PFD	Risk reduction factor
SIL 1	0.1–0.01	10–100
SIL 2	0.01–0.001	100–1,000
SIL 3	0.001–0.0001	1,000–10,000
SIL 4	0.0001–0.00001	10,000–100,000

defining and evaluating safety systems, which quantifies the risk of failure. As can be seen in Table 9.7, SIL levels range from 1 to 4, and it is significant to note that these values are for low-demand operations. In the following section, we will discuss these modes of operation in additional detail. A SIL 4 system has the lowest probability of failure or probability of failure on demand (PFD) and the highest reliability. PFD is a measure of the effectiveness of a safety function. The likelihood that the safety function will not work when needed is represented by this value. Consequently, SIL 1 has the highest probability of failure and the lowest level of safety. For ESD valves, SIL 2 or 3 is typically required, as described earlier.

In relation to a reliability analysis for a safety system, IEC 61508 describes two or three types of requirements. The IEC 61508 categorizes the frequency of demands into three categories, called demand modes of operation.

Low-demand: Safety system requirements or functions are occurring less frequently than once per year. Some characteristics of low-demand systems include: a safety system's response is rarely needed. It may also be possible for a failure to remain hidden for some time, so the safety system is not functioning for this reason.

High-demand: Safety system requirements or functions are occurring more frequently than once per year.

Continuous mode: The demands or functions of a safety system are always present.

As a general rule, high-demand and continuous modes of operation share two main characteristics. First, the safety system provides a response to prevent failure caused by risks occurring frequently or constantly. Furthermore, the failure of the safety system may directly result in a hazardous incident if the system malfunctions.

For a high-demand mode or continuous mode of operation, PFH is calculated as the average frequency of dangerous failure of the safety function. Table 9.8 provides the correlation between PFH and SIL. In Table 9.9, SILs are correlated with the associated average PFDs and PFHs.

Table 9.8 SIL Levels according to IEC 61508 for high-demand and continuous operation conditions.

SIL	PFH	Risk reduction factor
SIL 1	0.00001–0.000001	100,000–1,000,000
SIL 2	0.000001–0.0000001	1,000,000–10,000,000
SIL 3	0.0000001–0.00000001	10,000,000–100,000,000
SIL 4	0.00000001–0.000000001	100,000,000–1,000,000,000

Table 9.9 Levels of SIL according to IEC 61508 for low and high demand as well as continuous operation.

SIL	Average PFD (low-demand mode of operation)	Average PFH (high-demand or continuous mode of operation)
1	$\geq 10^{-2}$–$<10^{-1}$	$\geq 10^{-6}$–$<10^{-5}$
2	$\geq 10^{-3}$–$<10^{-2}$	$\geq 10^{-7}$–$<10^{-6}$
3	$\geq 10^{-4}$–$<10^{-3}$	$\geq 10^{-8}$–$<10^{-7}$
4	$\geq 10^{-5}$–$<10^{-4}$	$\geq 10^{-9}$–$<10^{-8}$

9.5.2 Probability of Failure on Demand (PFD)

PFD stands for probability of failure on demand. In many cases, safety systems are designed to work in the background, monitoring a process. However, they don't take any action until a safety limit has been exceeded, at which point they must take action to keep the process safe. As with reliability, PFD is also a probability value ranging from 0 to 1, inclusive. For example, a PFD of zero means there is no likelihood of failure. There is no question that a system designed for high reliability must exhibit a small PFD value (very close to zero). There's only one way to determine how low the PFD has to be, and that's based on how necessary the component or system is to the fulfilment of our human needs.

IEC 61508 presents a simplified formula for the calculation of PFD that is given based on Eq. (9.51).

Calculation of PFD Based on IEC 61508

$$\mathrm{PFD}_{\mathrm{avg}}^{\mathrm{G}} = \lambda_{\mathrm{D,G}}\, t_{\mathrm{GE}} \tag{9.51}$$

where:

$\lambda_{D,G}$: Group failure frequency of dangerous failures;
t_{GE}: Group equivalent mean downtime.

Note: The next section will explain the concept of mean downtime.

In addition, PFD can be calculated according to Eq. (9.52) as follows:

$$\mathrm{PFD_{avg}} = \frac{\text{Frequency of tolerable accidents}}{\text{Frequency of accidents without protection}} \tag{9.52}$$

As a result of Eq. (9.53), there is a relationship between availability and PFD.

Relationship Between Availability and PFD

$$\text{Availability} = 1 - \mathrm{PFD_{avg}} \tag{9.53}$$

According to Eq. (9.54), there is a correlation between *risk reduction factor (RRF)* and probability of loss (PFD).

Correlations Between PFD and RRF

$$\mathrm{RRF} = \frac{1}{\mathrm{PFD_{avg}}} \tag{9.54}$$

Example 9.19 The safe PFD for a SIS is 0.001/year, while the dangerous probability is 0.0001/year. Calculate the values of the PFD, availability, and risk reduction factor.

Answer

$$\mathrm{PFD_{avgS}} = 0.001/\text{year, and } \mathrm{PFD_{avgD}} = 0.0001/\text{year}$$

$$\mathrm{PFD_{avg}} = \mathrm{PFD_{avgS}} + \mathrm{PFD_{avgD}} = 0.001 + 0.0001 = 0.0011/\text{year}$$

Using Eq. (9.53) → $\text{Availability} = 1 - \mathrm{PFD_{avg}} = 1 - 0.0011 = 0.9989 = 99.89\%$

Using Eq. (9.54) → $\mathrm{RRF} = \frac{1}{\mathrm{PFD_{avg}}} = \frac{1}{0.0011} = 909.09$

In this regard, we observe that the values of the PFD, the availability and the risk reduction factor are 0.0011/year, 99.89%, and 90.09%, respectively.

9.5.3 Mean Downtime

The mean downtime of a system due to both dangerous detected and undetected failures t_{CE} includes the mean downtime of the system due to dangerous

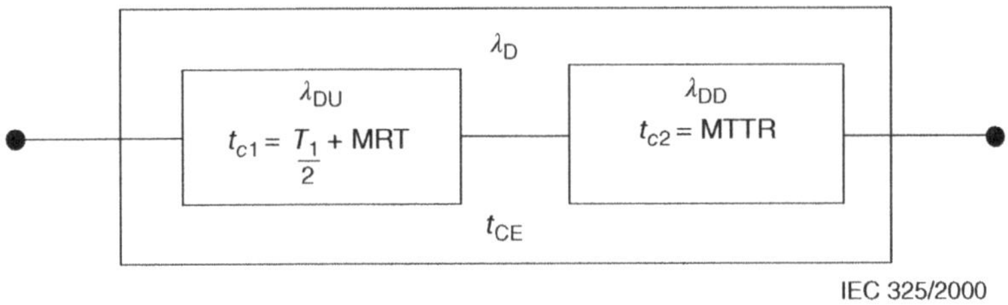

Figure 9.14 1oo1 reliability block diagram.

undetected failures t_{c1} plus the mean downtime of the system due to dangerous detected failures t_{c2} as shown in Figure 9.14. t_{c1} and t_{c2} are calculated using Eqs. (9.55) and (9.56):

Calculation of the Mean Downtime of the System Due to Dangerous Undetected Failures As Well As Dangerous Detected Failures (Refer to IEC 61508)

$$t_{c1} = \frac{T_1}{2} + \text{MRT} \tag{9.55}$$

$$t_{c2} = \text{MTTR} \tag{9.56}$$

where:

T_1: Proof test interval (hour);
MTTR: Mean time to restoration (hour);
MRT: Mean repair time (hour).

$$\begin{aligned}\text{MTTR} = {} & \text{Time to detect the failure} + \text{Time spent before starting the repair} \\ & + \text{The effective time to complete the repair} \\ & + \text{The time before the component is put back into operation}\end{aligned} \tag{9.57}$$

$$\text{MRT} = \text{MTTR} - \text{Time to detect the failure} \tag{9.58}$$

Figure 9.14 shows the reliability block diagram for architecture 1oo1:

Note 1
There is a distinction to be made between how MTTR is defined and calculated in Eq. (9.57) and what was explained earlier in this Chapters 9–17.

Note 2

A typical industrial valve that is a component of SIS is composed of three parts; the valve, the actuator, and the control panel. Generally speaking, an actuator is a mechanical device that is installed on the top of the valve to operate (open and close) the valve and control the valve movement. In order to control the flow of hydraulic fluid through the actuator, a control panel that consists of valves, a pressure gauge and tubes is used. As a side note, we should mention that the majority of actuators that are part of SIS work with hydraulic oil. It is true that these actuators convert hydraulic power to mechanical motion to operate valves through hydraulic power. As part of our SIL analysis, we use the 1oo1 architecture since these three components are considered as a single unit. Based on Figure 9.15, the architecture of the system is composed of a single channel in which any failure of any part compromises the safe operation of the system. The reliability equations are simplified when using the 1oo1 architecture, as opposed to scenarios where other architectures are employed.

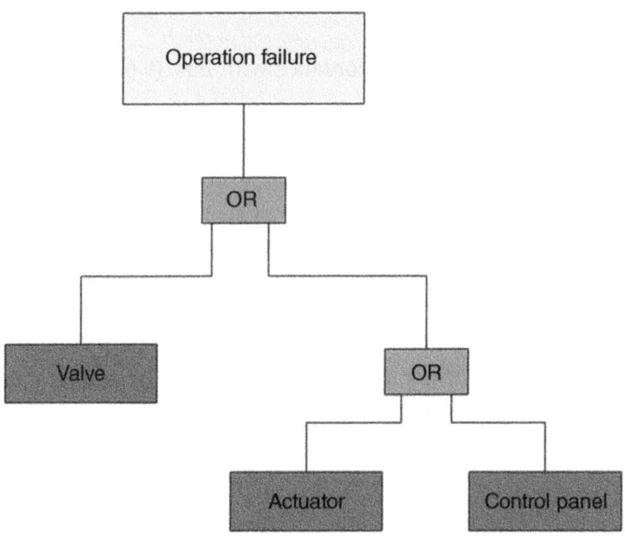

Figure 9.15 Fault tree for safety critical valves connected to the SIS.

Placing values of t_{c1} and t_{c2} from Eqs. (9.55) and (9.56) into Eq. (9.59):

$$t_{CE} = \frac{\lambda_{DU}}{\lambda_D}\left(\frac{T_1}{2} + \text{MRT}\right) + \frac{\lambda_{DD}}{\lambda_D}\text{MTTR} \tag{9.59}$$

9.5.4 Diagnostic Coverage

The number of dangerous failures detected by automatic online diagnostic tests is called *diagnostic coverage (DC)*. DC is the detected dangerous failure rate divided by the total rate of the dangerous failures, using Eq. (9.60).

Calculation of Diagnosis Coverage (DC)

$$\text{DC} = \frac{\lambda_{DD}}{\lambda_D} \tag{9.60}$$

$$\text{Therefore}, \lambda_{DD} = \lambda_D \times \text{DC} \tag{9.61}$$

$$\lambda_{DU} = \lambda_D - \lambda_{DD} = \lambda_D - (\lambda_D \times \text{DC}) \rightarrow \lambda_{DU} = \lambda_D(1 - \text{DC}) \tag{9.62}$$

9.5.5 Safe Failure Fraction (SFF)

The average PFD (PFD_G) is calculated in accordance with Eqs. (9.63) and (9.64):

$$\text{PFD}_G = (\lambda_{DU} + \lambda_{DD})\, t_{CE} \tag{9.63}$$

$$\rightarrow \text{PFD}_G = \lambda_D \times t_{CE} = \lambda_{DU}\left(\frac{T_1}{2} + \text{MRT}\right) + \lambda_{DD} \times \text{MTTR} \tag{9.64}$$

Another key safety parameter is called the *safe failure fraction (SFF)*, which verifies the suitability of a device for use in a system for a particular SIL. In fact, SFF is a term used to describe the proportion of failures that are safe (not hazardous) plus hazardous failures that are possible to detect by applying a test. On the basis of Eq. (9.65), the SFF is calculated as the ratio of the average safe failure rate plus the average dangerous detection rate to the average total failure rate:

Calculation of Diagnosis Safe Failure Fraction (SFF)

$$\text{SFF} = \left(\frac{\lambda_S + \lambda_{DD}}{\lambda_S + \lambda_D}\right) = \left(\frac{\lambda_{total} - \lambda_{DU}}{\lambda_{total}}\right) \tag{9.65}$$

SFF and SIL can be correlated based on the table for different levels of hardware fault tolerance. In terms of hardware fault tolerance, it can either be $N = 0$, 1, or 2 and can either be type A or B. The type B model provides a higher level of safety for

Table 9.10 Correlation between SIL and SFF for hardware fault tolerance – type A (IEC).

	Hardware fault tolerance – type A		
Safe failure fraction (SFF)	**0**	**1**	**2**
SFF < 60%	SIL 1	SIL 2	SIL 3
60% < SFF < 90%	SIL 2	SIL 3	SIL 4
90% < SFF < 99%	SIL 3	SIL 4	SIL 4
SFF ≥ 99%	SIL 3	SIL 4	SIL 4

more complex diagrams. Type A can be used with a single block diagram and simple 1oo1 architecture, including all safety critical valves. In Table 9.10, the maximum allowable SIL for a safety function implemented by a type A safety is illustrated for the different hardware types used. The hardware fault tolerance of N indicates that the loss of safety functions could be caused by a fault of $N + 1$. Because architecture-type 1oo1 has been selected in this chapter, which means that failure of any component may result in the failure of the entire system, the hardware fault tolerance value has been set to 0 in order to achieve the highest level of safety and reliability possible.

Example 9.20 On an offshore platform, a 20″ hydraulically actuated ball valve is connected to an ESD. This valve is used for the transportation of the produced gas from offshore to the final user. This valve is composed of a carbon steel body, and it can handle a maximum pressure of 250 bar. There are three failure rates for the valve, hydraulic actuator, and control panel summarized in Table 9.11. The following parameters need to be calculated:

1) DC for valves, actuators, control panels, and the entire system.
2) Assuming that the proof test interval is 9760 hours, the mean repair time is 24 hours, and the time needed to detect the problem is zero, calculate the average time that the whole system is unable to run.

Table 9.11 The failure rate values for the 20″ valve, its actuator, and its control panel.

Component	λ_S	λ_D	λ_{DU}	λ_{DD}
20″ valve	0	1.02×10^{-7}	3.3×10^{-8}	6.9×10^{-8}
Actuator	0	4×10^{-8}	1.7×10^{-8}	2.3×10^{-8}
Control panel	1.3×10^{-9}	1.3×10^{-7}	1.3×10^{-7}	0

3) The PFD for the entire system.
4) Safe failure fracture (SFF) for the entire system.

A minimum SIL 3 is also required by the end user for this valve that is actuated. Does the actuated valve meet the expectation that is required?

Answer

In the first step, we should calculate the safe failure rate λ_S, dangerous failure rate λ_D, dangerous detected λ_{DD}, and undetected failure rate λ_{DU}, for the entire system.

$$\begin{aligned}\lambda_S(\text{Total}) &= \lambda_S(\text{Valve}) + \lambda_S(\text{Actuator}) + \lambda_S(\text{Control panel})\\ &= 0 + 0 + 1.3 \times 10^{-9} = 1.3 \times 10^{-9}\\ \lambda_D(\text{Total}) &= \lambda_D(\text{Valve}) + \lambda_D(\text{Actuator}) + \lambda_D(\text{Control panel})\\ &= 1.02 \times 10^{-7} + 4 \times 10^{-8} + 1.3 \times 10^{-7} = 2.72 \times 10^{-7}\\ \lambda_{DU}(\text{Total}) &= \lambda_{DU}(\text{Valve}) + \lambda_{DU}(\text{Actuator}) + \lambda_{DU}(\text{Control panel})\\ &= 3.3 \times 10^{-8} + 1.7 \times 10^{-8} + 1.3 \times 10^{-7} = 1.8 \times 10^{-7}\\ \lambda_{DD}(\text{Total}) &= \lambda_{DD}(\text{valve}) + \lambda_{DD}(\text{Actuator}) + \lambda_{DD}(\text{Control panel})\\ &= 6.9 \times 10^{-8} + 2.3 \times 10^{-8} + 0 = 9.2 \times 10^{-8}\end{aligned}$$

Now, it is possible to calculate diagnosis coverage for the valve, actuator, control panel, and the entire system separately by using Eq. (9.60) as follows:

$$\text{For valve} \rightarrow \text{DC} = \frac{\lambda_{DD}}{\lambda_D} = \frac{6.9 \times 10^{-8}}{1.02 \times 10^{-7}} = 67.65\%$$

$$\text{For actuator} \rightarrow \text{DC} = \frac{\lambda_{DD}}{\lambda_D} = \frac{2.3 \times 10^{-8}}{4 \times 10^{-8}} = 57.5\%$$

$$\text{For control panel} \rightarrow \text{DC} = \frac{\lambda_{DD}}{\lambda_D} = \frac{0}{1.3 \times 10^{-7}} = 0$$

$$\text{For entire system} \rightarrow \text{DC} = \frac{\lambda_{DD}}{\lambda_D} = \frac{9.2 \times 10^{-8}}{2.72 \times 10^{-7}} = 33.82\%$$

The next step would be to conduct a calculation to obtain the mean downtime of a system due to both dangerous detected and undetected failures t_{CE} for the entire system using Eq. (9.59).

T_1: Proof test interval (hour) is 9760 hours
MTR: Maintenance repair time = 24 hours
Mean repair time (MRT) = MTTR – time to detect the failure

Time to detect the failure is zero and thus, MTTR = MTR = 24 hours

$$t_{CE} = \frac{\lambda_{DU}}{\lambda_D}\left(\frac{T_1}{2} + \text{MRT}\right) + \frac{\lambda_{DD}}{\lambda_D}\text{MTTR}$$

$$t_{CE} = \frac{1.8 \times 10^{-7}}{2.72 \times 10^{-7}}\left(\frac{9760}{2} + 24\right) + \frac{9.2 \times 10^{-8}}{2.72 \times 10^{-7}} \times 24$$

$$= 3245 + 8.12 = 3253 \text{ hours}$$

Thus, the average time that the whole system is unable to run is 3253 hours.

Now, it is possible to calculate the PFD for the entire system based on Eq. (9.64) as follows:

$$PFD_G = \lambda_D \times t_{CE} = 2.72 \times 10^{-7} \times 3253 = 8.85 \times 10^{-4}$$

The next step is to calculate the SFF for the whole system using Eq. (9.65):

$$SFF = \left(\frac{\lambda_S + \lambda_{DD}}{\lambda_S + \lambda_D}\right) = \left(\frac{1.3 \times 10^{-9} + 9.2 \times 10^{-8}}{1.3 \times 10^{-9} + 2.72 \times 10^{-7}}\right) = 34.13\%$$

According to Table 9.10, the SFF is less than 60%, considering zero fault tolerance and type A. This means that the level of safety achieved by the system is not sufficient to meet the requirements of the project.

Example 9.21 By applying more restrictive tests and accurate sensors to monitor the failures on the 20″ valve considered in the previous example, the failure rate values are modified to the values provided in Table 9.12.

There are a number of parameters that need to be calculated:

1) DC for valves, actuators, control panels, and the entire system.
2) Based on the assumption that the proof test interval, the mean repair time, and the time to detect the problem remain unchanged, calculate the average period of time that the whole system cannot be operated.
3) For a given system, the PFD is calculated.
4) SFF for the entire system.

There is also a requirement from the end user for this valve that is actuated to have a minimum SIL 3. How well does this valve fulfil the requirements set by the end user?

Table 9.12 Modified failure rate values for the 20″ valve, its actuator, and its control panel.

Component	λ_S	λ_D	λ_{DU}	λ_{DD}
20″ valve	0	8.65×10^{-8}	2.57×10^{-8}	5.99×10^{-8}
Actuator	0	2.58×10^{-8}	2.3×10^{-9}	2.35×10^{-8}
Control panel	1.05×10^{-6}	4.95×10^{-7}	4.95×10^{-7}	0

Answer

In the first step same as the previous example, we should calculate the safe failure rate λ_S, dangerous failure rate λ_D, dangerous detected λ_{DD}, and undetected failure rate λ_{DU}, for the entire system.

$$\begin{aligned}\lambda_S(\text{Total}) &= \lambda_S(\text{Valve}) + \lambda_S(\text{Actuator}) + \lambda_S(\text{Control panel})\\ &= 0 + 0 + 1.05 \times 10^{-6} = 1.05 \times 10^{-6}\end{aligned}$$

$$\begin{aligned}\lambda_D(\text{Total}) &= \lambda_D(\text{Valve}) + \lambda_D(\text{Actuator}) + \lambda_D(\text{Control panel})\\ &= 8.55 \times 10^{-8} + 2.58 \times 10^{-8} + 4.95 \times 10^{-7} = 6.06 \times 10^{-7}\end{aligned}$$

$$\begin{aligned}\lambda_{DU}(\text{Total}) &= \lambda_{DU}(\text{Valve}) + \lambda_{DU}(\text{Actuator}) + \lambda_{DU}(\text{Control panel})\\ &= 2.57 \times 10^{-8} + 2.3 \times 10^{-9} + 4.95 \times 10^{-7} = 5.23 \times 10^{-7}\end{aligned}$$

$$\begin{aligned}\lambda_{DD}(\text{Total}) &= \lambda_{DD}(\text{Valve}) + \lambda_{DD}(\text{Actuator}) + \lambda_{DD}(\text{Control panel})\\ &= 5.99 \times 10^{-8} + 2.35 \times 10^{-8} + 0 = 8.34 \times 10^{-8}\end{aligned}$$

Now, it is possible to calculate diagnosis coverage for the valve, actuator, control panel, and the entire system separately by using Eq. (9.60) as follows:

$$\text{For valve} \rightarrow \text{DC} = \frac{\lambda_{DD}}{\lambda_D} = \frac{5.99 \times 10^{-8}}{8.65 \times 10^{-8}} = 69.25\%$$

$$\text{For actuator} \rightarrow \text{DC} = \frac{\lambda_{DD}}{\lambda_D} = \frac{2.35 \times 10^{-8}}{2.58 \times 10^{-8}} = 91.09\%$$

$$\text{For control panel} \rightarrow \text{DC} = \frac{\lambda_{DD}}{\lambda_D} = \frac{0}{4.95 \times 10^{-7}} = 0$$

$$\text{For entire system} \rightarrow \text{DC} = \frac{\lambda_{DD}}{\lambda_D} = \frac{8.34 \times 10^{-8}}{6.06 \times 10^{-7}} = 13.76\%$$

$$t_{CE} = \frac{\lambda_{DU}}{\lambda_D}\left(\frac{T_1}{2} + \text{MRT}\right) + \frac{\lambda_{DD}}{\lambda_D}\text{MTTR}$$

$$t_{CE} = \frac{5.23 \times 10^{-7}}{6.06 \times 10^{-7}}\left(\frac{9760}{2} + 24\right) + \frac{8.34 \times 10^{-8}}{6.06 \times 10^{-7}} \times 24 = 4235.30 \text{ hours}$$

Thus, the average time that the whole system is unable to run is 4235.30 hours. Now, it is possible to calculate PFD_G using Eq. (9.64) as follows:

$$\text{PFD}_G = \lambda_D \times t_{CE} = 6.06 \times 10^{-7} \times 4235.30 = 2.56 \times 10^{-3}$$

The next step is to calculate the SFF using Eq. (9.65):

$$\text{SFF} = \left(\frac{\lambda_S + \lambda_{DD}}{\lambda_S + \lambda_D}\right) = \left(\frac{1.05 \times 10^{-6} + 8.34 \times 10^{-8}}{1.05 \times 10^{-6} + 6.06 \times 10^{-7}}\right) = 68.44\%$$

Based on the results shown in Table 9.10, the SFF is between 60 and 90%. As a result, the system's safety level is SIL 2, which is not acceptable to the end user.

Hence, in order to improve the safety and reliability of the system, various techniques need to be used such as more stringent tests and/or using sensors as part of condition monitoring or ValveWatch.

9.6 Condition Monitoring (ValveWatch)

Probably, the first literature about the use of online condition monitoring or ValveWatch on the Norwegian continental shelf was written in 2002. As shown in Figure 9.16, a complete set of pressure sensors is installed on a safety-critical valve in order to illustrate the type of hardware included in the ValveWatch system. The following is a brief explanation of each of the sensors:

1) **Actuator pressure sensor:** It is a sensor attached to the pressurized part of the actuator and measures the air or hydraulic pressure applied to the actuators. In case of a reduction in actuator pressure compared to normal circumstances (e.g. 160–200 bar hydraulic oil pressure), there could be various reasons for this, such as external or internal leakage in the actuator, or perhaps a problem with the actuator control system.
2) **Strain gauge:** There is a sensor installed on top of the valve which measures the torque or force that is exerted between the valve and the actuator. In physics, torque is a measure of the force that can cause an object to rotate around its axis. A practical way to calculate the torque is to multiply the force by distance according to Eq. (9.66). It has been observed that when the operating force detected by this sensor is higher than normal, it is likely caused by problems with stem packing such as misalignment or overtightening of the stem packing

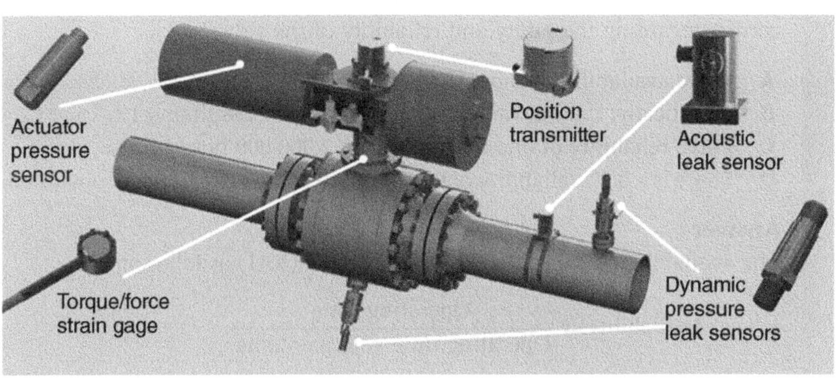

Figure 9.16 ValveWatch sensors by MRC Global. *Source:* MRC Global Inc.

which causes extra friction with the stem. In order to prevent leakage from the stem to the external environment, valve packings are used as stem sealings. The strain gauge is able to detect the bending of the stem, which causes packing wear and external leakage, to occur. If an operator applies extra torque on the stem higher than the maximum stem load capacity, also known as the MAST (Maximum Allowable Stem Torque), the operator may damage and bend the stem as well as wear it out. The bending and damage to the valve stem lead to an external leakage from the valve.

3) The valves use dynamic pressure sensors in order to detect leakage inside the valves. Figure 9.16 shows two dynamic pressure leak sensors, one located on the pipe and the other located in the valve cavity. Each of these sensors is sensitive to noise levels and can convert them into the value of pressure.

Calculation of the Valve Torque

$$\tau = F \times r \tag{9.66}$$

where:

τ: the magnitude of the torque to open or close the valve (Nm);
F: the force applied on the valve stem (N);
r: the distance from the stem to the center of closure member (m).

Questions and Answers

9.1 In an offshore oil rig, an industrial valve is installed on a piping system that carries crude oil. It is estimated that the MTTF is 2,000,000 hours and the mean time to restore (MTTR) is 10 hours. What is the sentence that is not correct regarding the safety and reliability of the valve?

A Valve availability is more than 99%.
B Statistically, the unavailability of the valve is less than 0.1%.
C Valve reliability of one during its first 2,000,000 hours of use.
D There is a 0.5% reliability of the valve after 2,000,000 hours of operation.

Answer

The availability is calculated according to Eq. (9.41) as follows:

$$\text{Availability} = \frac{\text{Operating time}}{\text{Operating time} + \text{Repair time}}$$
$$= \frac{2{,}000{,}000}{2{,}000{,}000 + 10} = 99.9995\%$$

Therefore, option A is the correct choice. The next step will be to calculate unavailability as follows:

$$\text{Unavailability} = 1 - \text{Availability} = 1 - 0.999995 = 0.000005 = 0.0005\%$$

Therefore, option B is correct as well. It should be noted that Option C is the correct answer because no failures occurred during the first 2,000,000 hours of the valve's operation. Option D is incorrect because we have no information regarding the operational reliability of the valve after 2,000,000 hours of use. So we cannot provide any reliability data for the valve.

9.2 There are 2000 industrial valves that have been working in an industrial plant for 10 years. During this time, only 20 of them have failed. In which of the following options would you find the correct failure rate and MTTF values for this group of valves?

A $\lambda = 1.14 \times 10^{-7}$, MTTF $= 0.88 \times 10^{7}$
B $\lambda = 1.14 \times 10^{-6}$, MTTF $= 0.88 \times 10^{7}$
C $\lambda = 1.14 \times 10^{-7}$, MTTF $= 0.88 \times 10^{6}$
D $\lambda = 1.14 \times 10^{-6}$, MTTF $= 0.88 \times 10^{6}$

Answer

The average failure rate for the group of valves is calculated as follows:

$$\lambda = \frac{\text{Number of failures}}{\text{Total time of the valve in operation}}$$

$$= \frac{n}{\tau} = \frac{20}{2000 \times 10 \times 365 \times 24} = 1.14 \times 10^{-7} \text{ failures/h}$$

$$\text{MTTF} = \frac{1}{\lambda} = \frac{1}{1.14 \times 10^{-7}} = 0.88 \times 10^{7}$$

Therefore, option A is the correct answer.

9.3 Which of the following statements concerning SIL formulas and calculations is correct?

A The probability of dangerous failure is equal to the sum of the detected dangerous failure rate and the undetected dangerous failure rate.
B The correlation between the SFF and SIL cannot be established.
C DC is the detected dangerous failure rate divided by the total rate of the failures.
D If there is a dangerous failure, the safety function goes into its safe state, such as closing a valve in an emergency situation.

Answer

The correct choice is option A. It would be wrong to choose option B because there could be a correlation between the SFF and the SIL. Option C is incorrect because DC is the ratio between the number of dangerous failures detected by the number of dangerous failures as a whole. It is incorrect to choose option D since safe failures put the emergency shutdown valve in a safe position.

9.4 Consider a pipeline that contains four identical valves and assume that each valve has a reliability of 0.93. In the event of one of the valves failing, the whole pipeline system will cease to function. In terms of safety, how reliable is the pipeline system with regard to these four identical valves?

A 81.2%
B 73.4%
C 90.5%
D 92.8%

Answer

Four identical valves are arranged in series in the system. Equation (9.27) is a formula we can use to determine the reliability of a series system or network that consists of identical units. The answer is:

$$R_S = 1 - m(1 - R) = 1 - 4 \times (1 - 0.93) = 0.72$$

A second method to calculate the reliability of the entire system would be to multiply all reliability numbers together as shown in the following calculation:

$$\begin{aligned} R_S &= \prod_{i=1}^{4} R_i = R_1 \times R_2 \times R_3 \times R_4 \\ &= (0.93) \times (0.93) \times (0.93) \times (0.93) \\ &= 0.7480 \end{aligned}$$

There is an average reliability of 73.4% calculated by using two methods. So option B is the correct answer.

9.5 In a piping system, two valves are installed; the first is located in the main line, and the second is located in the bypass line. The two identical valves work independently, and the probability of failure is 0.05 for each. Assuming that the system will function successfully if only one valve is functioning, what is the reliability of the entire system?

A 99.75%
B 98.85%

C 97.60%

D 95.50%

Answer

The system consists of two parallel valve units. As a first step, we have to determine the reliability of each valve by identifying the probability of failure for each as follows:

$$R\text{ (Reliability)} = 1 - F\text{ (Failure probability)} = 1 - 0.05 = 0.95$$

To determine the reliability of a parallel system consisting of identical units, we can use Eq. (9.37). The answer is:

$$R_P = 1 - (1 - R)^m = 1 - (1 - 0.95)^2 = 0.9975$$

Accordingly, the reliability of the entire system including two valves is equal to 0.9975 = 99.75%. Thus, option A is the correct answer.

9.6 There are seven independent valves installed in series in a piping system. A failure rate of 0.009 failures/h is expected from the entire piping system, which includes seven valves. As a result of past experience, the failure rates for the first to seventh valve have been determined to be 0.001 failures/h, 0.002 failures/h, 0.003 failures/h, 0.004 failures/h, 0.005 failures/h, 0.006 failures/h, and 0.007 failures/h, respectively. Which of the following statements is correct?

A The allocated failure rate to the first valve is 6.4×10^{-5} failures/h.

B The allocated failure rate to the seventh valve is 1.93×10^{-4} failures/h.

C The allocated failure rate to the second valve is 1.29×10^{-4} failures/h.

D The allocated failure rate to the third valve is 9.64×10^{-5} failures/h.

Answer

In order to begin, the first step would be to calculate the failure rate for the entire system based on past experience as per Eq. (9.28).

$$\lambda_S = \sum_{i=1}^{N} \lambda_i = 0.0001 + 0.0002 + 0.0003 + 0.0004 + 0.0005 + 0.0006$$
$$+ 0.0007 = 0.028 \text{ failures/h}$$

Using Eqs. (9.30) and (9.32), the relative weights and allocated reliability values for five subsystems are calculated as follows:

$$\theta_1 = \frac{0.0001}{0.028} = 0.00357 \rightarrow \lambda_1^* = \theta_1 \times \lambda_{rs}$$
$$= (0.00357)\,(0.009) = 3.2 \times 10^{-5} \text{ failures/h}$$

$$\theta_2 = \frac{0.0002}{0.028} = 0.00714 \rightarrow \lambda_2^* = \theta_2 \times \lambda_{rs}$$
$$= (0.00714)\,(0.009) = 6.4 \times 10^{-5}\ \text{failures/h}$$

$$\theta_3 = \frac{0.0003}{0.028} = 0.010714 \rightarrow \lambda_3^* = \theta_3 \times \lambda_{rs}$$
$$= (0.010714)\,(0.009) = 9.64 \times 10^{-5}\ \text{failures/h}$$

$$\theta_4 = \frac{0.0004}{0.028} = 0.0142857 \rightarrow \lambda_4^* = \theta_4 \times \lambda_{rs}$$
$$= (0.0142857)\,(0.009) = 1.29 \times 10^{-4}\ \text{failures/h}$$

$$\theta_5 = \frac{0.0005}{0.028} = 0.017857 \rightarrow \lambda_5^*$$
$$= \theta_5 \times \lambda_{rs} = (0.017857)\,(0.009)$$
$$= 1.61 \times 10^{-4}\ \text{failures/h}$$

$$\theta_6 = \frac{0.0006}{0.028} = 0.02143 \rightarrow \lambda_4^*$$
$$= \theta_4 \times \lambda_{rs} = (0.02143)\,(0.009)$$
$$= 1.93 \times 10^{-4}\ \text{failures/h}$$

$$\theta_7 = \frac{0.0007}{0.028} = 0.025 \rightarrow \lambda_5^*$$
$$= \theta_5 \times \lambda_{rs} = (0.025)\,(0.009)$$
$$= 2.25 \times 10^{-4}\ \text{failures/h}$$

Thus, option D is the correct answer.

9.7 The probability of detecting a total failure for an actuated valve connected to SIS by a traditional method of inspection is 35%. In order to detect defects in the actuated valves more reliably, three sensors are added to the valves and actuators as part of a process called condition monitoring or ValveWatch. As a result, these sensors, which are labelled A, B, and C, can identify 66.66%, 15.83%, and 20.83% of system defects, respectively. Can you identify which of the following statements could be the outcome of using the ValveWatch for the valve in this case?

A Reducing the SIL
B Increasing the DC
C Reducing the SFF
D Reducing the mean time to failure (MTTF)

Answer

ValveWatch results can be applied to failure detection more easily, thereby increasing the likelihood of dangerous detected failures λ_{DD} being detected, thereby increasing the DC value. As a result, the correct answer is option B.

Both options A and C are incorrect because increasing DC will result in an increase in reliability, SIL and SFF. Option D is not correct either because the MTTF value is independent of the improvement in failure detection.

9.8 There are values of 0.1 and 0.01, respectively, for safe failure rates and dangerous failure rates for an actuated valve connected to SIS. What are the values of the total mean time to failure as well as the total mean time between failures?

A $MTTF_{total} = MTBF_{total} = 100$ years
B $MTTF_{total} = MTBF_{total} = 10$ years
C $MTTF_{total} = MTBF_{total} = 9.9$ years
D $MTTF_{total} = MTBF_{total} = 9.09$ years

Answer

$$\lambda_S = 0.1/\text{year and } \lambda_D = 0.01/\text{year}$$

$$\rightarrow \lambda_{total} = \lambda_D + \lambda_S = 0.11/\text{year}$$

$$\rightarrow MTTF_{total} = MTBF_{total} = \frac{1}{\lambda_{total}} = \frac{1}{0.11} = 9.09\text{ years}$$

Thus, option D is the right choice.

9.9 The following values are provided for the safe and dangerous failure rates:

$$\lambda_{DD} = 0.14/\text{year}, \lambda_{DU} = 0.04/\text{year}, \lambda_{SD} = 0.22/\text{year}, \lambda_{SU} = 0.5/\text{year}$$

For the second condition, all these values remain unchanged, except for the failure rate of dangerous undetected value, which is changed to 0.4/year. What is the difference between the calculated safe failure fraction (SFF) in two conditions?

A 20.2%
B 27.3%
C 32.4%
D 38.6%

Answer

Equation (9.65) is used to calculate the SFF in both conditions as follows:

$$\text{First condition SFF} = \left(\frac{\lambda_S + \lambda_{DD}}{\lambda_S + \lambda_D}\right) = \left(\frac{\lambda_{SD} + \lambda_{SU} + \lambda_{DD}}{\lambda_{SD} + \lambda_{SU} + \lambda_{DD} + \lambda_{DU}}\right)$$

$$= \left(\frac{0.22 + 0.5 + 0.14}{0.22 + 0.5 + 0.14 + 0.04}\right) = \left(\frac{0.86}{0.90}\right) = 0.9555 = 95.55\%$$

$$\text{Second condition SFF} = \left(\frac{\lambda_S + \lambda_{DD}}{\lambda_S + \lambda_D}\right) = \left(\frac{\lambda_{SD} + \lambda_{SU} + \lambda_{DD}}{\lambda_{SD} + \lambda_{SU} + \lambda_{DD} + \lambda_{DU}}\right)$$

$$= \left(\frac{0.22 + 0.5 + 0.14}{0.22 + 0.5 + 0.14 + 0.4}\right) = \left(\frac{0.86}{1.26}\right) = 0.6825 = 68.25\%$$

The difference between SFF in first and second conditions = 95.55% − 68.25% = 27.3%

Therefore, option B is the correct answer.

9.10 The failure rate of a safety component is 200 FIT. What is the likelihood of this device failing in ten years?

A 1.43×10^{-2}
B 1.77×10^{-2}
C 1.98×10^{-2}
D 1.75×10^{-2}

Answer

The failure rate of 200 FIT means that 200 failures have happened over a billion hours.

$$\lambda = \frac{\text{Number of failures}}{\text{Total time of the valve in operation}} = \frac{200}{10^9} = 2 \times 10^{-7}\ \text{failures/h}$$

The duration of operation is ten years which is equal to $365 \times 24 \times 10 =$ 87, 600 hours.

Using Eq. (9.5) → Probability of failure $(P) = \lambda \times t = 2 \times 10^{-7} \times$ 87, 600 $= 1.75 \times 10^{-2}$

Thus, option D is the correct answer.

Further Reading

Basilio, A., Landrini, G., Capelle, T.V. et al. (2010). *Plant Engineering and Maintenance According to IEC 61508 and IEC 61511 Standards*, 3e. International Technology for Safety.

Colombo, D., GBA, L., Pereira, D.R. et al. (2020). Regression-based finite element machines for reliability modeling of downhole safety valves. *Reliability Engineering & System Safety* 198: 106894.

Gross, R.E. (2008). Reliability testing of pressure relief valves. *Proceeding Paper: ASME Pressure Vessel and Piping Conference*, San Diego, CA, USA, 57–166. Paper number: PVP2004-2610 (25–29 July 2004).

International Electrotechnical Commission (IEC) (2010). *IEC 61508, Functional Safety of Electrical/Electronic/Programmable Electronic Safety-Related Systems*, 2e. Geneva: International Electrotechnical Commission (IEC).

International Electrotechnical Commission (IEC) (2016). *IEC 61511, Functional Safety – Safety Instrumented System for the Process Industry Sector*, 2e. Geneva: International Electrotechnical Commission (IEC).

Jeeves, S. (2016). Valve monitoring across disciplines and operations. *MRC Global. Valve World Conference*, Dusseldorf. Presentation # PV16035 (29 November–1 December 2016).

Juvik, T., Carr, R., Hermansen, T., and Hale, S. (2002). Online valve monitoring system used on off-shore platforms in the North Sea. *21st International Conference on Offshore Mechanics and Artic Engineering*. Oslo: ASME. Paper No. OMAE2002-28403 (23–28 June 2002).

Kim, J.H., Lee, J.H., and Jo, Y. (2018). A study on monitoring method for accident prevention of industrial valves. *Proceeding Paper: International Science and Technology Conference*, Paris, France (18–20 July 2018).

Lemme, A. and Furseth, H.J. (2019). *Expansion of the Condition Monitoring Strategy for ESD and PSD Valves on Johan Sverdrup Phase 1 – A Case Study in Equinor ASA*. University of Stavanger, Faculty of Science and Technology.

Lundteigen, M.A. and Rausand, M. (2007). *The Effect of Partial Stroke Testing on the Reliability of Safety Valves*. Department of Production and Quality Engineering, NTNU, ESREL.

Molnes, E., Sundet, I., and Vatan, J. (1992). Reliability of Surface Controlled Surface Safety Valve. Phase 4. ETDEWEB. Technical Report. Report Number STF-75A91038.

MRC Global (2020). ValveWatch-sensors. https://www.mrcglobal.com/Global-Region/Products/ValveWatch/Sensors (accessed 13 April 2022).

Nesbitt, B. (2007). *Handbook of Valves and Actuators: Valves Manual International*, 1e. Oxford: Elsevier.

Norskolje & Gass (2018). *Application of IEC 61508 and IEC 61511 in the Norwegian Petroleum Industry (Recommended SIL Requirement)*. Norwegian Oil and Gas Association. Rev. 03.

Rausand, M. and Vatn, J. (1998). Reliability modeling of surface controlled subsurface safety valves. *Reliability Engineering & System Safety* 61 (1–2): 159–166.

Reza, M. (2019). System reliability and availability calculation. https://www.bmc.com/blogs/system-reliability-availability-calculations/#:~:text=Reliability%20is%20calculated%20as%20an,be%20used%20to%20calculate%20Reliability (accessed 15 April 2022).

Sangolt, J.K., Bjoroy, M., and Erstad, T. (2019). *Application of Industry 4.0 Concepts for Increasing Reliability of Valves*. Department of Mechanical and Marine Engineering, Western Norway University of Applied Sciences.

SINTEF (2006). *Reliability Prediction Method for Safety Instrumented System – The PDS Method Handbook*. SINTEF. ISBN: 82-14-03899-5.

Skousen, P.L. (2011). *Valve Handbook*, 3e. New York: McGraw-Hill.

Smit, P. and Zappe, R.W. (2004). *Valve Selection Handbook*, 5e. New York: Elsevier.

Smith, D.J. (2011). *Reliability, Maintainability and Risk: Practical Methods for Engineers*. Butterworth-Heinemann. ISBN: 978-0-08-096902-2.

Sotoodeh, K. (2019). Safety integrity level in valves. *Journal of Failure Analysis and Prevention* 19: 832–837. https://doi.org/10.1007/s11668-019-00666-2.

Sotoodeh, K. (2020). Development of a numeric method to validate the reliability improvement of safety critical valves during operation through online monitoring implementation. *Journal of Sensing and Imaging* https://doi.org/10.1007/s11220-020-00323-1.

Sotoodeh, K. (2021). *Safety and Reliability Improvement of Valves and Actuators for the Offshore Oil and Gas Industry Through Optimized Design*. University of Stavanger. Doctor Philosophiae (Dr. Philos) UiS no.573. Faculty of Science and Technology. ISBN: 978-82-7644-987-7.

Sotoodeh, K. (2021). *Subsea Valves and Actuators for the Oil and Gas Industry*, 1e. Austin, TX: Elsevier (Gulf Professional Publishing).

10

Valve Operation

10.1 Introduction

A valve must have a means of operation in order to function properly. It may be possible to divide all valves operating in the same way, except for pressure relief and check valves, into two groups; those which are controlled by hand and those which are controlled with actuators. There are valves that are manually controlled or manual valves that are operated by either a lever or a handwheel and a gearbox. As seen in Figure 10.1, a manually operated valve with a handwheel and a gearbox that is being driven by an individual (a human) is shown.

The calculations in this section are focused primarily on the torque and forces generated by the manual operator or actuator to operate the valve. Additionally, the generated loads from the operator or actuator are transferred through the valve stem to the internals of the valve in order to operate (open or close) the valve. Therefore, the valve internals, especially the stem, must be designed robustly in order to withstand the forces applied by the operator or actuator. There are also several valves which are used in cryogenic or extremely cold service environments like liquefied natural gas (LNG), and nitrogen. In fact, some of these valves are capable of operating at −196 °C. It is no surprise that LNG has since become a very popular source of energy as a clean and environmentally friendly fuel. So many investments have been made in LNG plants, and more investments will be made in the future. Cryogenic valves are designed with an extended bonnet and stem. The length of the extension should be sufficient to keep the stem packing from coming into contact with the cryogenic material passing through the valve. When the valve packing or stem sealing is exposed to a freezing temperature, there is a high probability that the packing will freeze and leak. By increasing the length of the stem, parameter L, the risk of stem buckling and misalignment for gate and globe valves will increase. Buckling in engineering is the deflection or deformation of a component such as a bar or a pipe under the pressure of a compression load.

Industrial Valves: Calculations for Design, Manufacturing, Operation, and Safety Decisions, First Edition. Karan Sotoodeh.

Figure 10.1 An operator who operates a manual valve by using a handwheel and gearbox. *Source:* Pichitstocker/Adobe Stock.

With a rotating motion, the stems of ball valves and butterfly valves are subject to torsional deflection. In this chapter, we explain the necessary calculations to design the valve stems to be able to withstand the loads from the valve operator or actuator while preventing buckling and torsional deflections.

10.2 Valve Torque

In the technical world, torque is defined as the amount of force that can be exerted on a valve operator to rotate the valve closure member to be opened or closed. One Newton-meter of torque has the same meaning as one Newton force applied perpendicularly to a 1-m-long moment arm, which is the result of one Newton force applied perpendicularly to the end of the moment arm. In accordance with the definition provided, the torque on the valve lever or handwheel is calculated according to Eqs. (10.1) and (10.2), respectively.

Calculation of the Torque for the Lever-Operated Valves

$$\text{Torque} = \text{Length of the lever} \times \text{Operator single hand force} \tag{10.1}$$

Example 10.1 A single-hand force is required to operate a lever-operated valve by a human operator, which is 100 N. In the case of a 70 Nm torque produced by the operator, what is the length of the lever?

Answer

$$\begin{aligned}\text{Torque} &= \text{Length of the lever} \times \text{Operator single hand force} \rightarrow 70\\ &= \text{Length of lever} \times 100 \rightarrow \text{Length of lever} = 0.7\,\text{m} = 70\,\text{cm}\end{aligned}$$

Calculation of the Torque for the Gear-Operated Valves

$$\text{Torque} = \text{Diameter of the handwheel} \times \text{Operator single hand force} \qquad (10.2)$$

Example 10.2 Figure 10.2 depicts the hands of a human operator using a handwheel to operate a manual valve. The handwheel is 300 mm in diameter, as illustrated in the figure. The force produced by the operator's hand when holding the valve is 200 N. Calculate the required torque for the valve to be operated.

Answer

$$\begin{aligned}\text{Torque} &= \text{Diameter of the handwheel} \times \text{Operator single hand force}\\ &= 0.3 \times 200 = 60\ \text{Nm}\end{aligned}$$

In physics, torque can be calculated using Eq. (10.3) as follows:

Calculation of Torque in Physics

$$\tau = F \times r \qquad (10.3)$$

where:

τ = Magnitude of torque (Nm);
F = Force applied on the bare stem valve (N);
r = Distance of the force from the rotational axis (m).

Figure 10.2 An operator who operates a manual valve equipped with a handwheel.

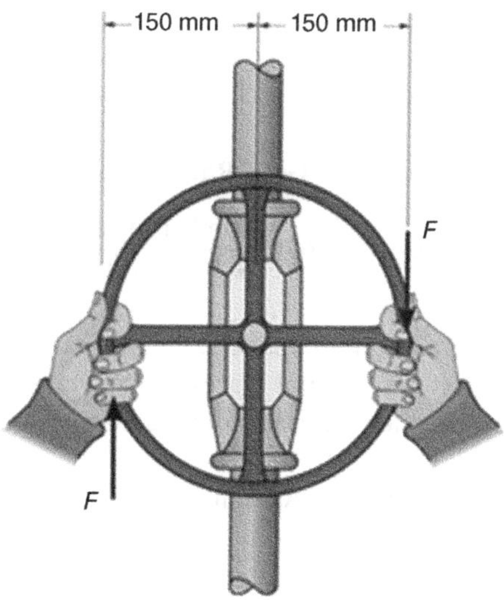

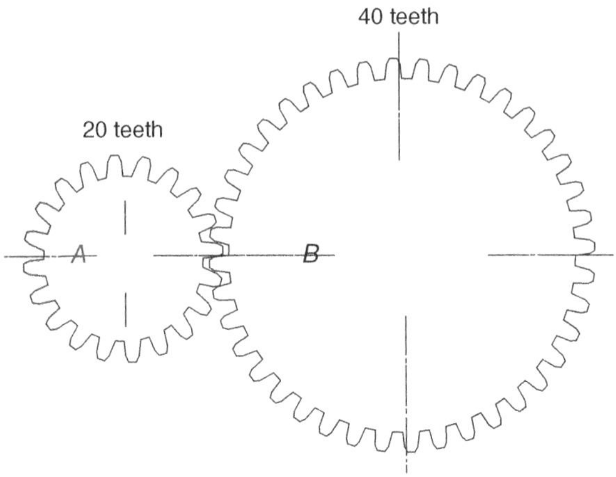

Figure 10.3 Gears in a gearbox for increasing input force.

In order to enhance the force and torque produced by the hand of the operator, the handwheel-operated valves can be fitted with a gearbox. Typically, gearboxes are used in conjunction with handwheels in order to facilitate valve operation. It can be seen as a simple and cheap method of valve operation in which gearing is used to increase the force and efficiency produced by the operator by moving the handwheel. There are gears in the gearbox, which are wheels with teeth that slot together. Assume that there is a gearbox made up of two gears, as illustrated in Figure 10.3; the first one on the ride, the "driving" gear, is smaller and has 20 teeth. With 40 teeth, the second gear is larger, referred to as the "driven" gear. Considering the number of teeth on the driving gear in this case, the gearbox, in this case, increases the input force applied by the operator through the handwheel. Depending on the gear ratio, which is calculated by Eq. (10.4), the amount of increase will vary.

Calculation of Gear Ratio

$$\text{Gear ratio} = \frac{\text{Number of teeth on driven gear}}{\text{Number of teeth on driver gear}} \tag{10.4}$$

As a result, in this case, the gear ratio is equal to two. In the example given earlier, if the operator places 300 Newtons (N) of force on the handwheel with two hands, then the gearbox in the example will increase the force to 600 Newtons (N).

Example 10.3 Assuming that a valve with a gear ratio of three is used in the previous example, how much torque is produced by the operator's one hand to operate the valve using the gear ratio of three?

Answer

The hand of the operator will have the ability to increase the produced torque and force three times by using a gearbox with a gear ratio of three. This is equivalent to an increase in torque from 60 to 180 Nm when using a gearbox with a ratio of three.

The force applied to the valve should overcome the unbalanced forces resulting from the pressure drop; impart sufficient force to the seat to prevent leakage; and overcome the packing friction force on a bare stem valve using Eq. (10.5) as follows:

Calculation of the Total Force Acting on the Valve Stem

$$F_{\text{total}} = F_{\text{pressure drop}} + F_{\text{seat}} + F_{\text{packing}} \tag{10.5}$$

As a result of the pressure drop across the valve, an unbalanced force is generated, which is calculated as follows using Eq. (10.6):

Calculation of the Required Force for the Valve Operation Due to the Pressure Drop

$$F_{\text{pressure drop}}\,(N) = \text{Seat area}\,(\text{m}^2) \times \text{Pressure drop}\,(\text{N/m}^2) \tag{10.6}$$

The seat of a ball valve is placed around the ball and its area is determined by Eq. (10.7).

Calculation of the Ball Valve Seat Area

$$\text{Seat}_{\text{area}} = \frac{\pi}{4}\left(\text{Seat external bore}^2 - \text{Ball or valve internal bore}^2\right) \tag{10.7}$$

Example 10.4 In an oil service, a 10″ ball valve with a pressure class of 300 (50 bar pressure nominal, refer to ASME B16.34 standard) is used. There is a 258 mm external bore on the seat, and a 252 mm minimum valve bore. There is a 0.5″ diameter stem on the valve, which is double reinforced with packing. There are various packing friction values with a stem which are based on the ASME pressure class and stem diameter which are listed in Table 10.1. Find the force required for this valve to operate. What would be the required torque to operate the valve if its operation is controlled by a handwheel with a diameter of 0.4 m?

Table 10.1 Packing friction forces.

Stem diameter (mm)	ASME pressure class	Packing type		
		Single PTFE	Double PTFE	Reinforced PTFE
12.5	150	170	250	284
	300			428
	600			563
	900			720
	1500			855

Answer

In the case of a ball valve, the seating force is provided by the fluid and the spring force behind the seat, so $F_{seat} = 0$ in this case. As a next step, the area of the ball valve seat needs to be calculated according to Eq. (10.7) as follows:

$$\text{Seat}_{\text{area}} = \frac{\pi}{4}\left(\text{Seat external bore}^2 - \text{Ball or valve internal bore}^2\right)$$
$$= \frac{\pi}{4}\left(258^2 - 252^2\right) = \frac{\pi}{4} \times 3060 = 2403.25\ \text{mm}^2 = 0.0024032\ \text{m}^2$$

The maximum pressure drop across a valve can be assumed equal to the maximum class pressure of the valve, which can be equated to the nominal pressure of 50 bar in this case. Each bar is equal to 100,000 (N/m^2). Thus, the pressure drop across the valve is 5,000,000 (N/m^2). Now it is possible to calculate the force of the pressure drop using Eq. (10.6) as follows:

$$F_{\text{pressure drop}}\ (\text{N}) = \text{Seat area}\left(\text{m}^2\right) \times \text{Pressure drop}\left(\text{N/m}^2\right)$$
$$= 0.0024032 \times 5{,}000{,}000 = 12{,}016\ \text{N}$$

The next step is to obtain the required force to overcome the packing friction from Table 10.1. $F_{packing} = 428$ and now it is possible to calculate the total force on the stem valve using Eq. (10.5) as follows:

$$F_{\text{total}} = 12{,}016 + 428 = 12{,}444\ \text{N}$$

Given that the valve is operated manually by a handwheel, it is possible to calculate the torque required for operation using Eq. (10.2) as follows:

$$\text{Torque} = \text{Diameter of the handwheel} \times \text{Operator single hand force}$$
$$= 0.4 \times 12{,}444 = 4977.6\ \text{Nm}$$

10.3 Stem Design

10.3.1 MAST Calculations

A valve manufacturer/supplier defines MAST as the maximum torque/thrust that can be applied to a valve train without causing damage. Engineering units for MAST are Nm or lb-in. The definition of valve drive train according to API 6D is "All parts of a valve drive between the operator and obturator (closure member, such as a disc or a ball), including the obturator but excluding the operator." In order to ensure torque delivered by the actuator is always less than the valve MAST during the engineering stage, it is the valve manufacturer's responsibility to calculate and provide valve MAST during the engineering stage. Previously, it was explained that the torque applied to the valve train comes from the valve operator, which can either be manual (e.g. a hand wheel with a gearbox) or actuated. The drive train in a ball valve includes a solid round stem, a stem key, and a ball, as well as a ball and stem joint. In order to make sure that the actuator's maximum output torque/thrust does not exceed the valve MAST at any point of travel, it is crucial to keep in mind that the actuator's maximum output torque/thrust is based upon the maximum supply voltage or operating pressure. In the case of manual valves, there is generally not enough torque generated by the gearbox to exceed MAST for the valve itself. In addition, the other reason for the high torque values for actuators is to have a safety factor (e.g. 2), calculated using Eq. (10.8), when designing and sizing the actuators.

Calculation of the Actuator Safety Factor

$$\text{Safety factor} = \frac{\text{Actuator torque value}}{\text{Valve torque value}} \tag{10.8}$$

Example 10.5 A 30-in. Class 1500-actuated top entry ball valve is installed on an oil export pipeline constructed of low-temperature carbon steel grade A352 LCC. The top entry design reduces the possibility of leakage from the valve because it is welded to the pipeline from both sides. This valve also includes an emergency shutoff function with a hydraulic actuator for quick action. High-pressure oil supplied to the actuator (e.g. 180 barg) along with high torque and a large spring in the actuator enable fast opening and closing of the valve. The safety factor of the actuator is two, and the torque values required from the valve manufacturer to open and close the valve are shown in Table 10.2. The meaning of each torque abbreviation is provided later. What is the minimum torque that the actuator should be able to generate?

Table 10.2 30″ class 1500 ball valve torque values.

Valve to open torques (Nm)			Valve to close torques (Nm)		
30″ class 1500 top entry ball valve, delta pressure = 250 barg					
BTO	Running	ETO	BTC	Running	ETC
110,016	16,215	36,852	110,016	16,215	36,852

There are six torque values associated with valves including ball valves:

1) **Break to open (BTO)**: The torque is measured when the valve is closed and the ball is in contact with just one seat under pressure. The torque, also known as breakaway torque, is the strongest torque in this case.
2) **Running torque (RT)**: It refers to the torque of the valve when the ball opens at approximately 35°–45°.
3) **End to open (ETO)**: The torque is applied when the ball is in an 80° position and closed to fully open the valve.
4) **Break to close (BTC)**: In the fully open position, the torque is required to break the open position of the valve and close it.
5) **End to close (ETC)**: When the valve is about to close, the torque is required to fully close the valve.

Answer

Considering the safety factor of two, the selected actuator must provide at least twice the torque of the valve at each position. The following formula determines the minimum actuator torque at different positions:

$$\text{Minimum actuator BTO torque} = 2 \times \text{Valve BTO} = 2 \times 110{,}016 = 220{,}032\,\text{Nm}$$
$$\text{Minimum actuator running torque} = 2 \times \text{Valve running} = 2 \times 16{,}215 = 32{,}430\,\text{Nm}$$
$$\text{Minimum actuator ETO torque} = 2 \times \text{Valve ETO} = 2 \times 36{,}852 = 73{,}704\,\text{Nm}$$
$$\text{Minimum actuator BTC torque} = 2 \times \text{Valve BTC} = 2 \times 110{,}016 = 220{,}032\,\text{Nm}$$
$$\text{Minimum actuator ETC torque} = 2 \times \text{Valve ETC} = 2 \times 36{,}852 = 73{,}704\,\text{Nm}$$

As such, the minimum torque an actuator must produce is 32,430 Nm, which is the running torque.

As illustrated in Figure 10.4, which includes four sections outlined later, the most complex shape of the stem is that of a ball valve.

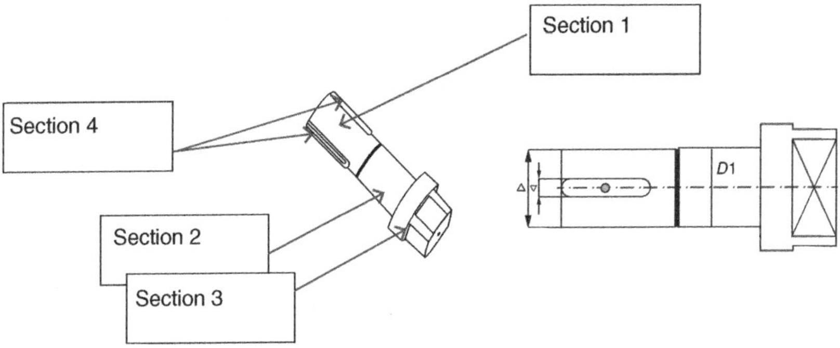

Figure 10.4 The stem of a ball valve consists of four sections.

1) The top section contains two keys
2) Circular part in the middle
3) The rectangular section at the bottom
4) Stem keys

It is important to note that only one stem key can be used on each valve stem in some cases as illustrated in Figure 10.5. In order to calculate the MAST for the stem section with one or two stem keys, we utilize Eq. (10.9).

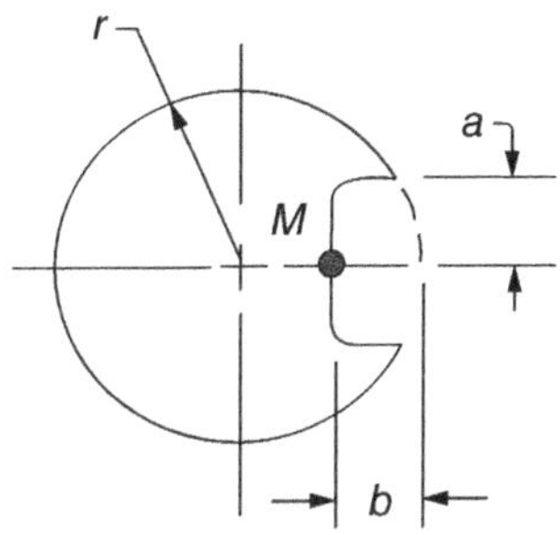

Figure 10.5 The portion of the stem containing one key.

Calculation of MAST for the Stem Section with One or Two Stem Keys

$$\text{MAST in a stem sectionwith one or two keys} = \text{MC1} = \frac{(0.536 \times Y_S \times r^3)}{B \times 1000} \tag{10.9}$$

where:

MAST is Nm;

Y_S = Yield strength $\left(\text{N/mm}^2\right)$

$$r = \frac{\text{Reduction of diameter at the top of the stem where the keys are located on the actuator side}}{2} \text{ (mm);}$$

B is calculated based on Roark's equation in accordance with Eq. (10.10) as follows:

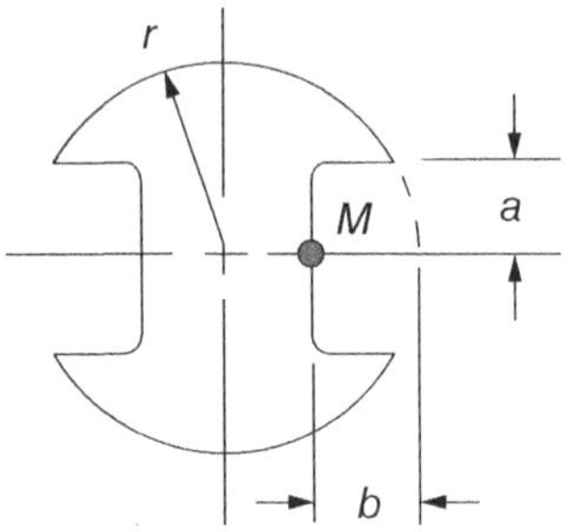

Figure 10.6 A section of the stem that contains two keys.

Calculation of Parameter *B* Based on Roark's Equation for the Stem Section with a Single Key

$$B = K_1 + K_2\frac{b}{r} + K_3\left(\frac{b}{r}\right)^2 + K_4\left(\frac{b}{r}\right)^3 \tag{10.10}$$

where for $0.5 \leq (a/b) \leq 1.5$

$$K_1 = 1.1690 - 0.3168\frac{a}{b} + 0.0490\left(\frac{a}{b}\right)^2$$

$$K_2 = 0.43490 - 1.5096\frac{a}{b} + 0.8687\left(\frac{a}{b}\right)^2$$

$$K_3 = -1.1830 + 4.2764\frac{a}{b} - 1.7024\left(\frac{a}{b}\right)^2$$

$$K_4 = +0.8812 - 0.2627\frac{a}{b} - 0.1897\left(\frac{a}{b}\right)^2$$

In the case of two stem keys on the top of the stem (see Figure 10.6), the MAST calculations are made in accordance with Eq. (10.9), but parameter B is calculated differently in accordance with Eq. (10.11).

Calculation of Parameter *B* Using Roark's Equation for the Stem Section with Two Keys

$$B = K_1 + K_2\frac{b}{r} + K_3\left(\frac{b}{r}\right)^2 + K_4\left(\frac{b}{r}\right)^3 \tag{10.11}$$

where for $0.5 \leq (a/b) \leq 1$

$$K_1 = 1.2512 - 0.5406\frac{a}{b} + 0.0387\left(\frac{a}{b}\right)^2$$

$$K_2 = -0.9385 + 2.3450\frac{a}{b} + 0.3256\left(\frac{a}{b}\right)^2$$

$$K_3 = 7.2650 - 15.338\frac{a}{b} + 3.1138\left(\frac{a}{b}\right)^2$$

$$K_4 = -11.152 + 33.710\frac{a}{b} - 10.007\left(\frac{a}{b}\right)^2$$

According to Eq. (10.12), the MAST in the circular section is calculated as follows:

Calculation of MAST for the Rounded Section of the Stem

$$\text{MAST in the rounded area of the stem} = \text{MC2} = \frac{\left(0.536 \times Y_S \times \pi \times d^3\right)}{16{,}000} \tag{10.12}$$

where:

MAST is Nm;
Y_S = Yield strength (N/mm^2);
d is the diameter of the stem at the rounded end.

Example 10.6 The stem of a 16″ ball valve is made of a super duplex material that has a yield strength of 80 Ksi. If the maximum torque of the valve is 2711 Nm, what is the diameter of the stem at the rounded end?

Answer
The yield strength of the stem in super duplex material is 80 Ksi, which is equal to 5.516×10^8 N/m^2

The stem of the valve is calculated as follows using Eq. (10.12):

$$\text{MAST in the rounded area of the stem} = \frac{\left(0.536 \times Y_S \times \pi \times d^3\right)}{16{,}000} \rightarrow 2711$$

$$= \frac{\left(0.536 \times 5.516 \times 10^8 \times \pi \times d^3\right)}{16{,}000} \rightarrow 0.43 \times 10^8$$

$$= 9.29 \times 10^8 \times d^3 \rightarrow d^3 = 0.0463\ \text{m}^3 \rightarrow d$$

$$= 0.359\ \text{m} = 359\ \text{mm}$$

Thus, the stem diameter in this case to withstand 2711 N.m is equal to 359 mm.

Equation (10.13) can be used to calculate the MAST at the rectangular section of the stem illustrated in Figure 10.7.

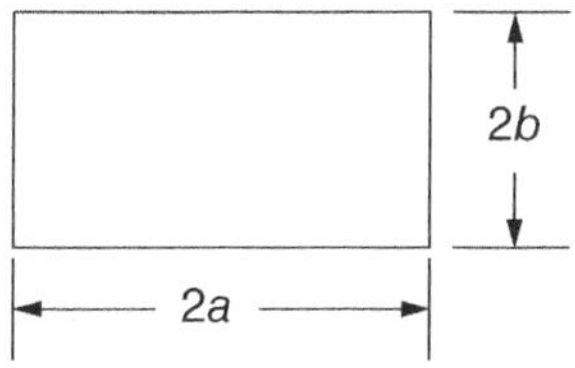

Figure 10.7 Section of the stem that is rectangular.

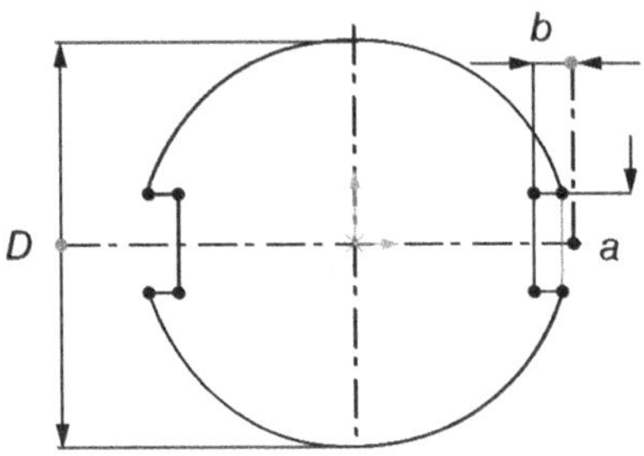

Figure 10.8 Part of the stem with keys.

Calculation of the Rectangular Section of the Stem Using MAST

$$\text{MAST in the rectangular sectionof the stem} = \text{MC3}$$

$$= \frac{\left[(0.536 \times Y_S) \times (2a \times 4b^2) / \left(3 + \frac{3.6b}{2a}\right)\right]}{1000}\ \text{Nm} \tag{10.13}$$

The next part of the stem for MAST is the part where the stem keys are installed, as illustrated in Figure 10.8. In accordance with Eq. (10.14), the calculation of MAST at the stem is performed with stem key(s) such as the one illustrated in the figure here.

Calculation of MAST at the Stem with Stem Key(s)

$$\text{MC4} = \left[\frac{(0.6 \times S_m \times D \times L \times a)}{2000}\right] \times n = \left[\frac{(0.6 \times 0.67 \times Y_S \times D \times L \times \text{a})}{2000}\right] \times n$$

$$= \left[\frac{(0.402 \times Y_S \times D \times L \times a)}{2000}\right] \times n \tag{10.14}$$

where:

a: width of key (mm);
L: length of key (mm);
n: number of keys;
D: stem diameter (mm);
b: key groove (mm).

The final valve MAST is calculated using Eq. (10.15) as follows:

Calculation of the Final MAST

$$\text{Valve final MAST} = \text{Min}(\text{MC1}, \text{MC2}, \text{MC3}, \text{MC4}) \tag{10.15}$$

Torque is typically greater as a function of force for valves with a larger size and a higher-pressure rating. Larger valves with a higher-pressure class usually possess bigger internals. So a higher force and consequently more torque is needed to operate the internals of the valve. By increasing the diameter of the valve stem or using a stronger material, it is possible to increase the torque that the valve stem can withstand.

10.3.2 Buckling Prevention

A cryogenic valve has an extended stem, meaning that its length is increased. The length of a stem, parameter L, increases the risk of stem buckling and misalignment. When applied to engineering, buckling refers to the deflection or deformation of a component, such as a bar or a pipe, that is subjected to a compression load. In Figure 10.9, a column is illustrated buckling under an axial compressive load (P). The same principle applies to the valve stem. The ratio of stem length to stem diameter, parameter L/D, has a direct impact on buckling: the higher the L/D ratio, the higher the chance of buckling. Thus, the stem extension of cryogenic valves increases the risk of buckling. So some measures should be taken as per the relevant cryogenic valve standards.

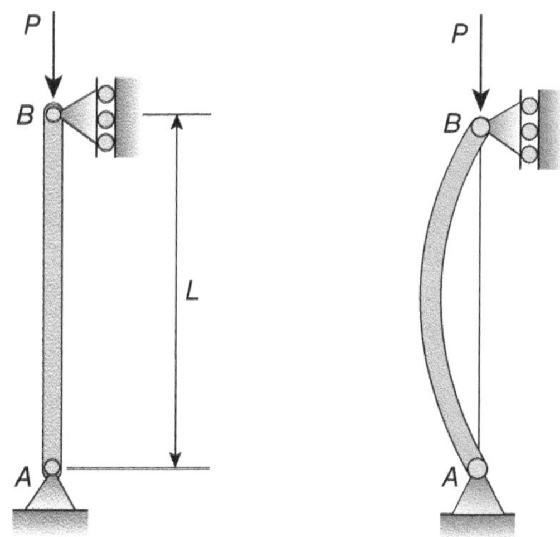

Figure 10.9 Buckling of a column under an axial compression load P.

There is a particular emphasis on stem design to prevent buckling for gate and globe valves with a linear stem motion in the MSS SP134 standard. In ball and butterfly valves, a rotary motion exposes the stem to torsional deflection, as explained further in this chapter. The MSS SP134 standard recommends that the ratio of stem length to diameter be limited for linear stem motion valves in order to prevent buckling. MSS SP 134 presents two equations for buckling prevention, depending on the disk/gate design: the first type of disk/gate is called a *stem-guided disk/gate*, and the second type is called a *body-guided disk/gate*. The stem-guided disk/gate and globe valves are those in which the movement of the disk/gate is guided and controlled by the stem.

Based on MSS SP134, gate and globe valves, which have a stem-disk connection through the stem nut, should have a limited stem length to diameter ratio when in the closed position as described by Eq. (10.16).

Ratio of Stem Length to Diameter in a Closed Position for Stem-Guided Disk Globe and Gate Valves

$$\frac{L}{d} \leq \frac{\pi}{2}\sqrt{\frac{E}{2S_{\text{pl}}N}} \tag{10.16}$$

where:

L: Unsupported length of the stem between the upper stem guide and the interface between the stem and the disk (m./in.);

d: Stem diameter (m/in.);

E: *Modulus of elasticity* of stem material (pound per square inch (psi)/Pa);

S_{pl}: *Material proportional limit* of the material, which is generally less than the material *yield strength* (psi/Pa). It is advisable to use the material yield strength instead of the material proportional limit in case of a worst-case scenario. On a stress-strain curve, Figure 10.10 illustrates the proportional limit for a material;

$N = 2$.

An alternative design, called a body-guided disk/gate, is more robust than the stem-guided disk/gate. It is an advantage of a body-guided disk design that the movement of the disk takes place entirely within the body enclosure and is guided by the valve body.

Example 10.7 According to the API 600 standard for gate valves, a 12-in. CL600 wedge-gate valve has a stem diameter of 2 in. or 50.80 mm. Due to the application of this valve in cryogenic environments, the bonnet and stem should be extended. The length of the stem without support is 45 in. or 1150 mm in the closed position,

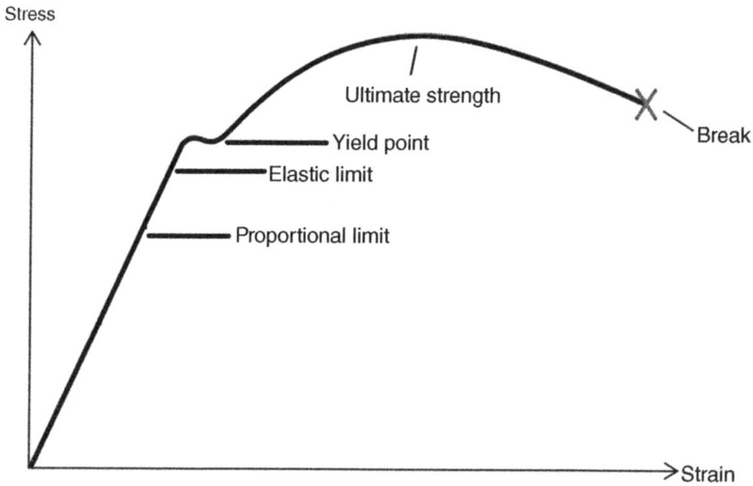

Figure 10.10 Proportional limit region on a stress-strain curve.

according to the MSS SP 134 standard. This valve is designed to operate at a minimum operating temperature of −129 °C or −200 °F. The stem of the valve contains stainless steel grade 304 austenitic stainless steel with a modulus of elasticity of 29.5×10^6 psi at the minimum operating temperature and a modulus of elasticity of 28×10^6 psi at ambient temperature, which is the maximum operating temperature. There is no change in the yield strength of SS304 at the different temperatures given earlier and it is equal to 30,000 psi. The gate of the valve has a stem guide. Is it possible that the stem of the valve can withstand a buckling axial force when the valve is in the closed position?

Answer

Since the gate valve is a stem-guided gate, the stem length-to-diameter conditions provided in Eq. (10.16) should be met.

$$\frac{L}{d} \leq \frac{\pi}{2}\sqrt{\frac{E}{2S_{\mathrm{pl}}N}}$$

$$\frac{L}{d} = \frac{45\text{ in.}}{2\text{ in.}} = 22.5\frac{\pi}{2}\sqrt{\frac{E}{2S_{\mathrm{pl}}N}}$$

$$= \frac{3.1415}{2}\sqrt{\frac{29.5 \times 10^6}{2 \times 3000 \times 2}} = 1.57\sqrt{245.83}$$

$$= 24.61 \text{ at minimum operating temperature}$$

Therefore, $(L/d) \leq (\pi/2)\sqrt{E/2S_{pl}N}$ at minimum operating temperature, since 22.5 < 24.61. So there is no risk of stem buckling for the wedge-gate valve at minimum operating temperature.

Now the equation should be checked for the maximum (normal) operating condition as follows:

$\pi/2\sqrt{E/(2S_{pl}N)} = 3.1415/2\sqrt{(28 \times 10^6)/(2 \times 3000 \times 2)} = 1.57 \times \sqrt{233.33} =$ 23.98 at a maximum operating temperature equal to room temperature.

Therefore, $L/d \leq \pi/2\sqrt{E/(2S_{pl}N)}$ at a minimum operating temperature since 22.5 < 23.98. So there is no risk of stem buckling for the wedge-gate valve at minimum operating temperature.

Due to the fact that the ratio of the stem length to a diameter at maximum and minimum operating temperatures is within the limits prescribed, the stem is not at risk of buckling during operation.

The *modulus of elasticity*, or elastic modulus, is a quantity that measures the resistance of an object to *elastic deformation*. According to Eq. (10.17), the modulus of elasticity is calculated as the ratio of stress to *strain*. In physics, strain is the amount of deformation or displacement of an object compared to its reference or initial length (refer to Eq. (10.18)). In addition to the effect of stress and strain on the modulus of elasticity, temperature also has an influence on the modulus of elasticity. As a matter of fact, the modulus of elasticity is reduced by increasing the temperature. In the ASME B31.3 process piping code, the modulus of elasticity values can be found in a table C-6, which contains the values for different materials at various temperature ranges. Steels such as 304 and 316, which are austenitic stainless steels, are commonly used for cryogenic valve parts, including the stem. According to ASME B31.3 table C-6, the modulus of elasticity of austenitic stainless steels is 28.3×10^6 psi at ambient temperature. There is a type of elastic deformation that results from stress, which is a nonpermanent deformation, which means that once the stress is released, the stressed object returns to its original form.

Modulus of Elasticity Calculation

$$E = \frac{\delta}{\varepsilon} \tag{10.17}$$

where:

E: Modulus of elasticity (psi/Pa);
δ: Stress (psi/Pa);
ε: Strain (dimensionless).

Strain Calculation

$$\varepsilon = \frac{\Delta L}{L} = \alpha \times \Delta T \tag{10.18}$$

where:

ε: Strain (dimensionless);
ΔL: Change in length of pipe (m/in./ft);
L: Initial length of pipe (m/in./ft).

Example 10.8 Under tensile stress, a steel bar has a cross-sectional area of 100 mm^{2}and a length of 50 mm. Calculate the value of the modulus of elasticity when the ratio of force to change in length for the bar is 410×10^3 N/mm.

Answer
The following parameters are given in this example:
As a result of the combination of Eqs. (10.17) and (10.18), the modulus of elasticity is given as follows:

$$E = \frac{\delta}{\varepsilon} = \frac{F/A}{\Delta L/L} = \frac{F}{\Delta L} \times \frac{L}{A} = 410 \times 10^3 \times \frac{50}{100} = 205 \times 10^3 \text{ N/mm}^2$$

Based on MSS SP134, gate and globe valves with a body-guided disk/gate connection, in the closed position, should be designed so that the stem length to the diameter ratio is determined by Eq. (10.19).

Ratio of Stem Length to Diameter in the Closed Position for Body-Guided Disk/Gate Globe and Gate Valves

$$\frac{L}{d} \leq \frac{\pi}{2}\sqrt{\frac{E}{S_{\text{pl}}N}} \tag{10.19}$$

where:

L: Unsupported length of the stem between the upper stem guide and the interface between the stem and the disk (m/in.).
d: Stem diameter (m/in.).
E: *Modulus of elasticity* of stem material (pound per square inch (psi)/Pa).
S_{pl}: *Material proportional limit* of the material, which is generally less than the *material yield strength* (psi/Pa). It is recommended that the material yield strength should be used in place of the material proportional limit in the

worst-case scenario. As shown in Figure 10.10, the proportional limits for a material are illustrated by a stress-strain curve.

$N = 2$.

In fact, a body-guided disk/gate connection is more robust than a stem-guided disk/gate connection. So a lower risk of buckling is expected for the stem with a more robust design.

Example 10.9 The size, pressure class, and minimum and maximum operating temperatures of the two gate valves are the same. In addition, the stems for these valves are of the same material and diameter. In this case, there are three assumptions: The first is that one valve has a body-guided disk and the other has a stem-guided disk. The second point is that the ratio of length to diameter should be as high as possible without buckling. According to the third assumption, the unsupported length of the stem of the body-guided disk is 500 mm. What would be the stem's unsupported length for the other design?

Answer

Because the material of the stem is identical in both valves, the values of modulus of elasticity and material proportional limit will also be identical. Given that both valve stems have the same diameter, the only difference between the stem of the first and second valves is their length. The stem length to diameter ratio should be the maximum without buckling, meaning that:

$$\frac{L}{d} = \frac{\pi}{2}\sqrt{\frac{E}{2S_{\text{pl}}N}}\text{(Applicable to the stem-guided disk.)}$$

$$\frac{500}{d} = \frac{\pi}{2}\sqrt{\frac{E}{S_{\text{pl}}N}}\text{(Applicable to the body-guided disk.)}$$

The following result is obtained by dividing both sides of the above equations:

$$\frac{L}{500} = \frac{1}{\sqrt{2}} \rightarrow L = 353.56\ \text{mm}$$

Consequently, the maximum unsupported length of the stem for stem-guided disks is 353.56 mm.

10.3.3 Torsional Deflection Prevention

Since *quarter-turn* valves do not experience axial (linear) loads, their stems are not at risk of buckling. This is because the stems of quarter-turn valves, i.e. ball,

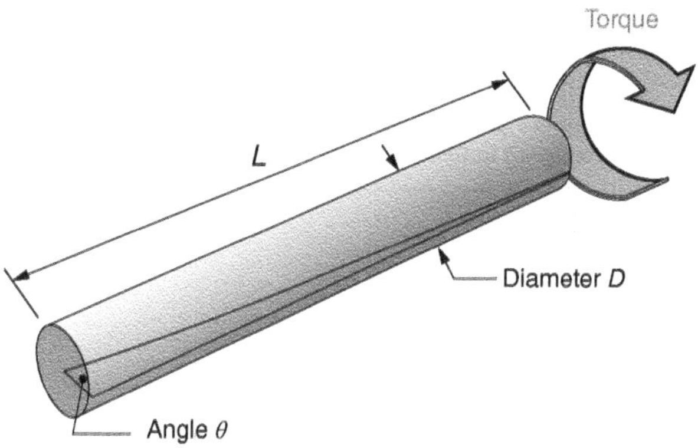

Figure 10.11 Twisting a piece of the valve stem.

butterfly, and plug types are not at risk of buckling. The length of the stem to the diameter of the valve must be as close to one as possible in order to prevent excessive *torsional deflection* of the stem. The term torsional deflection refers to the angular displacement that occurs when a twisting force is applied to a component. The *torsion* of a component is defined as the twisting of its one end, in this case, the stem of a quarter-turn valve, as shown in Figure 10.11. Torque is a measure of the force used to twist an object, in this case, the stem of a valve. A torque is applied to one end of the stem, which is called the rotating end. The *angle of twist*, parameter θ, of the shaft or stem of a valve that is under a torsional load is defined as the angle formed between the fixed and rotating ends as a result of torque.

According to an MSS SP 134 standard requirement, the stem length and diameter combination of a quarter-turn valve should limit the stem torsional deflection or angle of twist to $\pi/90$ or 0.0349 radian or 2°. The angle of the twist is determined by Eq. (10.20).

Angle of Twist

$$\theta = \frac{TL}{GJ} \leq \frac{\pi}{90} \tag{10.20}$$

where:

θ: Angle of twist (radian);
T: Maximum stem design torque (Nm);
L: Length of the stem from the point where torque is applied to the connection to the closure member (m);

Table 10.3 The values of the modulus of rigidity or modulus of shear for some important materials.

Material	Modulus of rigidity or shear modulus (parameter G) in GPa (10^9 Pa)
Carbon steel	77
Inconel 625	79
Monel	66
Stainless steel	77
Titanium	41

G: Modulus of rigidity or shear modulus (Pa or N/m^2). In general terms, the modulus of rigidity is defined as the ratio of *shear stress* to the displacement of the material or shear strain. Shear stress is the result of opposing forces applied by two separate objects to a cross-section that is on the same plane as the cross-section of the material. Shear forces tend to cause material deformation through slippage along the plane parallel to the imposed stresses. The values of the modulus of rigidity for some essential materials are provided in Table 10.3;
J: Polar moment of inertia of stem indicating the torsion deformation resistance of the shaft m^4.

The polar moment of inertia for a circular object such as a valve stem or shaft depends on the stem diameter and is calculated using Eq. (10.21):

Polar Moment of Inertia for a Valve Stem in a Circular Shape

$$J = \frac{\pi d^4}{64} \tag{10.21}$$

where:

J: Polar moment of inertia of stem indicating the torsion deformation resistance of the shaft (m^4);
d: Valve stem diameter (m).

10.3.4 MAST Limitation for Quarter-Turn Cryogenic Valves

In accordance with MSS SP 134, the MAST for a quarter-turn extended stem valve in cryogenic service shall not exceed the limit established by Eq. (10.22).

Limit of MAST in Quarter-Turn Valves

$$\text{MAST} < \frac{\pi d^3 \tau_{\text{pl}}}{12n} \tag{10.22}$$

where:

MAST: Maximum allowable stem torque (Nm).
d: Stem diameter (m).
τ_{pl}: The shear stress proportional limit for the stem material. It should be noted that this value is very close to the *shear strength* point of the material. Standard values for shear strength have not been published, but can be estimated to be 60% of the *tensile strength*. The unit of proportional shear stress is Pa or N/m^2.
n: Safety factor of 2.

Example 10.10 In a cryogenic application with a low temperature of −196 °C, a 20-in. CL300 ball valve requiring a maximum torque value of 6500 Nm is used. This valve can handle a maximum stem torque of 20,500 Nm. The valve stem is made of stainless steel 316 with a tensile strength of 480 MPa. To prevent stem failure caused by the loads applied by the valve operator, what is the minimum stem diameter for this valve?

Answer

There is a maximum operating torque for the valve of 6500 Nm, which is less than the MAST value, which is 20,500 Nm in this case. This is a crucial point to keep in mind because if the maximum operating torque is higher than the MAST, then the stem of the valve has a high risk of being damaged or failing during operation.

According to Eq. (10.22), the minimum stem diameter leads to the minimum value of $\pi d^3 \tau_{\text{pl}}/12n$ on the right-hand side. $\pi d^3 \tau_{\text{pl}}/12n$ must be at least slightly over the MAST to be considered a minimum value. In this case, we can assume that $\pi d^3 \tau_{\text{pl}}/12n$ is only one unit above the MAST. Therefore,

$$\frac{\pi d^3 \tau_{\text{pl}}}{12n} = 20{,}501\ \text{Nm, and}$$

$$\tau_{\text{pl}} = \text{Shear strength} = 0.6 \times \text{Tensile strength}$$

$$= 0.6 \times 480\ \text{MPa} = 288 \times 10^6\ \text{Pa or N/m}^2 \rightarrow \frac{3.1415 \times d^3 \times 288 \times 10^6}{12 \times 2}$$

$$= 20{,}501 \rightarrow d^3 = 543.7 \times 10^{-6} \rightarrow d$$

$$= 8.16 \times 10^{-2}\ \text{m} = 0.0816\ \text{m} = 81.6\ \text{mm}$$

Questions and Answers

10.1 The maximum torque required for the operation of a 4-in. cryogenic ball valve in pressure class 300 is 500 Nm. The handwheel's diameter should not exceed 700 mm. In accordance with the design standard of the valve, the maximum operator hand force is 360 N. Can an operator apply 360 N of force to provide the required torque? If not, what would be required to generate the torque necessary for valve operation?

A Yes. It is possible for the operator to apply the required torque in order to operate the valve.
B No. The operator is only able to produce 300 Nm torque in this case, which is insufficient.
C In this instance, a gearbox with a gear ratio of two is required.
D All of the answers are incorrect.

Answer
Maximum torque is produced when the diameter of the handwheel and the force applied by the operator are both at their maximum. The largest diameter of the handwheel is 700 mm, and the maximum torque produced by rotating the handwheel is calculated as follows:

$$\text{Torque} = 360\ \text{N} \times 0.7\ \text{m} = 252\ \text{Nm}$$

Due to the maximum amount of torque produced by the operator, option A is incorrect; option B states that the operator is only able to produce 300 Nm, whereas the actual torque produced is 252 Nm. As a result, option B is wrong as well. Option C is correct, as using a gearbox with a gear ratio of two could result in a torque of 504 Nm, which is more than the required torque. Hence, option D is incorrect since option C is correct.

10.2 For the stem of a ball valve with a rounded geometry, there are two material options: the first is super duplex which has a yield strength of 80 Ksi, and the second is Inconel 718 which has a yield strength of 150 Ksi. The required torque for the operation of the valve is 5000 Nm. In what number of millimeters is the diameter of the stem larger when a stem constructed of super duplex material is selected instead of an Inconel 718 material?

A 83.5 mm
B 85 mm
C 90 mm
D 95 mm

Answer

Y_S for a duplex material = 80 Ksi = 5.516×10^8 N/m^2

Y_S for an Inconel 718 material = 150 Ksi = 1.034×10^9 N/m^2

According to Eq. (10.12), the MAST of a rounded stem is calculated as follows:

MAST in the rounded area of the stem = $(0.536 \times Y_S \times \pi \times d^3)/16,000$

The super duplex stem material has been chosen → $5000 = (0.536 \times 5.516 \times 10^8 \times \pi \times d^3)/16,000 \rightarrow d^3 = 0.086 \rightarrow d = 0.4417$ m = 441.7 mm

The Inconel 718 material has been chosen → $5000 = (0.536 \times 1.034 \times 10^9 \times \pi \times d^3)/16{,}000 \rightarrow d^3 = 0.046 \rightarrow d = 0.3582$ m = 358.2 mm

The diameter difference between a super duplex and Inconel 718 stems = 441.7 − 358.2 = 83.5 mm. Thus, option A is the correct answer.

10.3 The stem of a large size ball valve that is installed on a large oil export pipeline is composed of four segments; two keys, a circular and rectangular section, and the area where the stem keys are installed. For these four sections, the calculated MAST values are 270,000 Nm, 1,450,000 Nm, 190,000 Nm, and 930,000 Nm, respectively. What is the final MAST value for the stem as a whole assembly of these sections?

A 270,000 Nm
B 1,450,000 Nm
C 190,000 Nm
D 930,000 Nm

Answer

According to Eq. (10.15), the final MAST must be selected from the lowest torque value given earlier, which is equal to 190,000 Nm. Therefore, option C is the correct response.

10.4 The below table lists the torque values for a valve and its actuator. What would be the correct number for the actuator's safety factor?

	BTC (N)	BTO (N)	ETO (N)	ETC (N)
8″ slab gate valve	8,145	27,450	8,322	23,241
Hydraulic fail-safe closed actuator	70,737	62,553	45,735	55,452
Safety factor	8.68	2.28	5.50	2.39

A 1
B 2
C 3
D 4

Answer

The most appropriate answer is option B with a safety factor of two. The choice of option A is incorrect since the safety factor of one implies that the actuator torque values should be close to the valve torque values, which is not common in the industry, including in this case. Neither option C nor D is correct, as safety factors 3 and 4 are not met for torque values of BTO and ETC.

10.5 The stem of a valve is made of a steel round bar that has been put under a stress of 280 MPa. The stem has a diameter of 80 mm and a length of 240 mm. A modulus of elasticity of 205 GPa is calculated for the bar. Which of the following statements is true?

- **A** The strain value is 0.0010.
- **B** The strain value is 0.002.
- **C** A force of 1.40 MN is applied to the stem.
- **D** All answers are incorrect.

Answer

E: Modulus of elasticity = 205 GPa = 205×10^9

Using Eq. (10.17) $\rightarrow E = (\delta/\varepsilon) \rightarrow \varepsilon = (\delta/E) = ((280 \times 10^6)/(205 \times 10^9)) = 0.00137$

As a result, both options A and B are incorrect. Therefore, the next step would be to calculate the applied force on the valve stem. By multiplying the stress by the cross-sectional area of the stem, it is easy to estimate the force applied to the stem.

The cross-sectional area of the stem $(A) = \pi(d^2/4)(d =$ Diameter of the stem$) = \pi(0.080^2/4) = 0.00503\ \text{m}^2$

$$F = \delta \times A = 280 \times 10^6 \times 0.00503 = 1.4084 \times 10^6$$

The calculated force is 1.40 MN. So option C is the correct choice. Option D is incorrect since option C is the correct choice.

10.6 A quarter-turn valve requires 200 N of force to overcome pressure drop, 160 N of force to overcome seating, and 140 N of force to overcome packing and stem friction. Suppose that the valve is to be manually operated by a gearbox and handwheel, that the size of the handwheel is 0.4 m, and that the torque to be produced by the handwheel operator is equal to 100 Nm, which of the following statements is correct?

- **A** The force produced by the operator is sufficient to move the internal mechanisms of the valve without the aid of a gearbox.

B It is necessary to apply a force of 500 N to the valve stem in order to operate it.
C A force of 350 N is applied by the operator using the handwheel.
D A gearbox is required, and its gear ratio should be two.

Answer
In this situation, the force generated by the operator is equal to the ratio of torque (100 Nm) to handwheel diameter (0.4 m) or 400 N. Therefore, option C is incorrect because it states that the force produced by the operator is 350 N. In order to determine the force required for valve operation, which is applied to the stem, we must add together the forces required to overcome friction, including a pressure drop (200 N), seating (160 N), and packing (140 N), which is 500 N. Therefore, option B is the correct choice. The first option is incorrect due to the fact that the force produced by the operator is 400 N, but a force of 500 N is necessary to operate the valve. Thus, in order to increase the force applied by the operator from 400 to 500 N, a gearbox is required. By this, we mean that the gear ratio should be less than two. As a result, option D is incorrect. The correct choice is, therefore, option B.

10.7 Stainless steel 304 is the material used for the construction of a body-guided wedge gate valve stem, which is 10 mm in diameter. The modulus of elasticity of the stem in this case should be considered to be 28×10^6 psi during the operation of the valve. It is noted that the yield strength of SS304 at the different operating temperature ranges of valves remains the same and is equal to 30,000 psi. How long can the stem of the valve be before it becomes bent or buckled?
A 0.16 m
B 0.22 m
C 0.34 m
D 0.47 m

Answer
According to MSS SP134, gate and globe valves with a body-guided disk/gate connection should have a stem length to diameter ratio in the closed position determined by Eq. (10.19) as follows:

$$\frac{L}{d} \leq \frac{\pi}{2}\sqrt{\frac{E}{S_{\text{pl}}N}}$$

where:

$d = 100$ mm
$E = 28 \times 10^6$ psi

S_{pl} = Proportional limit = Yield strength = 30,000 psi
$N = 2$ (Safety factor)

The length of the stem is maximum when both sides of the equation become identical.

$$\frac{\pi}{2}\sqrt{\frac{E}{S_{pl}N}} = \frac{3.14}{2} \times \sqrt{\frac{28 \times 10^6}{30{,}000 \times 2}} = 1.57 \times 21.60 = 33.92$$

$$\frac{L}{d} = 33.92 \rightarrow L = 33.92 \times 10 = 339.2\ \text{mm} = 0.34\ \text{m}$$

Thus, option C is the correct answer.

10.8 A 20-in. CL300 ball valve operates at a maximum torque of 4500 Nm in a cryogenic application at a low temperature of −196 °C. This valve is designed to handle a maximum stem torque of 18,500 Nm. The stem of the valve is made from stainless steel 316 with a tensile strength of 480 MPa. Considering that the diameter of the stem is 79 mm in this case, what would be the maximum length of the stem in order to prevent the stem from failing due to torsional deflection?

A 350 mm
B 276 mm
C 250 mm
D 220 mm

Answer

According to Eq. (10.20), increasing the stem's length increases the stem's angle of twist, and the maximum allowable angle of twist is 2°, $\pi/90$, or 0.0349 radian. This means that the maximum stem length should result in a maximum of 0.0349 radian angle of twist. For the purpose of calculating the maximum stem length using Eq. (10.20), it should be noted that parameter T, the maximum design stem torque value, is equal to 18,500 Nm. In addition to the modulus of rigidity, parameter G, is equal to 77×10^9 Pa or N/m^2 for stainless steel, as per Table 10.3. The polar moment of inertia, parameter J, is equal to $\pi d^4/64(d = 0.079\ \text{m})$ or $1.9 \times 10^{-6}\ \text{m}^4$.

$$\theta = \frac{TL}{GJ} \rightarrow 0.0349 = \frac{TL}{GJ} \rightarrow 0.0349 = \frac{18{,}500 \times L}{77 \times 10^9 \times 1.9 \times 10^{-6}}$$
$$\rightarrow L = 0.276\ \text{m} = 276\ \text{mm}$$

As a result, the maximum length of a stem that can prevent twisting by more than 2 degrees is 276 mm. So option B is correct.

10.9 There are two globe valves that have the same dimensions, pressure class, and minimum and maximum operating temperatures. The stems are made of the same material and have the same diameter. For this case, we have to make three assumptions: the first assumption is that the ratio of length to diameter should be the maximum value without the risk of buckling; second, one valve has a disk that is guided by the body, while the other has a disk that is guided by the stem; and as a third assumption, it is assumed that the unsupported length of the stem of the stem-guided disk is 1000 mm. If the stem of the other design were unsupported, what would be its length?

A 1414 mm

B 1560 mm

C 1710 mm

D 1800 mm

Answer

There is no difference in the parameters of Eqs. (10.16) and (10.19) between the two valves, with the exception of the length of the stem. There should be a maximum ratio of stem length to diameter in order to avoid buckling, meaning that:

$$\frac{1000}{d} = \frac{\pi}{2}\sqrt{\frac{E}{2S_{\text{pl}}N}} \text{(This applies to the stem-guided disk.)}$$

$$\frac{L}{d} = \frac{\pi}{2}\sqrt{\frac{E}{S_{\text{pl}}N}} \text{(This applies to the body-guided disk)}$$

By dividing both sides of the equations earlier $\rightarrow 1000/L = 1/\sqrt{2} \rightarrow L = 1414$ mm

Therefore, option A is the correct answer.

10.10 A cryogenic ball valve has a stem diameter of 200 mm and is constructed from stainless steel 316 material. What is the maximum amount of MAST that the valve can have without risk of stem failure?

A 10^6 Nm

B 0.2×10^6 Nm

C 0.3×10^6 Nm

D 0.4×10^6 Nm

Answer

In accordance with MSS SP 134, the measure of the MAST for a quarter-turn extended stem valve in cryogenic service should not exceed the measurements of Eq. (10.22).

$$\text{MAST} < \frac{\pi d^3 \tau_{\text{pl}}}{12n}$$

The stem of the valve is made of stainless steel 316 and has a tensile strength of 480 MPa. The stem material shear stress proportional limit is parameter τ_{pl}. This value is very close to the shear strength point of a material. This value can be estimated as follows:

$$\tau_{pl} = \text{Shear strength} = 0.6 \times \text{Tensile strength} = 0.6 \times 480\,\text{Mpa}$$
$$= 288 \times 10^6\ \text{Pa or N/m}^2$$

The value of $\pi d^3 \tau_{pl}/12n$ on the right side of the equation is calculated as follows:

$$\frac{\pi d^3 \tau_{pl}}{12n} = \frac{\pi(0.2)^3 \times 288 \times 10^6}{12 \times 2} = 0.3 \times 10^6$$

We can say that MAST should be almost as high as 0.3×10^6. Thus, option C is the correct answer.

Further Reading

American Petroleum Institute (API) (2012). *Standard for Actuator Sizing and Mounting Kits for Pipeline Valves. API 6DX*, 1e. Washington, DC: American Petroleum Institute (API).

American Petroleum Institute (API) (2014). *Specification for Pipeline and Piping Valves. API 6D*, 24e. Washington, DC: American Petroleum Institute (API).

American Society of Mechanical Engineers (ASME) (2004). *Valves – Flanged, Threaded, and Welding End*. New York: American Society of Mechanical Engineers (ASME).

Gokilakrishnan, G., Divya, S., Rajesh, R., and Selvakumar, V. (2014). Operating torque in ball valves: a review. *International Journal for Technological Research in Engineering* 2 (4): 2347–4718.

International Organization for Organization (ISO) (2007). *Pipeline Transportation Systems: Pipeline Valves. ISO 14313*, 2e. Geneva: International Organization for Organization (ISO).

International Organization for Organization (ISO) (2011). *Petroleum and Natural Gas Industries: Mechanical Integrity and Sizing of Actuators and Mounting Kits for Pipeline Valves. ISO 12490*, 1e. Geneva: International Organization for Organization (ISO).

Nesbitt, B. (2007). *Handbook of Valves and Actuators: Valves Manual International*, 1e. Oxford: Elsevier.

Skousen, P.L. (2011). *Valve Handbook*, 3e. New York: McGraw-Hill.

Smit, P. and Zappe, R.W. (2004). *Valve Selection Handbook*, 5e. New York: Elsevier.

Sotoodeh K. (2019). Actuator sizing and selection. *Springer Nature Applied Science*, 1, 1207. https://doi.org/10.1007/s42452-019-1248-z

Sotoodeh K. (2019). The importance of maximum allowable stem torque in valves. *Springer Nature Applied Science*, 1, 433. https://doi.org/10.1007/s42452-019-0445-0

Sotoodeh, K. (2022). *Cryogenic Valves for Liquified Natural Gas Plants*, 1e. Austin: Elsevier (Gulf Professional Publishing).

Young, C. and Budynas, R. (2002). *Roark's Formulas for Stress and Strain*, 7e. New York: McGraw-Hill.

11

Miscellaneous

11.1 Introduction

Equations and calculations related to joint efficiency, as well as valve sealing, are discussed in this chapter. Joint efficiency is important because it refers to the valve's ability to provide a secure seal between metallic components, such as the body and bonnet, to prevent leakage. Welded bodies and bonnets provide the tightest connections with the highest joint efficiency. However, the majority of valves are body pieces or body and bonnets that are bolted together. The second part of this chapter deals with the design formulas and calculations that are associated with valve stem sealing. The seals on valves, especially the stem seals, play an imperative role because they reduce the possibility of leaks and fugitive emissions from valves. In addition, they increase the life of the valves, improving performance, as well as reducing the amount of maintenance required. In view of this, it is essential to dedicate a separate chapter to valve seals. This is because a key aspect of valve seals is their ability to minimize the risk of emissions of gases and other environmental pollutants through the seal. It is very critical to select, design, and arrange valve seals in the most efficient way possible to prevent leakage and increase the life and performance of the valve. There are several parts of a valve that require sealing, such as the stem, the body and the bonnet, and the body of the valve pieces. Valve stem seals are known as the most important leakage points when it comes to fugitive emissions for two reasons. First, valve stem seals are classified as dynamic seals due to their movement. The seals are divided into two categories: static seals and dynamic seals. It should be noted that static seals are used between two nonmoving parts of the valve, while dynamic seals are used between a moving or rotating valve part and another part that may be in motion or not. Seals that are dynamic in nature are more critical since they are at a higher risk of wearing and tearing due to friction between one or two moving parts. Furthermore, unlike the valve seat or seat insert, leakage from the valve stem seals is released to the environment and is one of the major sources of fugitive emissions.

Industrial Valves: Calculations for Design, Manufacturing, Operation, and Safety Decisions, First Edition. Karan Sotoodeh.

Therefore, it is well worth paying more attention to valve stem seals in more detail. The term fugitive emission refers to the unintentional and undesirable release, leakage, or discharge of gases or vapors from equipment or facilities containing pressure, and from components inside industrial plants, such as valves, piping flanges, pumps, storage tanks, and compressors.

11.2 Joint Efficiency

In this section, we will focus on three joint efficiency conditions; welded joints, bolted joints, and threaded joints. Next, we will discuss the efficiency of welded joints.

11.2.1 Weld Joint Efficiency

It is crucial to understand that welded joints are defined by their efficiency, and therefore, in some literature, the term "welding joint efficiency" is used. American Petroleum Institute (API) standards and American Society of Mechanical Engineers (ASME) codes include references to joint efficiency. This is a numerical value expressed as the ratio between the strength of the welded part and the strength of the base material. This parameter is calculated as follows using Eq. (11.1).

$$E = \frac{S_w}{S_b} \tag{11.1}$$

where:

E: weld joint efficiency;
S_w: weld strength (Ksi);
S_b: base metal strength (Ksi).

According to paragraph 2.1.6 of ASME B16.34, the minimum acceptable welding joint efficiency is 0.8 or 80%. The ideal condition is when the weld joint strength is equal to the base material strength, in which case the joint efficiency would be 1.00 or 100%. Joint efficiency is also applicable between two welded components, such as two pieces of pipe or a pipe and a valve. Joint efficiency is determined by the type of welding and the type of post welding inspection. Nondestructive testing (NDT) refers to an inspection technique that is applied to materials, especially after welding or machining, in order to find possible defects and discontinuities in the material or the welding without destroying it or damaging it. On welding, NDTs are typically applied after a post weld heat treatment (PWHT). The primary purpose of PWHT is to reduce residual stress during welding. The residual stress refers to the stress that remains in the material after welding, leading to cracks, brittle fractures, and ultimately failure of the weld. NDT methods include visual inspection, liquid penetration test, magnetic particle test, ultrasonic test (UT), and radiography test (RT). In contrast to the first three methods, ultrasonic tests and RTs are volumetric NDTs, which are designed to detect defects below the surface. An RT is a type of NDT that increases the efficiency of the weld joint. It is possible to increase weld joint efficiency to 100% by applying an RT test. In Table 11.1,

Table 11.1 Longitudinal weld joint quality factor for piping according to ASME B31.3 code, table 302.3.4.

No.	Type of joint		Type of seam	Examination	Weld joint efficiency (E)
1	Furnace buttweld		Straight	As required by the specification or code	0.6
2	Electrical resistance weld		Straight or spiral	As required by the specification or code	0.8
				Additional spot radiography	0.9
				Additional 100% radiography	1.00
3	Electrical fusion weld (single butt weld)		Straight or spiral	As required by the specification or code	0.8
				Additional spot radiography	0.9
				Additional 100% radiography	1.00
4	Electrical fusion weld (double buttweld)	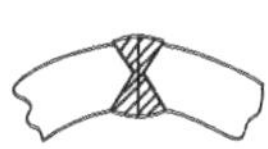	Straight or spiral	As required by the specification or code	0.8
				Additional spot radiography	0.9
				Additional 100% radiography	1.00
5	Submerged arc weld, gas metal arc weld	API 5L	Straight with one or two seams spiral	As required by the specification	0.95

Source: ASME B31.3/American Society of Mechanical Engineers.

we found that the longitudinal weld joint factors for different welding mechanisms were summarized based on the data in table 302.3.4 in ASME B31.3. It should be noted that electrical resistance welding and electrical fusion welding provide a higher level of weld joint efficiency than furnace buttwelding.

Furthermore, the joint efficiency of the fillet welds could be 0.45, 0.45, or 0.50 depending on whether or not the fillet weld is single or double. Welding in which two pieces of metal meet at an angle or perpendicularly is known as "fillet welding."

11.2.2 Bolted Joint Efficiency

This section discusses two types of bolted joints; the first is the bolted joint of the bonnet or the cover, and the second is the bolted joint of the body.

11.2.2.1 Bolted Bonnet or Cover Joints

With a bolted body and bonnet valve design, the body and bonnet are like flanges that are joined with bolts and nuts. A gasket is placed between the body and the bonnet to prevent leakage. The body and bonnet are connected by bolts, making it possible to maintain the valve. The bolting of bonnets and covers to valve bodies shall be threaded according to ASME B1.1 for inch-dimensioned bolting and, as a minimum, shall satisfy the following bolt cross-sectional area requirements as Eq. (11.2).

Bolted Bonnet or Cover Joint Requirement According to ASME B16.34

$$P_c \times \frac{A_g}{A_b} \leq K_1 \times S_a \leq 9000 \tag{11.2}$$

where:

A_b: Total effective bolt tensile stress area that is calculated either from Eq. (11.6) or (11.7);

A_g: Area enclosed by the seal effective outside periphery of a gasket, O-ring, or another seal, with the exception that in the case of a ring-joint gasket, the bounded area is determined by the pitch diameter of the ring-type joint gasket;

K_1: It is 65.26/MPa when the allowable bolt stress is expressed in MPa units and 0.45/psi when the allowable stress is expressed in psi unit;

P_c: Pressure rating class designation. In accordance with the mandatory appendix B of ASME B16.36, the value of the designation pressure rating is the same as the value of the pressure rating. There are several pressure ratings that are used for industrial valves, including 150, 300, 600, 900, 1500, and 2500. The pressure class designation number for a valve with a pressure class of 150 is equal to

150, for example. In the same way, if the valve pressure class is 300, the pressure class designation number is 300;

S_a: Allowable bolt stress at 38 °C or 100 °F. If the value is greater than 137.9 MPa or 20,000 psi, use 137.9 MPa (20,000 psi). The allowable bolt stress values shall be taken from the ASME boiler and pressure vessel code (BPVC), section II, part D.

When you select calculated tensile stress area in a bolt calculation program such as connector-bolt property manager in a bolt calculation program, the program calculates the tensile stress area (A_t), which is the minimum area of the threaded portion of the bolt according to Eq. (11.3) as follows:

Calculation of the Tensile Stress Area According to the Bolt Calculation Program

$$A_t = \left[\frac{d_3 + d_2}{2}\right]^2 \times \frac{\pi}{4} \tag{11.3}$$

where:

P: Pitch, axial distance between two adjacent threads;
n: $1/P$, number of bolt threads per inch or millimeter;
d: nominal bolt shank diameter.

The values of d_2 and d_3 are shown in Figure 11.1, and they are calculated as follows:

$$d_3 = d - 1.2268 \times P \tag{11.4}$$

$$d_2 = d - 0.6495 \times P \tag{11.5}$$

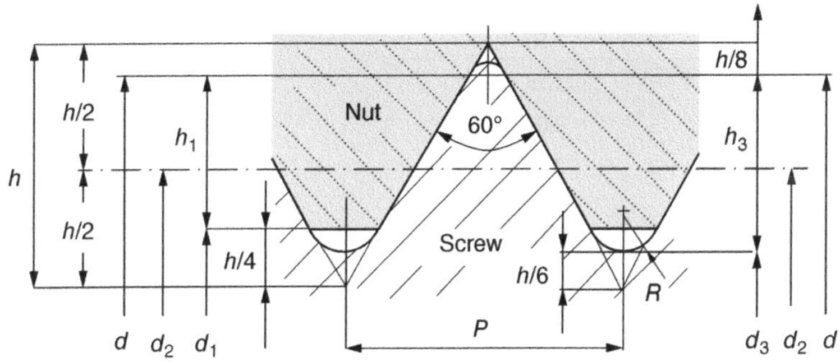

Figure 11.1 A schematic showing the contact between bolts and nuts. *Source:* Courtesy: Dassault systems.

For bolt dimensions in millimeters, the formula for the tensile stress area would be Eq. (11.6) and for bolt dimensions in inches, Eq. (11.7).

Calculation of the Bolt Tensile Stress Area Based on the Bolt Diameter and Thread Count

$$A_\text{t} = 0.7854 \times \left[d - \left(0.9382 \times \frac{1}{n}\right)\right]^2 \text{(The units are mm)} \tag{11.6}$$

$$A_t = 0.7854 \times \left[d - \left(0.9743 \times \frac{1}{n}\right)\right]^2 \text{(The units are in.)} \tag{11.7}$$

In Table 11.2, the nominal bolt diameter, pitch, and number of bolt threads per millimeter for ISO M large bolts are provided.

Table 11.2 Nominal bolt diameter, pitch, and number of bolt threads per millimeter for ISO M large bolts.

Nominal bolt diameter *d* (mm)	Pitch *P* (mm)	Number of bolt threads per millimeter (*n*)
M 1.60	0.35	2.86
M 2.00	0.40	2.50
M 2.20	0.45	2.22
M 3.00	0.50	2.00
M 4.00	0.70	1.43
M 4.50	0.75	1.33
M 5.00	0.80	1.25
M 6.00	1.00	1.00
M 8.00	1.25	0.80
M 10.00	1.50	0.67
M 12.00	1.75	0.57
M 16.00	2.00	0.50
M 20.00	2.50	0.40
M 24.00	3.00	0.33
M 30.00	3.50	0.29
M 36.00	4.00	0.25
M 42.00	4.50	0.22
M 48.00	5.00	0.20
M 56.00	5.50	0.18
M 64.00	6.00	0.17

Example 11.1 Calculate the bolt tensile stress area for a large ISO M bolt with a diameter of 2 mm. The area of tensile stress of the bolt is increased by how many square millimeters if the bolt diameter doubles?

Answer

Table 11.2 indicates a value of 2.5 for the number of threads per millimeter. Given the diameter of the bolt and the number of threads per millimeter, using Eq. (11.6), the area of tensile stress of the bolt can be calculated as follows:

$$A_t = 0.7854 \times \left[d - \left(0.9382 \times \frac{1}{n}\right)\right]^2 = 0.7854 \times \left[2 - \left(0.9382 \times \frac{1}{2.5}\right)\right]^2$$
$$= 0.7854 \times 2.63 = 2.07\ \text{mm}^2$$

According to Table 11.2, the number of threads per millimeter becomes 1.43 when the bolt diameter is doubled and increased to 4 mm. It is now possible to calculate the area of tensile stress for the bolt with a diameter of 4 mm as follows:

$$A'_t = 0.7854 \times \left[d' - \left(0.9382 \times \frac{1}{n'}\right)\right]^2 = 0.7854 \times \left[4 - \left(0.9382 \times \frac{1}{1.43}\right)\right]^2$$
$$= 0.7854 \times 11.18 = 8.78\ \text{mm}^2$$

$$8.78 - 2.07 = 6.71\ \text{mm}^2$$

As a result of doubling the bolt diameter from 2 to 4 mm in this example, 6.71 mm^2 of bolt tensile stress surface is created.

Example 11.2 There are eight bolts that are used to connect the body and bonnet of a valve, which is made of carbon steel material and has a pressure rating of 1500. The valve is operating at ambient temperature. As far as bolts and nuts are concerned, the bolts and nuts used are the American Society for Testing and Materials (ASTM) A193 B7 and A194 Gr.2H, respectively. In terms of the bolt's major diameter, it is 0.875 in., and there are 14 threads/in. on the bolt. If we take into consideration the gasket diameter between the body and the bonnet to have a sealing diameter of 4.5 in., what would be the total effective area of tensile stress? Can the bolts provide adequate force to secure the area between the body and the bonnet tightly without the possibility of leaks – as required by ASME B16.34?

Answer

In order to calculate the bolt stress area, Eq. (11.7) must be taken into account as follows:

$$A_t = 0.7854 \times \left[d - \left(0.9743 \times \frac{1}{n}\right)\right]^2 = 0.7854 \times \left[0.875 - \left(0.9743 \times \frac{1}{14}\right)\right]^2$$
$$= 0.7854 \times 0.649 = 0.51\ \text{in.}^2$$

Due to the eight bolts connecting the body and bonnet of the valve, the total bolt stress area can be calculated as follows:

$$A_b = 8 \times 0.51\ \text{in.}^2 = 4.076\ \text{in.}^2$$

Since the diameter of the seal is equal to 4.5 in., the area bounded by the effective outside of the gasket can be calculated as follows:

$$A_g : \text{Area enclosed by the seal} - \text{effective outside periphery of a gasket,}$$

$$\text{O-ring, or other seal} = \pi\left(\frac{d^2}{4}\right)(d : \text{Gasket diameter}) = \pi\left(\frac{4.5^2}{4}\right) = 15.904\ \text{in.}^2$$

The valve pressure class is 1500 meaning that $P_c = 1500$.

The maximum allowable stress for A193 B7 bolts refers to two ASME codes including ASME BPVC Sec. II and ASME B31.3, the process piping code, is equal to 20,000 psi at room temperature. It is critical to keep in mind that when the temperature increases, the value of the allowable stress is typically reduced. The value of K_1 is 0.45/psi when the allowable stress is expressed in psi unit. In order to ensure that the bolting (bolts and nuts) is able to provide enough force to ensure the body and bonnet joint remains tightly joined together with no chances for leaks, the following requirements must be met.

$$P_c \times \frac{A_g}{A_b} \leq K_1 \times S_a \leq 9000$$

The left side of the equation is calculated as follows:

$$P_c \times \frac{A_g}{A_b} = 1500 \times \frac{15.904}{4.076} = 5852.7$$

The middle section of the equation is calculated as follows:

$$K_1 \times S_a = 0.45 \times 20{,}000 = 9000$$

Therefore, $5852.7 \leq 9000 \leq 9000$ is met and the bolts used for the body and bonnet joint of the valve provide sufficient forces for the tight sealing of the valve.

11.2.2.2 Bolted Body Joints

The bolted body joints shall be threaded in accordance with ASME B1.1 for inch dimensional bolts and ASME B1.13M for metric bolts and, at a minimum, shall meet the following bolt cross-sectional area requirements given in Eq. (11.8).

Bolted body joint requirement according to ASME B16.34.

$$P_c \times \frac{A_g}{A_b} \leq K_2 \times S_a \leq 7000 \tag{11.8}$$

K_2: It is 50.76/MPa when the allowable bolt stress is expressed in MPa units and 0.35/psi when the allowable stress is expressed in psi unit.

Example 11.3 A ball valve is constructed from low-temperature carbon steel material and has a pressure rating of 300. There are eight bolts connecting its two body pieces. The valve operates at a temperature of 50 °C. In terms of bolts and nuts, the bolts and nuts used are ASTM A320 L7 and A194 Gr.4, respectively. There are 14 threads/in. on the bolt, and the bolt has a diameter of 0.875 in. Given that the gasket diameter between the body and the bonnet is 3.5 in., what is the total area of tensile stress applied to the bolts? Do the bolts provide adequate force to secure the area between two body pieces tightly without the possibility of leaks, as required by ASME B16.34?

Answer

The bolt stress area can be calculated using Eq. (11.7) as follows:

$$A_t = 0.7854 \times \left[d - \left(0.9743 \times \frac{1}{n}\right)\right]^2 = 0.7854 \times \left[0.875 - \left(0.9743 \times \frac{1}{14}\right)\right]^2$$
$$= 0.7854 \times 0.649 = 0.51\ \text{in.}^2$$

Due to the eight bolts connecting the body and bonnet of the valve, the total bolt stress area can be calculated as follows:

$$A_b = 8 \times 0.51\ \text{in.}^2 = 4.076\ \text{in.}^2$$

Given that the seal has a diameter of 3.5 in., the area bounded by the effective outside of the gasket can be calculated as follows:

$$A_g : \text{Area enclosed by the seal – effective outside periphery of a gasket,}$$
$$\text{O-ring, or other seal} = \pi\frac{d^2}{4}\ (d : \text{Gasket diameter}) = \pi\frac{3.5^2}{4} = 9.621\ \text{in.}^2$$

The maximum allowable stress for A320 L7 bolts refers to two ASME codes including ASME BPVC Sec. II and ASME B31.3, the process piping code, is equal to 25,000 psi at 50 °C. It is, however, required to use the allowable stress equal to 20,000 psi when the allowable stress exceeds 20,000, as in this case. The following requirements must be met in order to ensure that the bolting (bolts and nuts) is able to provide adequate forces to ensure that the body pieces remain tightly attached with no possibility of leaks.

$$P_c \times \frac{A_g}{A_b} \leq K_2 \times S_a \leq 9000$$

The left side of the equation is calculated as follows:

$$P_c \times \frac{A_g}{A_b} = 300 \times \frac{9.621}{4.076} = 708.11$$

The middle section of the equation is calculated as follows:

$$K_2 \times S_a = 0.35 \times 20{,}000 = 7000$$

Therefore, $708.11 \leq 7000 \leq 7000$ is met and the bolts used for the body pieces' connection of the valve provide sufficient force for tight sealing of the valve.

11.2.3 Threaded Joint Efficiency

This section discusses two types of threaded joints; the first is the threaded joint of the bonnet or the cover and the second is the threaded joint of the body. The following section discusses threaded bonnets or cover joints.

11.2.3.1 Threaded Bonnet or Cover Joints

In cases where bonnets or covers are attached to valve bodies by threads, the thread shear area must satisfy the following criteria as per Eq. (11.9):

Threaded Bonnet or Cover Joint Requirement According to ASME B16.34

$$P_c \times \frac{A_g}{A_s} \leq 4200 \tag{11.9}$$

where:

A_s: Total effective thread shear area.

Example 11.4 Imagine a valve that is in the pressure class 300 and has a sealing diameter of 4.5 in. There is a threaded contact between the body of the valve and the bonnet. In light of that, what would be the minimum effective threaded shear area necessary for a secure connection between the body and the bonnet?

Answer

$$P_c \times \frac{A_g}{A_s} \leq 4200 \rightarrow P_c \times \frac{A_g}{4200} \leq A_s$$

$$\rightarrow \text{Minimum } A_s = P_c \times \frac{A_g}{4200} = 300 \times \frac{4.5}{4200} = 0.32\ \text{in.}^2$$

The minimum effective threaded shear area is $0.32\ \text{in.}^2$.

11.2.3.2 Threaded Body Joints

As a minimum requirement, threaded body joints must satisfy the following thread shear area requirements according to Eq. (11.10):

Threaded Body Joint Requirement According to ASME B16.34

$$P_c \times \frac{A_g}{A_s} \leq 3300 \tag{11.10}$$

where:

A_s: Total effective thread shear area.

Example 11.5 The pressure class 150 valve has a minimum effective threaded shear area of 0.40 in.^{2}, which is in accordance with the minimum pressure class 150. Two pieces of the valve's body are attached together and there is a threaded contact between them. In light of this, what would be the maximum gasket diameter necessary for a secure connection between the threaded joint of the body and the valve?

Answer

Equation (11.10) is valid in this case because it describes the threaded joint between two body pieces of a valve.

$$P_c \times \frac{A_g}{A_s} \leq 3300 \rightarrow 150 \times \frac{A_g}{0.40} \leq 3300 \rightarrow A_g \leq 8.8$$

Therefore, the maximum gasket area in this case is 8.8 in.^{2}. It is now possible to determine the diameter of the gasket as follows:

A_g: Area enclosed by the seal-effective outside periphery of a gasket, O-ring, or other seal $= 8.8 = \pi(d^2/4)$ (d : Gasket diameter) $\rightarrow \pi(d^2/4) = 8.8$ in.$^2 \rightarrow d = 3.35$ in.

11.3 Stem Sealing

In order to prevent leakage from the valve stem area to the environment, stem sealing is critical. Furthermore, accuracy and tight sealing of valve stem seals are essential in fugitive emission applications where hazardous fluids, such as gases or vapors, can escape from valves that are not properly sealed. Hazardous fluid is defined as a fluid service in which the potential for personnel exposure is judged to be significant and in which a single exposure to a very small quantity of toxic

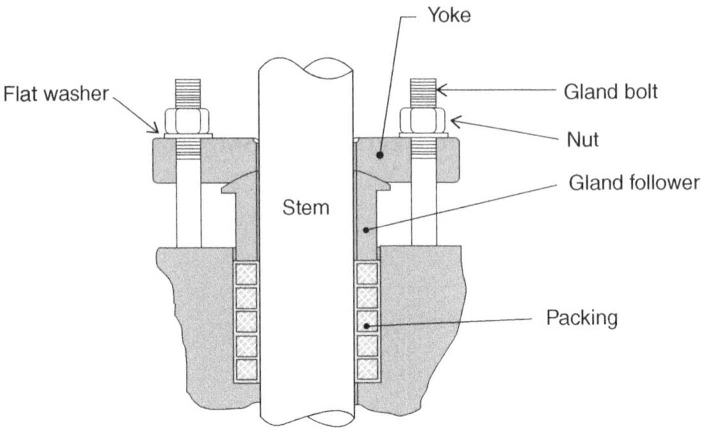

Figure 11.2 Stem packing arrangement including gland follower.

fluid can produce serious irreversible harm to someone upon breathing or bodily contact, even when prompt restorative measures are taken. The valve stem sealing can be performed with soft thermoplastics and elastomers, such as Teflon (PTFE), lip seals, V-packs, and Vitons, as well as graphite or graphoil. As illustrated in Figure 11.2, valve stem seals may consist of a compression packing that consists of soft materials placed in an area around the valve stem known as the "stuffing box" and compressed by a valve part known as the "gland or gland follower" to create stem seals.

The sealing capability of compression packing is defined as its ability to expand laterally against the stem and stuffing box wall when the gland is tightened. Based on Eq. (11.11), the relationship between stem seal lateral stress and stem seal axial stress can be expressed as follows:

Relationship Between Lateral and Axial Stresses in Stem Seals

$$\delta_{\mathrm{l}} = \delta_{\mathrm{a}}\left(\frac{1-\mu}{\mu}\right) \tag{11.11}$$

where:

δ_{l}: Lateral stress;
δ_{a}: Axial stress;
μ: Poisson's ratio (ratio of lateral expansion to axial compression of an elastic solid compressed between two faces calculated according to Eq. (11.12)).

Calculation of Poisson's Ratio

$$\mu = \frac{\text{Lateral strain}}{\text{Longitudinal strain}} \tag{11.12}$$

Strain is defined as deformation per unit of the original length, which is calculated using Eq. (11.13).

Calculation of Strain

$$\mathcal{E} = \frac{\Delta L}{L} \tag{11.13}$$

where:

ε: Strain;
ΔL: Length difference;
L: Length.

Example 11.6 If soft rubber with a Poisson's ratio of 0.5 is used for stem sealing, what would be the ratio of lateral to axial stress?

Answer
In order to express the relationship between the lateral stress on the stem seal and the axial stress on the stem seal, Eq. (11.11) can be used as follows:

$$\delta_\text{l} = \delta_\text{a}\left(\frac{1-\mu}{\mu}\right) \text{ and } \mu = 0.5 \rightarrow \delta_\text{l} = \delta_\text{a}\left(\frac{1-0.5}{0.5}\right) \rightarrow \delta_\text{l} = \delta_\text{a}$$

As the lateral and axial forces are equal, the ratio of lateral to axial stresses is one.

Torque is a measure of the force required to operate a valve between its open and closed positions. There are several factors that determine the torque required to operate a valve, such as the packing or stem sealing material, the number of stem sealing rings, the load of the gland flange, the stem surface finish, the temperature, and the cycle frequency. Equation (11.14) can be used to calculate the force of packing on the valve stem.

Calculation of the Packing Force on the Stem

$$F = \pi \times d \times H \times \text{GS} \times \mu \times Y \tag{11.14}$$

where:

F: Force required to overcome the packing friction (N);
d: Stem diameter (m);
H: Packing height (m);
GS: Compressive stress on the packing (axial gland stress on the packing) (N/m^2) that is calculated according to Eq. (11.15);
μ: Packing coefficient of friction (dimensionless);
Y: Ratio of lateral to axial load transference (for example, 0.5);
GS: Known as the axial gland stress on the packing, is calculated using Eq. (11.15).

Calculation of Axial Gland Stress on the Packing

$$\text{GS} = \frac{T.n/0.2.b}{\left(\pi\left(D^2 - d^2\right)\right)/4} \tag{11.15}$$

where:

T: the torque applied to the gland flange bolts calculated as per Eq. (11.16);
n: the number of bolts;
b: the nominal bolt diameter;
D: the packing box diameter;
d: the stem diameter.

Calculation of Axial Gland Stress on the Packing

$$T = \frac{kFBb}{12n} \tag{11.16}$$

where:

T: Required bolt torque (lb-ft);
k: Nut factor that is typically considered 0.2;
FB: bolt force (lb);
b: Nominal bolt diameter (in.);
n: Number of gland bolt.

Example 11.7 In the gland flange, there are two bolts with a diameter of 3/8″ that provide a torque value of 20 Nm. How much torque should be applied to the bolts to tighten the gland and the packing?

Answer

Equation (11.16) is applicable in this case. The value of torque equal to 20 Nm corresponds to 14.75 lb of force foot.

$$T = \frac{kFBb}{12n} \rightarrow \text{FB} = \frac{12T.n}{\text{kb}} = \frac{12 \times 14.75 \times 2}{0.2 \times (3/8)} = 4720\ \text{lb}$$

Example 11.8 In the previous example, if the stem diameter of the gland packing is 1/2 in. and the packing box diameter is 1 1/2 in., calculate the compressive stress on the gland packing.

Answer

For this case, Eq. (11.15) must be used as follows:

$$\text{GS} = \frac{(T.n/0.2.b)}{(\pi(D^2 - d^2))/4} = \frac{(14.75 \times .2)/0.2 \times (3/8)}{(\pi(1.5^2 - 0.5^2))/4}$$

$$= \frac{(14.75 \times .2)/(0.2 \times (3/8))}{(\pi(1.5^2 - 0.5^2))/4} = \frac{393.33}{1.57075} = 250.41\ \text{psi}$$

Example 11.9 How much force is required to overcome the packing friction on the valve stem if the diameter of the valve stem is 50 mm and the packing height is 20 mm? Axial gland pressure on the packing is 1.4×10^6 newtons per square meter and the packing coefficient of friction is 0.5. In addition, the lateral to axial force ratio is 0.5. If the valve stem diameter is 50 mm and the packing height is 20 mm, how much force is required to overcome the packing friction on the valve stem?

Answer

Equation (11.14) can be used to calculate the force of packing on the valve stem.

$$F = \pi \times d \times H \times \text{GS} \times \mu \times Y = \pi \times 0.05 \times 0.02 \times 1.4 \times 10^6 \times 0.5 \times 0.5 = 1099.5\ \text{N}$$

Questions and Answers

11.1 A ball valve is constructed of a body and a bonnet that are butt-welded together. It is estimated that the weld strength to base metal is 0.8. What is the correct answer?

- **A** It is more accurate to refer to the welding process used to join the valve body and bonnet as furnace buttweld.
- **B** Through the application of a 100% radiography test, the joint efficiency can be increased to 100%.

C The valve-welded joint between the body and bonnet has a lower strength compared with the fillet-welded joint.
D Weld joint efficiency can be increased by up to 85% by using spot radiography.

Answer
Option A is incorrect because the furnace buttweld is the weakest type of welding, based on Table 11.1, which can only provide a joint efficiency of 60% or 0.6, while the joint efficiency, in this case, would be 80% or 0.8. The correct answer is B. The choice of option C is incorrect because the joint efficiency through fillet welding is generally lower than through buttwelding, and it can range from 40 to 50%. Option D is incorrect since spot radiography, as shown in Table 11.1, can increase the joint's efficiency to 90%. Accordingly, option B is the correct answer.

11.2 Which of the following statements is true regarding the area of tensile stress value for various bolting diameters?
A For an M6 bolt, the area of tensile stress is 20.1 mm^2.
B The tensile stress area value for a bolt with a diameter of M12 is double that of a bolt with a diameter of M6.
C For an M12 bolt, the area of tensile stress is 70 mm^2.
D All of the options are incorrect.

Answer
There is only one thread per millimeter on an M6 bolt as per Table 11.2, which has a diameter of 6 mm. According to Eq. (11.6), the area of tensile stress for an M6 bolt is calculated as follows:

$$A_t = 0.7854 \times \left[d - \left(0.9382 \times \frac{1}{n}\right)\right]^2 = 0.7854 \times \left[6 - \left(0.9382 \times \frac{1}{1}\right)\right]^2$$

$$= 0.7854 \times 25.62 = 20.12\,\text{mm}^2$$

Table 11.2 indicates that the number of threads per millimeter for an M12 bolt is 0.57. In accordance with Eq. (11.6), the area of tensile stress for an M12 bolt is calculated as follows:

$$A_t = 0.7854 \times \left[12 - \left(0.9382 \times \frac{1}{0.57}\right)\right]^2$$

$$= 0.7854 \times \left[12 - \left(0.9382 \times \frac{1}{0.57}\right)\right]^2$$

$$= 0.7854 \times 107.20 = 84.20\ \text{mm}^2$$

There is no doubt that option A is the correct answer. M6 bolts have a tensile stress area almost four times greater than M12 bolts, which is the reason why option B is wrong. Option C is incorrect for the reason that an M20 bolt's area of tensile stress is 84.20 square millimeter. Option D is incorrect as well since option A is the correct answer.

11.3 A top entry ball valve is installed on an oil export pipeline constructed from low-temperature carbon steel material and has a pressure rating of 1500 psi. The body and bonnet of the valve are connected by 24 bolts. At all times, the valve operates at a temperature that is less than 70 °C. Bolts and nuts used are ASTM A320 L7 and A194 Gr.4, respectively. The bolt has 8 threads/in., and its diameter is 1 in. When the gasket diameter between the body and the bonnet is 10 in., what is the total area of tensile stress applied to the bolts? Do the bolts provide sufficient force to secure the area between two body pieces tightly without the possibility of leaks, as required by ASME B16.34?

A The bolt stress area is 10 in.^{2}, and the bolts provide adequate force to seal the joint.

B The bolt stress area is 11.5 in.^{2}, and the bolts do not provide adequate force to seal the joint.

C The bolt stress area is 14.5 in.^{2}, and the bolts provide adequate force to seal the joint.

D The bolt stress area is 17 in.^{2}, and the bolts do not provide adequate force to seal the joint.

Answer

The bolt stress area can be calculated using Eq. (11.7) as follows:

$$A_t = 0.7854 \times \left[d - \left(0.9743 \times \frac{1}{n}\right)\right]^2$$

$$= 0.7854 \times \left[1 - \left(0.9743 \times \frac{1}{8}\right)\right]^2$$

$$= 0.7854 \times 0.7712 = 0.6057 \text{ in.}^2$$

Due to the 24 bolts connecting the body and bonnet of the valve, the total bolt stress area can be calculated as follows:

$$A_b = 24 \times 0.6057 \text{ in.}^2 = 14.538 \text{ in.}^2$$

According to the following formula, since the seal has a diameter of 10 in., the area bounded by the actual outside of the gasket can be calculated.

A_g: Area enclosed by the seal-effective outside periphery of a gasket, O-ring, or other seal $= \pi(d^2/4)$ (d: Gasket diameter) $= \pi(10^2/4) = 78.54$ in.^{2}

According to two ASME codes including ASME BPVC Section II and ASME B31.3, the process piping code, the maximum stress for A320 L7 bolts is equal to 25,000 psi at a maximum temperature of 70 °C. However, it is required to use the allowable stress equal to 20,000 psi when the allowable stress exceeds 20,000, which is the case here. It is critical to meet the following requirements in order to ensure that the bolting (bolts and nuts) is strong enough to provide adequate forces to ensure that the body pieces remain tightly attached with no leakage potential.

$$P_c \times \frac{A_g}{A_b} \leq K_1 \times S_a \leq 9000$$

The left side of the equation is calculated as follows:

$$P_c \times \frac{A_g}{A_b} = 1500 \times \frac{78.54}{14.538} = 8103.6$$

The middle section of the equation is calculated as follows:

$$K_1 \times S_a = 0.45 \times 20{,}000 = 9000$$

Therefore, $8103.6 \leq 9000 \leq 9000$ is met and the bolts used for the body and bonnet joint of the oil export pipeline valve provide sufficient forces for tight sealing of the valve. Therefore, option C is the correct answer.

11.4 In order to connect two body pieces of a ball valve in a pressure class of 600, twelve 1/2″-in. bolts are installed with 13 coarse threads per inch. As far as bolts and nuts are concerned, the bolts and nuts used are the ASTM A193 B7 and A194 Gr.2H, respectively. The operating temperatures of the valve are ambient. What is the largest gasket diameter that can be used in order to keep body joints in place during operation?

A 14.10″
B 15.30″
C 15.90″
D 16.20″

Answer

The bolt stress area can be calculated using Eq. (11.7) as follows:

$$A_t = 0.7854 \times \left[d - \left(0.9743 \times \frac{1}{n}\right)\right]^2$$

$$= 0.7854 \times \left[0.5 - \left(0.9743 \times \frac{1}{13}\right)\right]^2$$

$$= 0.7854 \times 0.1807 = 0.1419\ \text{in.}^2$$

Due to the twelve bolts connecting the body pieces of the valve, the total bolt stress area can be calculated as follows:

$$A_b = 12 \times 0.1419\ \text{in.}^2 = 1.703\ \text{in.}^2$$

The allowable stress of the bolts in A193 B7 is 20,000 psi. Thus, the middle section of the equation is calculated as follows:

$$K_2 \times S_a = 0.35 \times 20{,}000 = 7000$$

The maximum gasket diameter results in a maximum value of the left side of the equation $P_c \times (A_g/A_b)$ and the maximum value of $P_c \times (A_g/A_b)$ is when it is equal to $K_2 \times S_a = 7000$.

$$P_c \times \frac{A_g}{A_b} = 7000 \rightarrow 600 \times \frac{A_g}{1.703} = 7000 \rightarrow A_g = 198.68\ \text{in.}^2$$

$$\rightarrow A_g = \pi\frac{d^2}{4} \rightarrow 198.68\ \text{in.}^2$$

$$= \pi\frac{d^2}{4} \rightarrow d^2 = 252. \rightarrow d = 15.90\ \text{in.}$$

Therefore, option C is the correct answer.

11.5 There is a threaded connection between the body pieces of a ball valve in pressure class 150. The gasket area is at its maximum. What is the ratio of the gasket area to the effective thread shear area of a ball valve?

A 20

B 22

C 24

D 26

Answer

In this case, Eq. (11.10) is valid since it describes the threaded joint between two body pieces of a valve.

$$P_c \times \frac{A_g}{A_s} \le 3300$$

Gaskets should not result in an area exceeding the left side of the equation by more than 3300.

$$P_c \times \frac{A_g}{A_s} \le 3300 \rightarrow 150 \times \frac{A_g}{A_s} = 3300 \rightarrow \frac{A_g}{A_s} = 22$$

Thus, option B is the correct answer.

11.6 The lateral force caused by a stem seal is double the axial force created by the gland on the sealing. What is Poisson's ratio of the stem seal?

A 0.13
B 0.23
C 0.33
D 0.43

Answer

Equation (11.11) can be used to express the relationship between the lateral stress on the stem seal and the axial stress on the stem seal as follows:

$$\delta_l = \delta_a \left(\frac{1-\mu}{\mu} \right) \rightarrow 2 = \left(\frac{1-\mu}{\mu} \right) \rightarrow 2\mu$$
$$= 1 - \mu \rightarrow 3\mu = 1 \rightarrow \mu = 0.33$$

Therefore, option C is the correct answer.

11.7 If the valve stem diameter is 2″ and the packing height is 1″, how much force is required to overcome the packing friction on the valve stem? The axial gland pressure on the packing is 200 psi and the packing coefficient of friction is 0.5. Additionally, the ratio of lateral to axial force is one.

A 570 lb
B 628 lb
C 750 lb
D 820 lb

Answer

Equation (11.14) can be used to calculate the force of packing on the valve stem.

$$F = \pi \times d \times H \times \mathrm{GS} \times m \times Y$$
$$= 3.1415 \times 2 \times 1 \times 200 \times 0.5 \times 1$$
$$= 628.3 - \mathrm{lb\ force}$$

Thus, option B is the correct answer.

11.8 When two bolts are used to fasten a gland flange with a diameter of 1/2″ and torque values of 12 lb-ft are applied, what is the value of compressive stress on a gland packing? It should be noted that the diameter of the gland packing is 1/2 in. and the diameter of the packing box is 1 1/2 in.

A 152.87
B 155.90
C 164.78
D 171.90

Answer

For this case, Eq. (11.15) must be used as follows:

$$\mathrm{GS} = \frac{(T.n/0.2.b)}{((\pi(D^2 - d^2))/4)} = \frac{((12 \times .2)/(0.2 \times 0.5))}{((\pi(1.5^2 - 0.5^2))/4)} = \frac{240}{1.57} = 152.87 \text{ psi}$$

Thus, option A is the correct answer.

Further Reading

Aksenov, A., Iliine, K., Luniewski, T. et al. (2006). Oil leakage through a valve stem seal. *ABAQUS Users' Conference 2005*, Stockholm, Sweden (18–20 May 2006).

American Petroleum Institute (API) 6D (2014). *Specification for Pipeline and Piping Valves*, 24e. Washington, DC: American Petroleum Institute (API).

American Society of Mechanical Engineers (ASME) (2016). *Process Piping. ASME B31.3*. New York: American Society of Mechanical Engineers (ASME).

American Society of Mechanical Engineers (ASME) (2020). *Boiler and Pressure Vessel Code (BPVC) VIII Div.02. Rules for Construction of Pressure Vessels Division 2 – Alternative Rules*. New York: American Society of Mechanical Engineers (ASME).

American Society of Mechanical Engineers (ASME) (2020). *Boiler and Pressure Vessel Code (BPVC) Section II. Materials*. New York: American Society of Mechanical Engineers (ASME).

American Society of Mechanical Engineers (ASME) B1.1 (2019). *Unified Inch Screw Threads (UN, UNR, and UNJ Threaded Forms)*. New York: American Society of Mechanical Engineers (ASME).

American Society of Mechanical Engineers (ASME) B16.34 (2020). *Valves – Flanged, Threaded, and Welding End*. New York: American Society of Mechanical Engineers (ASME).

Cox, J.C. (2010). Thwarting fluid-system leaks: saving energy, reducing leaks. *Digital Refining, Processing, Operations and Maintenance* 15 (9): 75–79.

Dassault Systems (2022). Tensile stress area of a bolt. https://help.solidworks.com/2021/english/SolidWorks/cworks/r_Tensile_Stress_Area_Bolt.htm (accessed 5 May 2022).

Drago, J. and Evans, W. (2017). Gauging the force effects of stem packing on valve stem actuation. *Proceedings ASME/NC Pump and Valve Symposium*, PVS2017-3518, Silver Spring, Maryland, USA (17–18 July 2017).

Engineering Edge (2019). Welded joint efficiency table recommendations. https://www.engineersedge.com/weld/welded_joint_efficiency_14419.htm (accessed 4 May 2022).

Fluid Sealing Association (2018). *Compression Packing Technical Manual*, 4e. Wayne, PA: Fluid Sealing Association.

Gobind, K. (2016). Valve fugitive emissions vs. current industry practices – part 2. *Valve World Magazine.* 21 (10): 121–122.

Nesbitt, B. (2007). *Handbook of Valves and Actuators: Valves Manual International*, 1e. Oxford: Elsevier.

Picard, D. (1999). Fugitive emission from oil and natural gas activities. *Journal of Good Practice Guideline and Uncertainty Management in National Greenhouse Gas Inventories.* 103–127.

Skousen, P.L. (2011). *Valve Handbook*, 3e. New York: McGraw-Hill.

Smit, P. and Zappe, R.W. (2004). *Valve Selection Handbook*, 5e. New York: Elsevier.

Sotoodeh, K. (2020). A review of valve stem sealing to prevent leakage from the valve and its effect on valve operation. *Journal of Failure Analysis and Prevention* https://doi.org/10.1007/s11668-020-01050-1.

Zulaika, L.C. (2016). Torque increase due to heavy packing tightening in valves. *AMPO. Valve World Conference 2016*, Dusseldorf, Germany (29 November–1 December 2016).

Index

Note: Page number followed by '*f*' and '*t*' refer to figures and tables respectively.

Industrial Valves: Calculations for Design, Manufacturing, Operation, and Safety Decisions, First Edition. Karan Sotoodeh.

d

Printed and bound by CPI Group (UK) Ltd, Croydon, CR0 4YY
17/09/2026
14964309-0004